Multidimensional Lithium-Ion Battery Status Monitoring

Multidimensional Lithium-Ion Battery Status Monitoring focuses on equivalent circuit modeling, parameter identification, and state estimation in lithium-ion battery power applications. It explores the requirements of high-power lithium-ion batteries for new energy vehicles and systematically describes the key technologies in core state estimation based on battery equivalent modeling and parameter identification methods of lithium-ion batteries, providing a technical reference for the design and application of power lithium-ion battery management systems.

- Reviews Li-ion battery characteristics and applications
- Covers battery equivalent modeling, including electrical circuit modeling and parameter identification theory
- Discusses battery state estimation methods, including state of charge estimation, state of energy prediction, state of power evaluation, state of health estimation, and cycle life estimation
- Introduces equivalent modeling and state estimation algorithms that can be applied to new energy measurement and control in large-scale energy storage
- Includes a large number of examples and case studies

This book has been developed as a reference for researchers and advanced students in energy and electrical engineering.

Emerging Materials and Technologies

Series Editor:
Boris I. Kharissov

Hybrid Polymeric Nanocomposites from Agricultural Waste
Sefiu Adekunle Bello

Photoelectrochemical Generation of Fuels
Edited by Anirban Das, Gyandshwar Kumar Rao, and Kasinath Ojha

Emergent Micro- and Nanomaterials for Optical, Infrared, and Terahertz Applications
Edited by Song Sun, Wei Tan, and Su-Huai Wei

Gas Sensors
Manufacturing, Materials, and Technologies
Edited by Ankur Gupta, Mahesh Kumar, Rajeev Kumar Singh, and Shantanu Bhattacharya

Environmental Biotechnology
Fundamentals to Modern Techniques
Sibi G.

Emerging Two Dimensional Materials and Applications
Edited by Arun Kumar Singh, Ram Sevak Singh, and Anar Singh

Advanced Porous Biomaterials for Drug Delivery Applications
Edited by Mahaveer Kurkuri, Dusan Losic, U.T. Uthappa, and Ho-Young Jung

Thermal Transport Characteristics of Phase Change Materials and Nanofluids
S. Harikrishnan and A.D. Dhass

Multidimensional Lithium-Ion Battery Status Monitoring
Shunli Wang, Kailong Liu, Yujie Wang, Daniel-Ioan Stroe, Carlos Fernandez, and Josep M. Guerrero

Scanning Probe Lithography
Fundamentals, Materials, and Applications
Yu Kyoung Ryu and Javier Martinez Rodrigo

Engineered Nanoparticles as Drug Delivery Systems
Nahid Rehman and Anjana Pandey

For more information about this series, please visit: www.routledge.com/Emerging-Materials-and-Technologies/book-series/CRCEMT

Multidimensional Lithium-Ion Battery Status Monitoring

Edited by

Shunli Wang

Kailong Liu

Yujie Wang

Daniel-Ioan Stroe

Carlos Fernandez

Josep M. Guerrero

CRC Press

Taylor & Francis Group

Boca Raton London New York

CRC Press is an imprint of the
Taylor & Francis Group, an **informa** business

First edition published 2023
by CRC Press
6000 Broken Sound Parkway NW, Suite 300, Boca Raton, FL 33487-2742

and by CRC Press
4 Park Square, Milton Park, Abingdon, Oxon, OX14 4RN

CRC Press is an imprint of Taylor & Francis Group, LLC

ISBN: 978-1-032-35602-0 (hbk)
ISBN: 978-1-032-36790-3 (pbk)
ISBN: 978-1-003-33379-1 (ebk)

DOI: 10.1201/9781003333791

Typeset in Times
by codeMantra

Contents

Preface

The development of new energy technology is a strategic measure to alleviate the world energy crisis and achieve the goal of carbon neutrality. Focusing on new energy technology based on the lithium-ion battery, this book examines concepts such as equivalent circuit modeling, parameter identification, and state estimation in lithium-ion battery power applications. This book systematically describes the key technologies in state estimation based on equivalent modeling and parameter identification methods of lithium-ion batteries. It mainly includes an overview of the lithium-ion batteries, equivalent modeling, parameter identification, state of charge estimation, state of health estimation, state of power estimation, state of energy estimation, and cycle life estimation. It also provides practical references for the design and application of lithium-ion battery management systems (BMSs), which are based on the technical requirements of lithium-ion battery applications compiled from the perspective of equivalent circuit modeling and state estimation of lithium-ion batteries. This book has distinctive features and employs the use of a variety of methods and techniques for state estimation and has rich examples that can serve as a guide and reference material. This book can be used as a textbook for control science and engineering, automation, electrical engineering, and other related majors in colleges and universities, as well as a resource in the field of new energy measurement and control. These measurement and control techniques can be used as a technical reference for researchers.

In recent years, power battery technology has rapidly developed and gradually matured, and lithium-ion batteries have become the main body of power batteries. During the energy storage and supply process of the lithium-ion battery pack, the BMS monitors its working status and manages the available energy. Due to the high safety requirements and complex working conditions of the lithium-ion battery packs, equivalent circuit modeling and state of charge estimation in applications have become the research hot spots.

Summarizing the battery pack application theory, experience, and design methods formed by the development of the high-power lithium-ion BMS for many years, the authors from the perspective of battery management combined the technical requirements of high-power lithium-ion batteries for new energy vehicles, the application of lithium-ion batteries, and battery power management as a starting point for the compilation of this book. Summarizing the understanding and experience of lithium-ion battery equivalent modeling and state estimation, we hope that this book can provide some technical references for the design, modeling, and application of lithium-ion BMSs. It also proposes some contributions to the global development of new energy technology and application business.

This book is written and compiled by the new energy measurement and control research team at the Southwest University of Science and Technology. The research team focuses on the field of new energy detection and control, with rich experience in teaching and research, with a fine tradition of close integration of production, education, and research. This book is compiled based on the research results regarding related materials and divided into six chapters. Professor Shunli Wang constructed the overall framework and carried out the unified supplements, revisions, and finalization. The research team participated in the content compilation of this book.

Chapter 1 provides an overview of lithium-ion batteries to give readers a comprehensive understanding and lays the foundation for subsequent equivalent circuit modeling and state estimation analyses of lithium-ion batteries, led by Chunmei Liu, Xiao Yang, Pu Ren, Shunli Wang, Yi Chai, Yujie Wang, Yunlong Shang, Zonghai Chen, and Paul Takyi-Aninakwa. Chapter 2 describes the battery equivalent circuit modeling and parameter identification methods, most of which are mainly written by Chunmei Yu, Haotian Shi, Xiao Yang, Shunli Wang, Yongcun Fan, Yuhong Jin, Xin Xiong, Weijia Xiao, Jie Cao, and Wenhua Xu. Chapter 3 presents the methods and analysis of battery state of charge estimation, completed by Peng Yu, Dan Deng, Daniel-Ioan Stroe, Jinsong Qiu, Yanxin Xie, Qingyun Ma, Shunli Wang, Jialu Qiao, and Xiao Yang. Chapter 4 mainly elaborates on the estimation of battery health status, completed by Pu Ren, Mingfang He, Shunli Wang, Yangtao Wang, Carlos Fernandez, Ji Wu, Lei Chen, Peng Yu, XIao Yang and Siyu Jin. Chapter 5 explains battery power state estimation, led by Lili Xia, Bowen Li, Yuyang Liu, Liying Xiong, Wen Cao, Jinhao Meng, Li Zhang, Shunli Wang, Xiao Yang and Jiawei Peng. Chapter 6 describes the estimation of battery energy state and cycle life, completed by Weihao Shi, Siyu Jin, Shunli Wang, Junhan Huang, Josep M. Guerrero, Kailong Liu, Wu Tang, Xiao Yang, Long Zhou, and Ran Xiong.

Professor Zonghai Chen of the University of Science and Technology of China reviewed the entire book. Professor Wu Tang of the University of Electronic Science and Technology of China provided a wealth of reference materials for this book. Professor Yi Chai of Chongqing University provided a lot of constructive comments on the publication of this book. We are grateful to them for their hard work and dedication to the success of this book. This book passed inspection from the Mianyang Product Quality Supervision and Inspection Institute (National Electrical Safety and Quality Supervision and Inspection Center), Deyang Product Quality Supervision and Inspection Institute, Mianyang Weibo Electronics Co. Ltd., Shenzhen Yakeyuan Technology Co. Ltd., Dongguan Bell Experimental Equipment Co. Ltd., and Shenzhen Xinwei New Energy Technology Co. Ltd. Thanks a lot for their scientific help and technical support.

Lithium-ion battery condition monitoring involves a wide range of areas and is limited by the knowledge level of the authors. Inevitably, there are improprieties in the organization and compilation of this book, and the readers are therefore welcome to criticize and correct constructively through the website https://www.researchgate.net/lab/DTlab-Shunli-Wang. We hope that this book serves as a platform for knowledge exchange and establishment of contact with readers, promoting the progress of key technologies for monitoring the battery states of lithium-ion batteries ultimately.

Editor
May 2022

The MathWorks, Inc.
3 Apple Hill Drive
Natick, MA 01760-2098 USA
Tel: 508-647-7000
Fax: 508-647-7001
E-mail: info@mathworks.com
Web: www.mathworks.com

Editors

Shunli Wang is a full professor at the Southwest University of Science and Technology, China. He is an authoritative expert in the field of new energy research and the head of NELab, conducting modeling and state estimation strategy research for lithium-ion batteries. He has undertaken more than 50 projects and 30 patents, published more than 150 research papers as well as won 20 awards such as the Young Scholar, and Science & Technology Progress Awards.

Kailong Liu is an assistant professor at the University of Warwick, United Kingdom. His research experience lies at the intersection of AI and electrochemical energy storage applications, especially data science in battery management. His current research focuses on the development of AI strategies for battery applications.

Yujie Wang is an associate professor at the Department of Automation, University of Science and Technology, China. He received his Ph.D. in control science and engineering from the University of Science and Technology, China in 2017. He has co-authored over 60 SCI journal papers on battery-related topics. His research interests include energy saving and new energy vehicle technology, complex system modeling, simulation and control, fuel cell system management, and optimal control.

Daniel-Ioan Stroe is an associate professor at AAU Energy, Aalborg University, Denmark, and the leader of the Batteries Research Group. He received his Ph.D. in lifetime modeling of lithium-ion batteries from Aalborg University in 2010. He has co-authored one book and over 150 scientific peer-reviewed publications on battery performance, modeling, and state estimation. His research interests include energy storage systems for grid and e-mobility, lithium-based batteries' testing, modeling, lifetime estimation, and their diagnostics.

Carlos Fernandez is a senior lecturer at Robert Gordon University, Scotland. He received his Ph.D. in electrocatalytic reactions from the University of Hull and then worked as a consultant technologist in Hull and in a postdoctoral position in Manchester. His research interests include analytical chemistry, sensors and materials, and renewable energy.

Josep M. Guerrero is a full professor at AAU Energy, Aalborg University, Denmark. He is the director of the Center for Research on Microgrids (CROM). He has published more than 800 journal articles in the fields of microgrids and renewable energy systems, which have been cited more than 80,000 times. His research interests focus on different microgrid aspects, including hierarchical and cooperative control, and energy management systems.

Contributors

Jie Cao
School of Information Engineering
Southwest University of Science and Technology
Mianyang, China

Wen Cao
School of Information Engineering
Southwest University of Science and Technology
Mianyang, China

Yi Chai
School of Automation
Chongqing University
Chongqing, China

Lei Chen
School of Information Engineering
Southwest University of Science and Technology
Mianyang, China

Zonghai Chen
School of Information Science and Technology Department of Automation
University of Science and Technology of China
Hefei, China

Dan Deng
School of Automation
Guangxi University of Science and Technology
Liuzhou, China

Yongcun Fan
School of Information Engineering
Southwest University of Science and Technology
Mianyang, China

Carlos Fernandez
School of Pharmacy and Life Sciences
Robert Gordon University
Aberdeen, United Kingdom

Josep M. Guerrero
Department of Energy and Technology
Aalborg University
Pontoppidanstraede Aalborg East, Denmark

Mingfang He
School of Information Engineering
Southwest University of Science and Technology
Mianyang, China

Junhan Huang
School of Information Engineering
Southwest University of Science and Technology
Mianyang, China

Siyu Jin
Department of Energy and Technology
Aalborg University
Pontoppidanstraede Aalborg East, Denmark

Yuhong Jin
School of Information Engineering
Southwest University of Science and Technology
Mianyang, China

Bowen Li
School of Information Engineering
Southwest University of Science and Technology
Mianyang, China

Chunmei Liu
School of Information Engineering
Southwest University of Science and Technology
Mianyang, China

Kailong Liu
WMG
University of Warwick
Coventry, United Kingdom

Yuyang Liu
School of Information Engineering
Southwest University of Science and Technology
Mianyang, China

Qingyun Ma
School of Information Engineering
Southwest University of Science and Technology
Mianyang, China

Jinhao Meng
College of Electrical Engineering
Sichuan University
Chengdu, China

Jiawei Peng
School of Information Engineering
Southwest University of Science and Technology
Mianyang, China

Jialu Qiao
School of Information Engineering
Southwest University of Science and Technology
Mianyang, China

Jingsong Qiu
School of Information Engineering
Southwest University of Science and Technology
Mianyang, China

Pu Ren
School of Information Engineering
Southwest University of Science and Technology
Mianyang, China

Yunlong Shang
School of Control Science and Engineering
Shandong University
Jinan, China

Haotian Shi
School of Information Engineering
Southwest University of Science and Technology
Mianyang, China

Weihao Shi
School of Information Engineering
Southwest University of Science and Technology
Mianyang, China

Daniel-Ioan Stroe
Department of Energy and Technology
Aalborg University
Pontoppidanstraede Aalborg East, Denmark

Paul Takyi-Aninakwa
School of Information Engineering
Southwest University of Science and Technology
Mianyang, China

Shunli Wang
College of Electrical Engineering
Sichuan University
Chengdu, China
and
School of Information Engineering
Southwest University of Science and Technology
Mianyang, China

Yangtao Wang
School of Information Engineering
Southwest University of Science and Technology
Mianyang, China

Yujie Wang
School of Information Science and Technology, Department of Automation
University of Science and Technology of China
Hefei, China

Ji Wu
School of Automotive and Transportation Engineering
Hefei University of Technology
Hefei, China

Wu Tang
School of Materials and Energy
University of Electronic Science and Technology of China
Chengdu, China

Lili Xia
School of Information Engineering
Southwest University of Science and Technology
Mianyang, China

Weijia Xiao
School of Information Engineering
Southwest University of Science and Technology
Mianyang, China

Yanxin Xie
School of Information Engineering
Southwest University of Science and Technology
Mianyang, China

Liying Xiong
School of Information Engineering
Southwest University of Science and Technology
Mianyang, China

Ran Xiong
School of Information Engineering
Southwest University of Science and Technology
Mianyang, China

Xin Xiong
School of Information Engineering
Southwest University of Science and Technology
Mianyang, China

Wenhua Xu
School of Information Engineering
Southwest University of Science and Technology
Mianyang, China

Xiao Yang
School of Information Engineering
Southwest University of Science and Technology
Mianyang, China

Chunmei Yu
School of Information Engineering
Southwest University of Science and Technology
Mianyang, China

Peng Yu
School of Information Engineering
Southwest University of Science and Technology
Mianyang, China

Li Zhang
School of Information Engineering
Southwest University of Science and Technology
Mianyang, China

Long Zhou
School of Information Science and Technology
University of Shanghai for Science and Technology
Shanghai, China

1 Battery Characteristics and Parameters

Chunmei Liu, Paul Takyi-Aninakwa, and Pu Ren
Southwest University of Science and Technology

Shunli Wang
Sichuan University
Southwest University of Science and Technology

Xiao Yang
Southwest University of Science and Technology

Yi Chai
Chongqing University

Yujie Wang
University of Science and Technology of China

Yunlong Shang
Shandong University

Zonghai Chen
University of Science and Technology of China

CONTENTS

DOI: 10.1201/9781003333791-1

Introduction: As an indispensable renewable energy source, lithium-ion batteries have the outstanding advantages of being lightweight, environmentally friendly, and portable. They are used in various fields of society, such as automobiles, mobile phones, aerospace, and other high-tech industries [1]. Therefore, more attention needs to be paid to the consumption and pollution caused by lithium-ion batteries. According to the current technological situation, the technologies applied for lithium-ion batteries are not enough, and there are still many problems waiting to be solved [2,3]. Accidents resulting from out of control such as battery leakage, battery explosion, and battery fire still occur from time to time. Therefore, the efficient application and maintenance technology of lithium-ion batteries have become a research hot spot. The improvement in the reliability, safety, and service life of lithium-ion batteries has also become an urgent problem to be solved.

1.1 OVERVIEW OF LITHIUM-ION BATTERIES

1.1.1 State of the Art

The battery industry develops rapidly, promoting the progress of human lives. On the other hand, it has also caused environmental pollution and energy crises. In various fields, the massive consumption of crude oil has caused serious air pollution and damage to the earth and atmosphere. It has also brought about an energy crisis, which has triggered a series of economic and political problems [4]. The resource crisis affects the rapid development of science and technology and even seriously affected human survival [5]. Severe weather conditions caused by air pollution have affected most cities seriously [6]. According to a survey, there are hundreds of large cities, but only less than one percent of these cities have air quality that meets the required safety standard. The pollution caused by the consumption of oil and other resources destroys the environment and has terrible consequences, which have negatively affected the lives of humans and plants. More terrible problems such as rising global temperatures and melting glaciers are threatening humanity [7]. The scarcity of resources and environmental pollution have gradually gained world attention. In the environment, new energy technology has been continuously developed with the attention and strong support of various countries. China has already raised resources to combat these environmental issues to the height that it is a national strategy.

During the working process of the lithium-ion batteries, the battery management system (BMS) is used to monitor the working status and related parameters of the battery. Under abnormal conditions, the BMS can alarm and cut off the power supply in time to avoid damage and accidents. It makes the battery safer and more reliable during the application, which extends the service life of the batteries [8–11]. In recent years, BMS technology has been regarded as an important development direction and research objective by universities, scientific research institutions, and companies.

The battery state of charge (SOC) estimation has always been a very important task of the BMS. The SOC is also called the remaining capacity, which is the ratio of the remaining dischargeable capacity after the battery has been used for some time or left unused for a long time compared to its fully charged state. The SOC value of the battery can reflect the endurance and the remaining use time [12–16]. During the working process, the BMS reflects the state of the battery, and the SOC of the battery is one of the most important state parameters. The state estimation of the battery is an important technology, but it is difficult to carry out the effective equivalent circuit modeling and accurate SOC estimation of lithium-ion batteries [17,18]. It is impossible to directly measure the SOC value of the battery because it is a dynamic parameter. To obtain the SOC value of the battery, it is necessary to measure its voltage, current, temperature, internal resistance, and other parameters and estimate the battery SOC value through a relational expression. Due to the interference from external conditions such as temperature in the process, it is difficult to accurately estimate the SOC. Therefore, research on different methods and technologies to estimate the SOC accurately is necessary [19,20].

Accurate SOC estimation of lithium-ion batteries has great theoretical value and practical significance for its development. Specifically, it can make the lithium-ion battery pack safer from usage and improve the performance of batteries[7,21]. If the SOC can be estimated accurately, the BMS can detect and display the status of the battery better, and the data can guide users to maintain the batteries. It highly reduces the maintenance cost and time, prolongs the battery life, thereby maintaining high performance and improving the user experience in the process.

Among all kinds of power batteries, lithium-ion batteries are widely used in various fields because of their outstanding advantages over other batteries. Advanced technology is a prerequisite for people to study lithium-ion batteries, and advances in science and technology have also enabled people to develop rapidly in the research of lithium-ion batteries and have achieved good results [22,23]. As research progresses, the various characteristics of lithium-ion batteries are gradually being understood. Lithium-ion batteries have outstanding advantages, such as high voltage, lightweight, and high energy density [24]. Nowadays, researchers have developed a variety of lithium-ion batteries suitable for different application scenarios for high specific energy and fast charging conditions, which have expanded the application field of lithium-ion batteries and accounted for its rapid development in modern society. The earliest use of lithium-ion batteries enabled pacemakers to be implanted or operated on the human body for a long time without recharging. With the rapid development and wide application of digital products, lithium-ion batteries are used in various digital products and power storage systems. Nowadays, countries all over the world are actively investing in the research and development of lithium-ion batteries.

For lithium-ion battery research, Japan's Sony Corporation is one of the pioneer companies to research and develop lithium-ion batteries compared to other countries and regions. In 2010, the Japanese government decided to increase investment in lithium-ion battery research and development. The government increased investment in electric vehicle research, by directly organizing relevant professional members to implement special national plans to research, develop, advance, and apply lithium-ion battery technology to a new generation of electric vehicles. Currently, the lithium-ion battery produced in Japan has reached an international leading level. Therefore, Japan attaches great importance to the development of new energy sources. The resource crisis is a huge predicament that cannot be ignored [25]. As the first country to study electric vehicles, Japan has been in the leading position in electric vehicles.

The research and production of lithium-ion batteries have rapidly grown with the development of society. As far as the current situation is concerned, the research and production of lithium-ion batteries are still growing, and their position in the global ranking continues to expand. In an environment-friendly world, new energy and environmental protection issues are important research topics, and lithium-ion batteries are relatively environmentally friendly in use. At present, lithium-ion batteries have gradually replaced those that have a greater impact on environmental pollution. It has also become an indispensable energy storage component in social development [26]. The development forecast of the battery industry points out that the frequency of use of lithium-ion batteries will increase in the future, the application fields will expand, and the market share will continue to increase. At this stage, the number of manufacturers producing lithium-ion batteries is gradually increasing, including companies of Wanxiang Group and AVIC lithium-ion batteries, and they serve as a broad platform and space for lithium-ion battery research and production. The lithium-ion battery system itself is a complex system integrating chemical, electrical, and mechanical characteristics [27–29]. Consequently, the requirements of various characteristics must be considered in the design.

In particular, the safety and life attenuation characteristics contained in the chemical characteristics of battery cells cannot be evaluated intuitively, and it is not easy to predict in a short time [30]. Therefore, when designing the battery system, it is necessary to adopt battery technology, group technology, BMS technology and also take into account the safety, reliability, and durability of the battery. Lithium-ion batteries generally come in cylindrical and rectangular shapes [31–34]. The interior of the cylindrical battery adopts a spiral winding structure, which is made of extremely fine and highly permeable positive and negative electrode material, mainly including polyethylene, polypropylene, polyethylene, and polypropylene composite materials. The rectangular lithium-ion battery is formed by lamination, with a diaphragm placed on the positive electrode, then a diaphragm on the negative electrode, and so on. The positive electrode includes a lithium-ion collector composed of lithium-containing materials, such as one or more mixed materials of lithium cobalt oxide, lithium manganate, nickel-cobalt lithium manganate, and a current collector composed of the aluminum film. The current collecting electrode is composed of a layered carbon ion collecting electrode and a copper ion collecting electrode. The battery is filled with organic electrolyte solution and equipped with a safety valve and positive temperature coefficient (PTC) resistor element, which has the advantages of small thermal resistance, high heat exchange efficiency, non-combustion, safety, and reliability [35–37]. It can effectively prevent the battery from being damaged in an abnormal state or output short circuit.

Under the influence of abnormal factors such as short circuits, high heat, and overcharge, it is easy to generate high-pressure gas in the battery, and it causes the deformation of the battery shell and even the risk of explosion. To use it safely, a safety valve must be installed in the battery to discharge the generated gas in case of abnormality to avoid explosion. When the internal pressure of the battery container rises to an abnormal state, the safety valve can quickly open and discharge the gas, playing the role of safety protection under abnormal conditions [38]. Since the PTC element is in the low resistance state under normal temperature, the danger of battery overheating can be prevented by adjusting the current in case of abnormality. For example, when an abnormally large current generated by a short circuit or overcharge causes the battery to overheat, the PTC element switches to a very high resistance state to reduce the current in the circuit. Therefore, the PTC element is usually used to prevent the battery overcurrent and the resulting overheating and protect the battery [39,40]. The voltage of a single lithium-ion battery is generally 2.80~4.20 V, and its capacity is generally 1.5 Ah, while the single capacity of a lithium-ion battery is generally in the range of 2–200 Ah. For new energy vehicles, a voltage of hundreds of volts is required to meet the requirements of endurance mileage, and a single battery cannot provide high voltage and energy. Therefore, the single lithium-ion battery is often processed in series and parallel to form a battery pack to meet the requirements of voltage and electricity.

1.1.2 Lithium-Ion Battery Composition

Lithium-ion batteries are mainly composed of four parts: positive and negative electrodes, diaphragm, and electrolyte. The cathode material determines the capacity and voltage of the batteries, which provides lithium ions for the batteries and determines the capacity and voltage [41,42]. Commonly used materials of positive electrodes are lithium manganese oxide, and conductive fluid collection using 10–20-micron thickness of electrolytic aluminum foil [43]. The anode material is mainly graphite, or carbon with a graphitic structure, and the conductive collector uses electrolytic copper foil with a thickness of 7–15 microns. The anode material at the anode acts as a current through an external circuit while allowing reversible absorption or emission of lithium ions released from the cathode. The main function of the anode is to store lithium ions and realize the insertion and intercalation reaction to lithium ions during the charge–discharge process.

The diaphragm is a special composite mold that prevents electrons from free shuttle between the positive and negative electrodes, while the lithium ions in the electrolyte can pass freely [5]. The electrolytes are mainly materials with high ionic conductivity, which can make it easy for lithium ions to move back and forth, it is generally composed of lithium salt and organic solvent, and it has the function of conducting ions.

The diaphragm is essentially an insulator, free electrons cannot pass through it, so it does not conduct electricity. In the battery, the element exists in the form of ions that can easily pass through the separator, and electrons leave the element on a new carrier, positive or negative electrode material. When it is in contact with the separator, the separator cannot absorb the free electrons on the electrode, thereby preventing the passage of electrons [44]. The common materials of the separator are single-layer PP film, PE film, and PP/PE/PP three-layer composite film. The electrolyte realizes the conduction of lithium ions between the positive and negative electrodes of the batteries. The cathode and anode determine the basic performance of the battery, while the electrolyte and diaphragm determine the safety of the battery [45,46]. Currently, $LiPF_6$ is the most widely used electrolyte. There are lithium ions, metal ions, oxygen ions, and carbon layers in the batteries. The composition of the lithium-ion battery is made up of some compounds. The reaction is done by the movement of ions inside the cell. The cell's diaphragm acts as a barrier, separating the battery's poles [47–49]. The internal structure of the battery is objective, as shown in Figure 1.1.

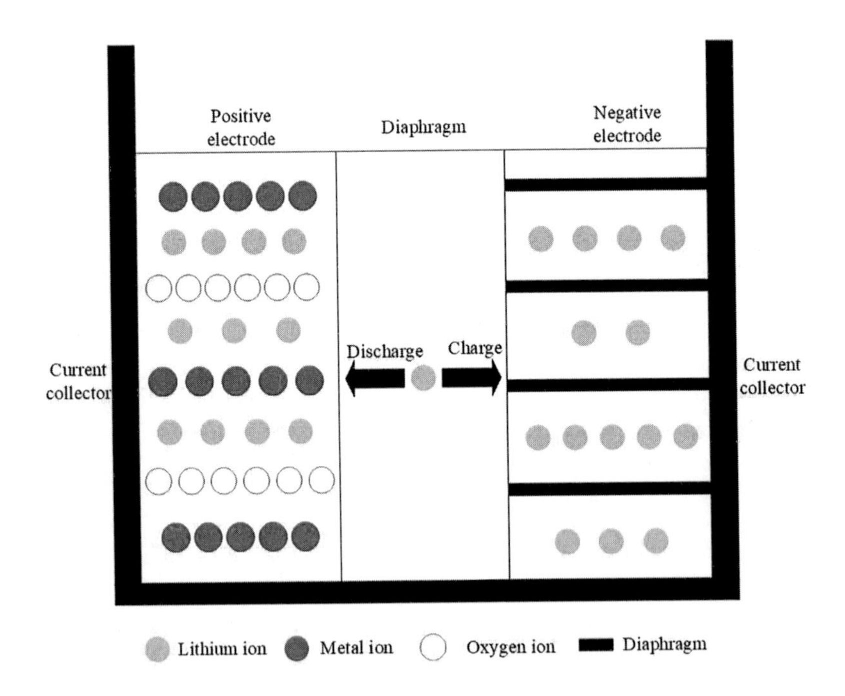

FIGURE 1.1 Schematic diagram of the lithium-ion battery structure.

In summary, the reason why lithium-ion batteries are used more than other batteries is that lithium-ion batteries have many advantages compared with other batteries [50]. (1) The energy density of lithium-ion batteries is very high; that is, in the unit cubature, the heat released is very high. (2) Lithium-ion battery is more environmentally friendly and meets the requirements of green social development. (3) Long service life. Under normal conditions, lithium-ion batteries can handle hundreds of times repeated charge–discharge processes, so they can be used for a long time. (4) No memory effect. Ordinary batteries lose their capacities in the process of working, resulting in less and less capacity, but lithium batteries do not have this problem. Also, there are many other advantages of lithium-ion batteries, such as good safety performance, low self-discharge, fast charging, and a wide range of operating temperatures accounting for its wide application.

1.1.3 Battery Working Principle

The internal chemical reaction of the lithium-ion battery is a basic redox reaction, which is also the working principle of a lithium-ion battery in the actual application process. It converts electric energy into heat energy through a chemical reaction. It can be observed from the chemical reaction equation that the charge–discharge process of lithium-ion batteries is the intercalation and deintercalation process of lithium ions. When a lithium-ion battery is charged, the lithium atoms in the positive electrode oxidize to lose electrons, thus becoming lithium ions [51]. A large number of lithium ions produced by the oxidation reaction to the positive electrode pass through the electrolyte solution to the carbon layer of the negative electrode of the batteries. The capacity of the battery is related to the number of lithium ions produced by the reaction to the positive electrode on the one hand, and the number of lithium ions exchanged with the negative electrode by the electrolyte on the other hand [52–55].

During the discharging process, an oxidation reaction occurs in the negative electrode, and the lithium ions embedded in the carbon layer of the negative electrode come out and move back to the positive electrode. The more lithium ions return to the positive electrode, the higher the discharge capacity is. Similarly, during charging, lithium ions are generated in the positive electrode of the battery, and the generated lithium ions move to the negative electrode through the electrolyte [56]. The lithium ions are embedded in the negative electrode and the micropores of the carbon layer. The more lithium ions embedded in the carbon layer, the higher the charging capacity. The internal chemical reaction process in the lithium-ion battery is shown in Figure 1.2.

Most of the cathode materials of lithium-ion batteries are lithium compounds, such as lithium cobalt oxide, lithium manganate, lithium iron phosphate, and ternary materials. Because of this, they are called lithium-ion batteries [57,58]. The anode materials of lithium-ion batteries were originally made of alloys and metal lithium, but eventually became graphite for its better performance. The electrolyte is an organic solution that dissolves lithium salts. In general, the internal electrochemical reaction process of a lithium-ion battery is the exchange of lithium ions between

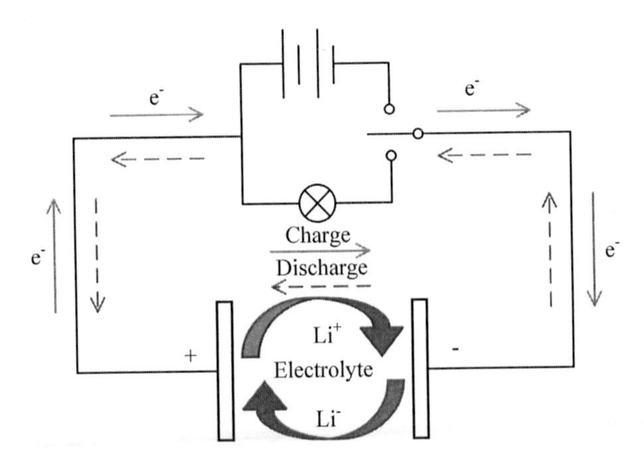

FIGURE 1.2 The working principle of lithium-ion batteries.

the positive and negative poles. The positive and negative reactions and the overall reaction equations are described as follows.

The positive electrode reaction is shown in Equation (1.1).

$$LiM_xO_y = Li_{(1-x)}M_xO_y + xLi^+ + xe^- \tag{1.1}$$

The negative electrode reaction is shown in Equation (1.2).

$$nC + xLi^+ + xe^- = Li_xC_n \tag{1.2}$$

The total reaction of the battery is shown in Equation (1.3).

$$LiM_xO_y + nC = Li_{(1-x)}M_xO_y + Li_xC_n \tag{1.3}$$

In Equation (1.3), M can be C_0, Mn, Fe, or Ni, which represent lithium cobalt oxide, lithium manganate, lithium iron phosphate, and lithium nickel oxide batteries, respectively. In the process of charging and discharging lithium-ion batteries, there is no metal lithium, only lithium ions. When the battery is charged, lithium ions are generated on the positive electrode of the battery, and the resulting lithium ions travel through the electrolyte to the negative electrode. The carbon as the negative electrode is layered and has many micropores [44]. The lithium ions that reach the negative electrode are embedded in the pores of the carbon layer. The more lithium ions embedded in the carbon layer, the higher the charging capacity. Similarly, when a battery is discharged, the lithium ions embedded in the carbon layer of the negative electrode break out and move back to the positive electrode [44]. The more lithium ions re-polarize, the higher the discharge capacity.

The battery capacity refers to the discharge capacity. In the charge–discharge process of lithium-ion batteries, it is in a state of motion from positive electrode to negative electrode to positive electrode. The charging – discharging process of a lithium-ion battery is the process of intercalation and deintercalation of lithium ions inside the battery [59]. That is, they can be reversibly inserted or extracted from the host material. In the two stages of charge–discharge, intercalation and intercalation are carried out back and forth between the positive and negative electrodes: During charging, lithium ions are first intercalated from the positive electrode, reach the negative electrode through the electrolyte, and insert lithium ions in the negative electrode.

The negative electrode of the battery is rich in lithium and lithium-rich is a positive electrode material made with a small amount of lithium doped, which is combined with positive electrode active materials, such as $LiMn_2O_4$. The lithium-rich condition can cause the unit cell to shrink and change from the cubature of charging–discharging and improve the structural stability of the material cycle performance. The process of discharging and charging is opposite to each other. The cathode material of a lithium-ion battery is composed of a lithium-intercalation compound. If there is an external electric field, the Li^+ in the cathode material can be extracted and inserted into the crystal lattice under the action of the electric field. Taking $LiCoO_2$ as an example, its positive electrode is shown in Equation (1.4).

$$LiCoO_2 \rightarrow xLi^+ + Li_{1-x}CoO_2 + xe^- \tag{1.4}$$

The negative electrode is shown in Equation (1.5).

$$xe^- + xLi^* + 6C \rightarrow Li_xC_6 \tag{1.5}$$

The general equation of battery reaction is shown in Equation (1.6).

$$LiCoO_2 + 6C \Leftrightarrow Li_{1-x}CoO_2 + Li_xC_6 \tag{1.6}$$

The lithium-ion battery has three important components: diaphragm, positive and negative electrodes. Its operation mainly depends on the back-and-forth movement of ions between the positive and negative electrodes. The working principle of the lithium-ion battery mainly depends on the lithium-ion concentration difference between ions at both ends. During charging, Li^+ is de-embedded from the positive electrode and embedded into the negative electrode through the corresponding electrolyte. After this series of chemical reactions, the positive electrode is in the low lithium state and the negative electrode is in the multi-lithium state. Also, its compensation charge is supplied to the negative electrode from the external circuit. During discharge, Li^+ is removed from the negative electrode and embedded into the positive electrode again through the action of the electrolyte.

1.1.4 Development Prospects

As a high-performance secondary battery, lithium-ion batteries are widely used in daily life and account for a large proportion of the market. At present, lithium-ion batteries are mainly used in various portable electronic products, such as mobile phones and notebook computers [60,61]. With the progress of battery material performance and design technology, the application scope of lithium-ion batteries has been expanding, and the main application fields are as follows.

1. Application in electronic products
   Lithium-ion batteries are widely used in portable electronic products because of their high volume-specific energy, small volume, and lightweight. With the popularity of mobile phones, digital cameras, video cameras, laptops, and handheld game consoles, the lithium-ion battery market has maintained rapid growth and occupied most of the market share. With the shortage of oil resources, the aggravation of environmental pollution, and the improvement of human environmental awareness, the development of green energy has been promoted, and the global electric bicycle industry has become a boom [62,63].
2. Application on transport vehicles
   1. Electric bicycle
      As the main mode of urban transportation development in the future, public transportation has been recognized by all walks of life. In addition, based on the analysis of the objective factors, electric bicycles have more advantages and practicability. Especially in recent years, with the shortage of oil resources and the intensification of environmental pollution to promote the development of green energy, the development of the electric bicycle industry has ushered in a boom [64,65].
   2. Electric vehicle
      The continuous development of electric vehicles can reduce greenhouse gas emissions, which is in line with the scientific outlook on development and becomes a strategic opportunity for the automotive industry. Compared with developed countries, lithium-ion batteries have the same technical advantages of resources and market. Therefore, the key technologies of electric vehicles are the focus of key research. Rapidly promoting the industrialization of electric vehicles is not only a strategic choice for China's national conditions but also an important way to ensure energy security [66,67]. In the 2008 Beijing Olympic Games, 50 pure electric buses miraculously created zero anchors and zero fault records, which showed the charm and style of Green Olympics effectively. The 2010 Shanghai World Expo used more than 1000 new energy vehicles for the first time, including fuel cell vehicles, hybrid electric vehicles, supercapacitor vehicles, and pure electric vehicles. During the Expo, it is expected to save about 10000 tons of traditional fuels, 118 tons of harmful gas emissions are reduced and 28400 tons of greenhouse gas emissions are reduced. In addition, electric vehicle charging stations and other related facilities have been completed and put into use, and the development of the electric vehicle industry is becoming more mature.
3. Application in aerospace
   In 2004, lithium-ion batteries have been used in Mars landers and rovers, and they are used in future exploration missions. In addition, NASA's space

exploration agency and other space agencies also consider the application of lithium-ion batteries in space missions [68–70]. At present, the main function of lithium-ion batteries in the aviation field is to provide support for the launch, flight correction, and ground operation, improving the efficiency of the battery and supporting night operation.

4. Application in the energy storage device
The power storage plant stores the power (valley power) generated by the power plant during the low peak period, and the electricity price is cheap at this time. The stored electric energy is sold at the peak of power consumption. At this time, the electricity price of the storage farm has certain advantages over the gradient electricity price (peak electricity) of the power plant. It is a difficult problem to adopt peak valley power regulation. Usually, to ensure peak power consumption, more power plants need to be built, but this method increases the investment cost. In low peak power consumption, the power plants need to operate, as usual, resulting in a waste of energy. Therefore, some enterprises have put forward the idea of investing in the construction of power storage plants, changed the procurement of large and medium-sized energy storage equipment, charged low fees during the peak period, stored electric energy during the low peak period, and peak valley electricity price, forming a win-win situation [71,72]. As green energy, lithium-ion batteries have a high energy density, good cycle performance, and high charge retention performance. It is recognized as an ideal choice for high-capacity and high-power batteries.

With the rapid development of the above fields, digital entertainment products, such as cameras, multimedia playback equipment, and electronic book equipment, have placed higher requirements on secondary batteries. The main indicators of lithium-ion batteries include energy density, power density, circularity, temperature characteristics, and price. Currently, research is being developed in the following five directions.

1. High-energy-density battery: It is mainly used in wireless communication office products and digital entertainment products.
2. High-power battery: It is mainly used in power tools, transportation, and other high-power devices.
3. Long-life energy storage battery: It is mainly used in backup power supply, solar power stations, wind power stations, and other decentralized and independent power supply systems of energy storage batteries.
4. Micro-sized lithium-ion batteries: It is used for wireless sensors, micro-unmanned aircraft, implantable medical devices, micro-robots, integrated chips, etc.
5. High-energy-density and high-power-density lithium-ion batteries: It has a wide range of applications, such as electric vehicles and multimedia information processing systems.

As a relatively mature and advanced battery at present, lithium-ion batteries have been favored by the market because of their lightweight and large power storage. Especially in the development of mobile phones, smart wearable devices, and new energy vehicles, lithium-ion batteries are in short supply. The entire industry is in a hot state, and the assessment of many listed companies in the capital market is constantly increasing. The upstream industry of lithium-ion batteries mainly includes lithium raw materials, cathode materials, anode materials, electrolytes, separators, and production equipment [73]. The downstream applications of lithium-ion batteries mainly include power batteries, 3C batteries, and energy storage batteries. Although the current energy storage battery market has not fully opened, its scale has a good development prospect.

Therefore, the lithium-ion battery industry seeks to show a good high-speed development trend in the future, mainly because the demand for lithium-ion batteries in the main application markets is still increasing. This is because the demand for lithium-ion batteries is currently very high in major markets. Unless the fuel cell can be widely used by breaking through the limitations of high assembly process and high cost, the status and market share of lithium-ion batteries will not be greatly affected. As the demand for lithium-ion batteries continues to increase, most technological companies see opportunities to carry out lithium-ion battery projects one after another. With the influx of major companies into the industry, the competitiveness of the market has become increasingly fierce. However, whether in nature or the market, the law of survival of the fittest has always been followed as well as the trend of social development. Ternary lithium-ion batteries have high energy density and low prices, making them the first choice of manufacturers. In addition, the Energy-saving and New Energy Automobile Industry Development Plan puts forward the requirement that the energy density of battery modules must be greater than 150 Wh/kg, but lithium iron phosphate batteries cannot compete with it.

However, ternary lithium-ion batteries also have some drawbacks. They decompose at 200°C, and the electrolyte burns rapidly. The cooling of the battery has become a key research direction because if the temperature gets too high and the environment is airtight, there is a greater risk of spontaneous combustion and explosion in the 'stove city'. In addition, if there is improper recycling or indecent disposal, the cathode material in the lithium-ion battery causes heavy metal pollution, which becomes an environmental problem that must be solved in the development process. However, as far as the Chinese market is concerned, the battery is still mainstream. Both the disposable lithium-ion battery and the ternary lithium-ion battery contain a variety of metals and organic solvents such as electrolytes, which are quite corrosive and chemically toxic. When the battery explodes and burns, it also produces toxic chemical gases, which are harmful to the environment. It is believed that due to the fierce competition in the lithium-ion battery industry, some relatively strong companies survive the competition. Some emerging small companies may be on the verge of bankruptcy or may undergo mergers and acquisitions. Therefore, this is not only a test but also an opportunity for some lithium-ion battery companies. As an important part of the new energy field, the lithium-ion battery industry develops rapidly and has become the focus of new investment in the manufacturing field. Lithium-ion battery companies have increased the construction of new factories in hopes of increasing production capacity and winning with the help of the scale effect.

With the expansion of the new energy vehicle market, the demand for power lithium-ion batteries increases significantly. At the same time, the rapid development of mobile phones, electric vehicles, power tools, digital cameras, and other industries continues to increase the demand for lithium-ion batteries. In addition, due to the technological innovation of lithium-ion battery manufacturers, the demand for lithium-ion batteries continues to grow, and the lithium-ion battery industry has a good development prospect. According to the forecast of the Foresight Industry Research Institute, the potential scale of China's new annual cascade utilization will reach 33.6 GWh in 2025, and the market size of China's lithium-ion battery industry will exceed ¥260 billion in 2026. The national grid's demand for peak-shaving and valley-filling economy has led to the explosive growth of distributed energy storage devices [74]. The acceleration of the deployment of electric vehicle energy storage and charging stations has boosted the demand for cascading utilization. In the future, the growth of consumer electronics demands for lithium-ion batteries is stabilized at a relatively low level, and the increase in installed capacity of the energy storage field is small. Therefore, the demand for high-power batteries is the main driving force for the high growth rate of the lithium-ion battery industry.

1.2 THE TYPES AND CHARACTERISTICS OF LITHIUM-ION BATTERIES

There are many types of lithium-ion batteries. According to the state of the electrolyte used in the battery, it can be divided into liquid lithium-ion battery, polymer lithium-ion battery, and all-solid lithium-ion batteries. According to the temperature variation, the battery types can be divided into high-temperature lithium-ion batteries and room-temperature lithium-ion batteries [75]. According to the classification of cathode materials, lithium-ion batteries can be divided into lithium iron phosphate batteries [76–78], lithium cobalt acid batteries, lithium-manganese acid batteries, lithium nickel acid batteries, and ternary material batteries.

1.2.1 Lithium Iron Phosphate Battery

Lithium iron phosphate batteries refer to lithium-ion batteries that use lithium iron phosphate compound as cathode material and have the following advantages and characteristics.

1. High safety performance
   The P–O bond in the lithium iron phosphate crystal is stable and difficult to decompose [79–81]. Even in the case of high temperature or overcharge, the structure does not collapse and generate heat or form strong oxidizing substances. Therefore, it has good safety. Even if the battery is damaged

internally or externally, the battery does not burn or explode, which is the best safety. It has been reported that a small number of samples of lithium iron phosphate batteries burn in practical operation or short-circuit experiments, but no explosion occurred. However, in the overcharge experiment, when the battery is charged at a high voltage that is several times beyond its discharge voltage, the explosion still occurs. Nevertheless, its overcharge safety has been greatly improved compared with other batteries.

2. High life expectancy
   The cycle life of the lead–acid battery is about 300 times, the highest is only 500 times, the cycle life of a lithium iron phosphate battery is more than 2000 times. The life span of lead–acid batteries of the same quality is at most 1–1.5 years. Under the same conditions, the theoretical life of lithium iron phosphate batteries can reach 7–8 years, and in theory, the performance ratio is 4 times of lead–acid batteries [82–84].
3. Good high-temperature performance
   The electric thermal peak of lithium iron phosphate can reach 350°C–500°C, while the thermal peak of lithium manganese acid and lithium cobalt acid is only about 200°C. Lithium iron phosphate battery operating temperature range is wide (–20°C to +75°C) and has high-temperature resistance characteristics, which expands the battery's applicable range. When the external temperature is 65°C, the internal temperature is described as high as 95°C, and the temperature can reach 160°C at the end of battery discharge. The structure of the battery is safe and intact [85,86].
4. Large capacity
   It has a larger capacity than ordinary batteries, ranging from 5 to 1000 Ah.
5. High-efficiency output
   Standard discharge of 2~5 C, continuous high current discharge of 10 C, instantaneous pulse discharge of 20 C for 10 seconds.
6. No memory effect
   If the battery is constantly being charged, its capacity quickly falls below its rated capacity, a phenomenon known as the memory effect. For example, nickel-metal hydride and nickel-cadmium batteries have memory, while lithium iron phosphate batteries do not have this phenomenon [87]. No matter what state the battery is in, it can be used at any time without first discharging and then charging.
7. Small size and lightweight
   The cubature of a lithium iron phosphate battery with the same capacity is 2/3 of that of a lead–acid battery, and its weight is 1/3 of that of lead–acid batteries.
8. Environmental protection and pollution-free
   Lithium iron phosphate as lithium-ion battery material is a matter of recent years. Its safety performance and cycle life are incomparable to other materials, which are also the most important technical indicators of power batteries. The large-capacity lithium-ion battery made of lithium iron phosphate material can easily increase the cruising range and meet the frequent charging and discharging needs of new energy vehicles. It has the advantages of

being non-toxic, pollution-free, good safety performance, wide source of raw materials, low price, long life, etc. It is an ideal cathode material for the new generation of lithium-ion batteries. This project belongs to the development of functional energy materials in high and new technology projects. It is a key field supported by the national high and new technology industry development plan.

1.2.2 Lithium Cobaltate Battery

Lithium cobaltate battery has a stable structure, high capacity ratio, and outstanding comprehensive performance, but its safety is poor and the cost is very high. It is mainly used in small and medium-sized electric cells and is widely used in notebook computers, mobile phones, and other small electronic devices with a nominal voltage of 3.7 V [88,89].

1. Characteristics of lithium cobaltate
   1. Excellent electrochemical performance.
      - A. Capacity attenuation per cycle is less than 0.05% on average.
      - B. The specific capacity of the first discharge is >135 mWh/g.
      - C. 3.6 V initial discharge platform ratio >85%.
   2. Excellent machining performance.
   3. The high vibration density is helpful to improve the cubature-specific capacity of the battery.
   4. Stable product performance and good consistency.
2. The use of lithium cobaltate

   It mainly used in the manufacture of battery cathodes for portable electronic devices such as mobile phones and notebook computers.
3. Technical standard for lithium cobaltate
   1. Name: lithium cobaltate; molecular equation: $LiCoO_2$; molecular weight: 97.88.
   2. Main application: lithium-ion battery.
   3. Appearance requirements: grayish-black powder, no agglomeration.

   The performance of lithium cobalt oxide batteries is stable, and the technology currently used in mobile phones is the most mature. However, due to the high cost of lithium cobalt oxide, there are fewer applications of lithium cobalt oxide batteries.

1.2.3 Lithium-Manganese Battery

Lithium manganate battery refers to the battery that uses lithium-manganese oxide material as the positive electrode. It uses the special treatment of manganese dioxide as the active substance of the positive electrode the high potential, high specific energy of lithium metal as the active substance of the negative electrode, and the organic electrolyte with good conductivity is used as the electrolyte [90]. The battery structure is divided into fully sealed and semi-sealed, and the nominal voltage of the lithium manganate battery is 2.50–4.20 V. Manganese acid lithium has the advantages of low cost, safety, and good low temperature performance, but it is not so stable, easy to decompose to produce gas, so much for mixed with other materials used, to reduce the cost of batteries, but the attenuation of cycle life quickly, prone to bulge, high-temperature performance is a poorer, relatively short life, it is mainly used for large and medium-sized batteries.

1.3 BASIC PARAMETERS

1.3.1 Voltage

1. Electromotive force
   The electromotive force is one of the ways to calculate the battery's theoretical output energy. If other conditions remain the same, the higher the electromotive force, the greater the output energy [91]. The electromotive force of a battery is the difference between the potentials of the two electrodes, and its mathematical expression is shown in Equation (1.7).

$$E = \phi_+ - \phi_- \tag{1.7}$$

   In Equation (1.7), E is the battery electromotive force, ϕ_+ is the positive balance potential, and ϕ_- is the negative balance potential.
2. Open-circuit voltage
   Open-circuit voltage refers to the potential difference between the two poles of the battery in an open-circuit state with no current flows. The open-circuit voltage of the battery depends on the activity of the battery's positive and negative electrodes, electrolyte and temperature conditions, etc. [92]. It has nothing to do with the battery geometry and size. In general, the open-circuit voltage of the battery is less than its electromotive force.
3. Rated voltage
   The rated voltage is also called the nominal voltage, which refers to the standard voltage at which the battery works under specified conditions. In general, the nominal voltage is the central voltage of the platform voltage, around which the voltage changes very slowly during charge–discharge. The chemical system of a battery can be distinguished by its rated voltage.
4. Working voltage
   Working voltage refers to the voltage displayed by the battery during the discharging process after it is connected to a load. It is also known as the load voltage or the discharge voltage, and its mathematical expression is shown in Equation (1.8). After the battery is connected to the load, due to the existence of ohmic resistance and polarization internal resistance, the operating voltage of the battery is lower than the open-circuit voltage and the electromotive force.

$$V = E - IR_i = E - I\left(R_\Omega + R_f\right) \tag{1.8}$$

In Equation (1.8), I is the battery working current, R_Ω is the ohmic resistance, and R_f is the polarized internal resistance.

5. End-of-discharge voltage
   For all lithium-ion batteries, the end-of-discharge voltage is an important indicator that must be strictly regulated. The discharge termination voltage is also called the discharge cutoff voltage, which refers to the lowest operating voltage limited to at which the battery discharges when the voltage drops to the point where it is no longer suitable for discharge. According to different types of batteries and discharge conditions, the requirements for the capacity and life of the battery are also changed, as well as the specified discharge end voltage. When discharging at a low temperature or high current, the final voltage is regulated lower. When discharging with a small current for a long time or intermittently, the final voltage is regulated higher.
6. End-of-charge voltage
   The end-of-charge voltage refers to the voltage at which the battery reaches full charge during the specified constant current charging period. If it continues to charge after reaching the end-of-charge voltage, it is overcharged, which generally affects battery performance and life.
7. The average voltage
   The average voltage refers to the value obtained by dividing the number of watt-hours by the number of ampere-hours during the specified charge–discharge process. In the case of constant current, it is the average voltage over a certain period.

1.3.2 Capacity

Battery capacity: The total amount of electricity that the battery can release under certain discharge conditions is called battery capacity, which is represented by the symbol C. The unit is usually expressed in Ah or mAh. Battery capacity is an important unit to measure the quality of a battery, which is related to the use time of the battery. Typically, the laboratory uses the constant current or constant voltage method to discharge lithium-ion batteries [93]. The storage capacity of the lithium-ion battery can be calculated by measuring the energy released by the battery during the whole discharge process or the product of the discharge current and time.

1. Theoretical capacity
   The theoretical capacity is the amount of electricity that can be provided assuming that all active materials participate in the electrochemical reaction of the batteries. The theoretical capacity can be accurately calculated according to the amount of electrode active material in the battery reaction equation and the electrochemical equivalent of the active material calculated according to Faraday's law. The calculation processing equation is shown in Equation (1.9).

$$Q = \frac{zmF}{M} \tag{1.9}$$

Faraday law states that when current passes through the electrolyte solution, the amount of chemically reacted substance on the electrode is proportional to the amount of electricity passed. In Equation (1.9), Q is the amount of electricity passed in the electrode reaction (Ah), z is the electronic measurement coefficient in the electrode reaction equation, m is the mass of the reactive material (g), M is the molar mass of the active substance (g/mol), and F is the Faraday constant of about 965000 C/mol or 26.8 Ah/mol. It can also be understood that the active material with mass m can release electricity Q after a complete reaction. The electric quantity Q is the theoretical capacity (C_0) of the electrode active material, so it can also be written in the form of Equation (1.10).

$$C_0 = 26.8z\frac{m}{M} = \frac{1}{K}m \tag{1.10}$$

In Equation (1.10), K is the electrochemical equivalent of the active material, and $K = M/26.8z$ (g/Ah) is the mass of active material required to obtain 1 Ah of electricity. Equation (1.10) is the calculation equation for the theoretical capacity of the electrode active material.

2. The rated capacity

   The rated capacity (C_g) is the minimum capacity that the battery should discharge under certain discharge conditions (temperature, discharge rate, termination voltage, etc.) according to the standards stipulated by the state or relevant departments, from the fully charged state to the stop voltage.

3. The actual capacity

   The actual capacity (C) refers to the total amount of electricity released by the battery under actual application working conditions. It is equal to the integral of the discharge current and the discharge time, which is in the lower right part of the capacity C and is expressed in Arabic numerals, such as C_5=50 Ah, which means that the discharge rate is completed in 5 hours. And the obtained capacity is 50 Ah. The actual capacity calculation method is described as follows.

   When discharging with constant current, the calculation process is shown in Equation (1.11).

$$C = It \tag{1.11}$$

When discharging with variable current, the calculation process is shown in Equation (1.12).

$$C = \int_0^T I(t)\,dt \tag{1.12}$$

In Equation (1.12), I is the discharge current as a function of the discharge time point t. T is the total discharge time. Due to the existence of internal resistance and other various reasons, the active material cannot be fully utilized; that is, the utilization rate of the active material is always less than 1.

Therefore, the actual capacity and the rated capacity of the chemical power supply are always lower than the theoretical capacity [94]. The actual capacity of the battery is closely related to the discharge current. In the process of high current discharge, the electrode polarization and internal resistance increase, but the discharge voltage drops rapidly, which is related to the energy efficiency of the battery, so the actual discharge capacity is always lower. Correspondingly, under low current discharge conditions, the discharge voltage drops slowly, and the actual capacity of the battery is often higher than the rated capacity.

4. The remaining capacity

   The remaining capacity refers to the remaining usable capacity of the battery after being discharged at a certain discharge rate. The remaining capacity estimation is affected by factors such as the discharge rate and discharge time of the battery, the degree of battery aging, and the application environment. Therefore, there are certain difficulties in the accurate estimation.

1.3.3 Internal Resistance

Inside a lithium-ion battery, lithium ions move from one electrode to another through the electrolyte, and the factors that hinder the movement of ions in the process constitute the internal resistance. Due to the internal resistance of the battery, the terminal voltage of the battery is lower than the electromotive force and open-circuit voltage during discharge, and the terminal voltage of the battery is higher than the electromotive force and open-circuit voltage during charging. The essence of electric current is the directional movement of electric charges, and the motion of the electrons is opposed by the material itself and the magnetic field. Internal resistance is one of the important parameters of lithium-ion batteries that are not always static. The magnitude of the internal resistance directly affects the battery operating voltage, operating current, and battery capacity [95,96]. In the process of battery charge–discharge, changes in temperature, charge–discharge rate, charge–discharge time, electrolyte concentration, and active material quality cause changes in the internal resistance of the battery.

The internal resistance of the battery includes the ohmic resistance R_0 and the polarization of internal resistance R_p. The ohmic resistance comes from the resistance of the electrolyte, diaphragm, and the contact resistance of various parts. During the internal oxidation–reduction reaction of the battery, an electric field is generated. Under the action of the electric field, the dielectric generates polarization charges due to the polarization effect. The polarization resistance is the resistance of the polarization charge to the current. During the charge–discharge process of lithium-ion batteries, the internal resistance generated by the polarization reaction is mainly electrochemical and concentration polarization [97]. The nature of the active material, the structure of the electrode, the manufacturing process of the battery, and the operating conditions of the battery lead to the difference in the polarization internal resistance. The main influencing factor of the polarization internal resistance is the operating conditions of the batteries. The total internal resistance R of the battery

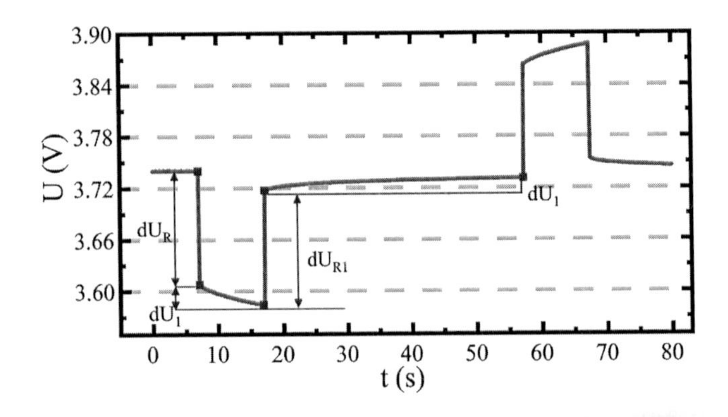

FIGURE 1.3 The voltage change curve of the cyclic discharge-shelving experiment.

is equal to the ohmic resistance R_0 plus the polarization internal resistance R_p. The calculation equation is shown in Equation (1.13).

$$R = R_0 + R_p \tag{1.13}$$

In Equation (1.13), different types of lithium-ion batteries show different internal resistance characteristics. Even the same type of lithium-ion battery, due to the difference in internal material composition, operating environment, and aging degree, also shows different internal resistances. The internal resistance is an important indicator to measure the battery performance. Therefore, the research on the change characteristics of the internal resistance of the battery is of great significance for equivalent modeling and SOC estimation. During the battery cycle discharge experiment, the partial voltage change curve is shown in Figure 1.3.

In Figure 1.3, when the battery is connected to or disconnected from the external circuit, the terminal voltage of the battery drops or rises rapidly for dU_R and dU_{R1}. It is due to the ohmic resistance of the lithium-ion battery itself, causing the voltage to change drastically at the beginning or end of discharge, which is called the internal resistance effect. In the first stage, the terminal voltage drops rapidly and then slowly decreases, or when the discharge ends, the voltage slowly rises (dU_1 and dU_2 in the figure), which is caused by the polarization effect of the lithium-ion batteries. When the battery is put aside for a while, the internal balance of the battery is reached, and the polarization effect disappears. The polarization effect produces polarized internal resistance and internal structure. The working conditions affect the polarization internal resistance of the batteries. However, the discharge current and the operating temperature have a particularly obvious effect on the polarization internal resistance of the batteries. It is due to its effect on the movement of lithium ions inside the battery during operation.

The internal resistance is most sensitive to temperature, and the internal resistance value can vary greatly at different temperature conditions. An important reason for the degradation of the battery performance at low temperatures is that the internal resistance increases drastically at low temperatures. Under normal circumstances, lithium-ion batteries are used as power sources. From the external analysis, the internal resistance is usually as small as possible. Especially in the case of high-power applications, small internal resistance is a necessary condition.

1.3.4 Polarization Characteristics

The polarization effect of the battery refers to the effect that the electrode of the lithium-ion battery deviates from the original equilibrium point when the battery is charged and discharged. It can be represented by resistance and capacitance. Generally, the polarization of lithium-ion batteries can be divided into three types: ohm polarization, electrochemical polarization, and concentration polarization [98]. The response speed of each polarization is also different. There are many factors affecting the degree of polarization, but in general, the higher the charge–discharge current density, the greater the polarization. Its relationship with SOC is the polarization characteristic of the lithium-ion battery pack. The polarization characteristics of the battery can be obtained through the hybrid pulse power characterization (HPPC) test, as shown in Figure 1.4.

In Figure 1.4, it can be observed that R_p has a relatively large value at SOC [(0%, 20%), (80%, 100%)] and a relatively small value at $\text{SOC}\left[20\%, 80\%\right]$. However, R_p and SOC have a nonlinear relationship, and there is no obvious correlation in the whole process. The polarization capacitance C_p has a relatively large value in the $\text{SOC}\left[0\%, 20\%\right]$ range and is relatively small and stable in other SOC value ranges. Generally, C_p and SOC have a positive correlation.

1.3.5 Energy and Power Density

1. Energy density
   The energy density of a battery refers to the amount of energy stored in a given space unit or material mass unit. The battery energy density is generally divided into two dimensions: weight energy density (Wh/kg) and cubature energy density (Wh/L), also known as mass-specific energy or cubature ratio energy. Specific energy is an important indicator for

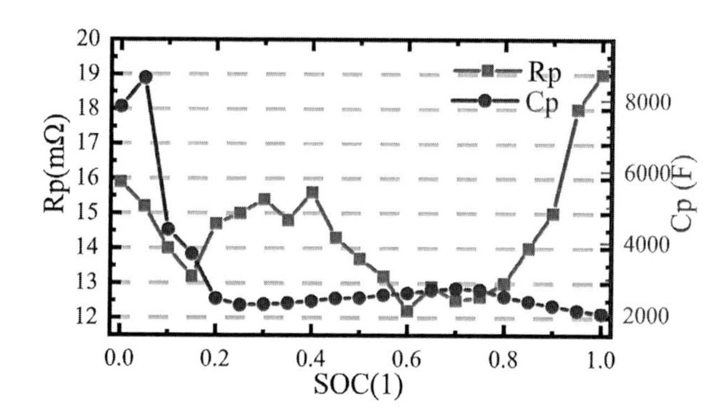

FIGURE 1.4 Relationship between R_p and C_p corresponding to SOC variation.

evaluating whether high-power batteries can meet the application requirements of new energy vehicles [99,100]. It is also an important indicator for comparing the performance of different types of batteries.

In the application process of new energy vehicles, the installation of battery packs requires corresponding components such as battery boxes, connecting wires, and current and voltage protection devices. The actual specific energy of the battery pack is less than the rated specific energy of the batteries. The smaller the gap between the actual specific energy and the rated specific energy, the higher the design level and integration degree of the battery pack. Therefore, the specific energy serves as an important parameter used to evaluate the performance of the battery pack for new energy vehicles.

2. Power density

   The power output per unit mass or cubature of the battery is called power density, also known as specific power, and the unit is W/kg or W/L. The specific power is the magnitude of the working current that the battery can withstand. Higher specific power means that the battery can withstand high current discharge [101]. Specific power is an important indicator for evaluating whether batteries and battery packs meet the rapid acceleration and climbing capabilities of new energy vehicles.

1.4 CORE BATTERY STATE FACTORS

1.4.1 State of Charge

The SOC is the percentage of available capacity and rated capacity of the lithium-ion battery, which describes the remaining power of the battery and is an important parameter in the use of lithium-ion batteries [102–106]. The parameter is related to the historical charge–discharge capacity and charge–discharge current of the batteries. The SOC value is a relative quantity, which is generally expressed as a percentage, and the value range of SOC is 0%–100%. When SOC = 0, the battery is fully discharged, and when SOC = 1, the battery is fully charged. The SOC of the battery cannot be measured directly, but can only be estimated by the battery terminal voltage, charge–discharge current, and internal resistance. The battery's SOC is affected by the basic characteristic parameters of the power battery, including aging degree, terminal voltage, working current, temperature, capacity, internal pressure, internal resistance, and the number of charge–discharge cycles.

The nonlinear characteristics of power batteries make it very difficult to measure the SOC value of the battery pack. At present, the simpler method for the research on the power of the battery pack is to equate the battery pack with a cell [107]. The relationship between the SOC and these parameters is obtained by measuring the current, voltage, internal resistance, and other external parameters of the battery pack. In the application process, to ensure the safety and service life of the battery pack, the SOC estimation of the cells with poor energy is often used to define the overall SOC estimation of the battery pack. Common SOC estimation methods include the open-circuit voltage method, ampere-hour accumulation method, electrochemical test method, battery model method, neural network method, impedance frequency advance method, and Kalman filter method.

1.4.2 State of Health

The state of health (SOH) characterizes the ability of current lithium-ion batteries to store electrical energy relative to new lithium-ion batteries [108–111]. It is the state of the battery from the beginning to the end of its life as a percentage, which is used to quantitatively describe the performance state of the lithium-ion batteries. China automotive industry standard 'QC/T743-2006' stipulates the end-of-life conditions for lithium-ion batteries: The available capacity decays to 80% of the standard capacity. That is, the SOH of the lithium-ion battery is reduced from 100% to 80%, which is the battery's life span. Presently, the definition of SOH is mainly embodied in several aspects such as capacity, power, internal resistance, number of cycles, and peak power. The SOH, which is generally recognized in the industry, is defined in terms of capacity. The expression is shown in Equation (1.14).

$$\mathrm{SOH} = \frac{Q_a}{Q_r} \times 100\% \tag{1.14}$$

In Equation (1.14), Q_r is the rated capacity of the battery when it leaves the factory, and Q_a is the available capacity of the battery after it is put into application. If $Q_{\max}$ is obtained directly through complete discharge, although it is more accurate, the method takes a long time and cannot be estimated in real time. The calculation method of the current time point capacity can be directly derived, as shown in Equation (1.15).

$$Q_{t+1} = \frac{\int_t^{t+1} I(t)\eta\, dt}{\mathrm{SOC}_{t+1} - \mathrm{SOC}_t} \tag{1.15}$$

In Equation (1.15), the SOC estimation of the battery has no direct internal connection with the health state. The above equation only establishes the numerical relationship between SOC and the current capacity of the battery at the time point. Since the SOC does not mutate into the experiment, the capacity obtained by the equation does not mutate in theory. However, it depends on the accuracy of the denominator, which requires a higher estimation accuracy of SOC value.

1. Forgetting factor dual particle filtering algorithm construction
   The particle filtering algorithm is a kind of random tracking prediction algorithm, which approximates the probability density function of system variables through a series of random samples and is suitable for nonlinear and non-Gaussian distributions [112,113]. The traditional particle filtering algorithm is commonly used to realize SOC estimation, but its accuracy immensely depends on the accuracy of parameter identification. In the research, the forgetting factor algorithm is introduced to improve the

accuracy of parameter identification and then improve the estimation accuracy. The state-space equation of the system is shown in Equation (1.16).

$$\begin{cases} x_{k+1} = A_k x_k + B_x u_k + w_k \\ y_k = C_k x_k + D_k u_k + v_k \end{cases} \tag{1.16}$$

In Equation (1.16), x_k is the state vector, and the first element is the SOC of the batteries. w_k is the process noise, v_k is the expression of the observation noise, u_k is the input variable, which is the terminal current I_L, and y_k is the output variable, namely the terminal voltage U_L. The matrices of A_k, B_k, C_k, and D_k are defined, as shown in Equation (1.17).

$$\begin{cases} A_k = \begin{bmatrix} 1 & 0 & 0 \\ 0 & e^{-T/(R_1C_1)} & 0 \\ 0 & 0 & e^{-T/(R_2C_2)} \end{bmatrix}; B_k = \begin{bmatrix} -\eta T/C \\ R_1\left(1-e^{-T/(R_1C_1)}\right) \\ R_2(1-e^{-T/(R_2C_2)}) \end{bmatrix} \\ C_k = \begin{bmatrix} U'_{OC} \\ -1 \\ -1 \end{bmatrix}; D_k = R_k \end{cases} \tag{1.17}$$

In Equation (1.17), R_k is the internal resistance of the battery at time point k.

The extended particle filtering algorithm mainly includes six steps: particle initialization, particle state update, weight normalization, resampling, state estimation, capacity, and SOH prediction. To reduce the influence of experimental measurement errors, this research introduces the particle voltage junction, current, and voltage measurement noises, respectively. The flowchart of the extended particle filtering algorithm is shown in Figure 1.5.

1. Particle initialization: The N particle is selected in each sampling interval to form the n-dimensional particle set $\left\{\mathrm{SOC}_0^i\right\}_{i=1}^N$. To ensure accuracy, the number of particles sampled in the research is 1000. The initial value of N particles is set as 0.9 to test the convergence of the algorithm, as shown in Equation (1.18).

$$\left\{L_0^i\right\}_{i=1}^N = \left\{\mathrm{SOC}_0^i\right\}_{i=1}^N \tag{1.18}$$

2. Particle state update: The samples are randomly sampled to produce a new set of n-dimensional particles $\left\{\mathrm{SOC}_k^i\right\}_{i=1}^N$. According to the characteristic that the weight satisfies the Gaussian distribution, the recursive relation between weights can be utilized, the weights of all the particles

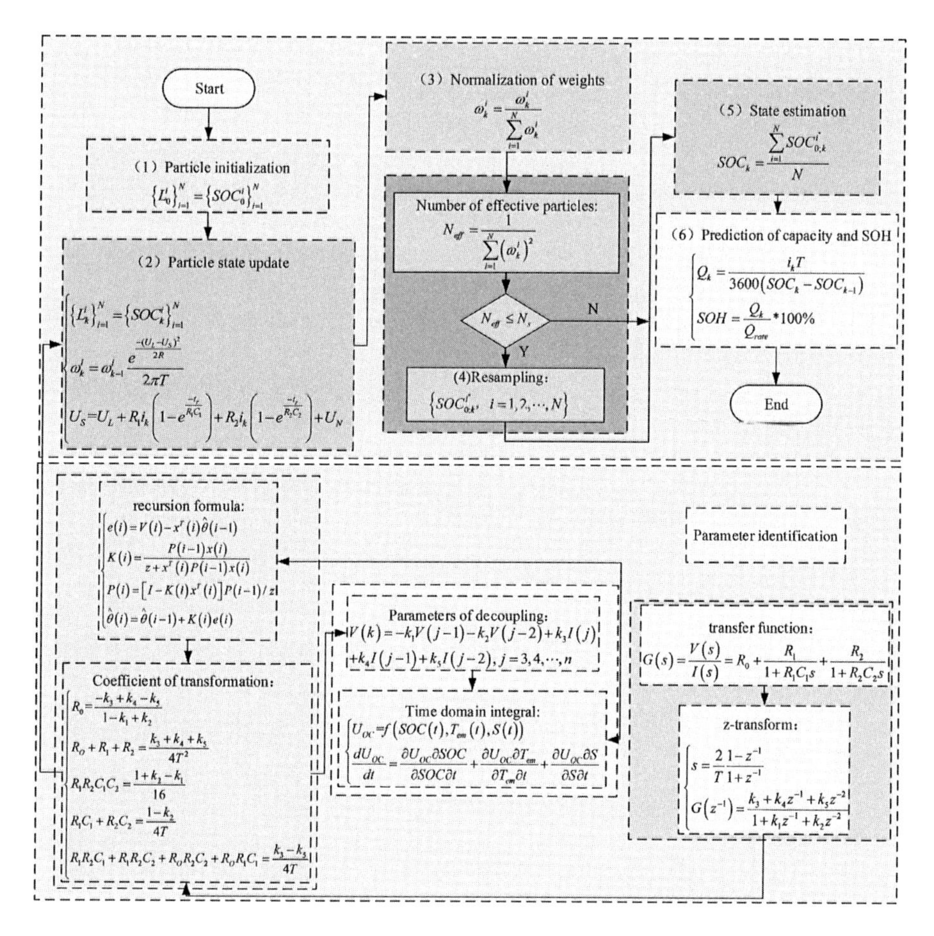

FIGURE 1.5 Flowchart of FFDPF algorithm.

ω_k^i are obtained, R is the observed noise covariance, U_S is the analog voltage, U_N is the voltage noise, and t_r is the zero-state response duration, as shown in Equation (1.19).

$$\begin{cases} \left\{L_k^i\right\}_{i=1}^N = \left\{\mathrm{SOC}_k^i\right\}_{i=1}^N, \ \omega_k^i = \omega_{k-1}^i \dfrac{e^{\frac{-(U_L-U_S)^2}{2R}}}{2\pi T} \\ U_S = U_L + R_1 i_k\left(1-e^{\frac{-t_r}{R_1C_1}}\right) + R_2 i_k\left(1-e^{\frac{-t_r}{R_2C_2}}\right) + U_N \end{cases} \tag{1.19}$$

3. Normalization of weight: After the weights are updated, they need to be normalized again to ensure that the sum of the ownership values is 1, which satisfies the Gaussian distribution and provides an important prerequisite for subsequent resampling, as shown in Equation (1.20).

$$\omega_k^i = \frac{\omega_k^i}{\sum_{i=1}^{N} \omega_k^i} \tag{1.20}$$

4. Resampling: In the filtration process, after several iterations, the weight of many particles becomes too small to be considered, and only a few particles have a large weight, the phenomenon known as particle degradation. To suppress this phenomenon, it is necessary to resample the particles. A new n-dimensional particle set $\left\{\mathrm{SOC}_{0;k}^{i^*},\ i = 1,2,\ldots,N\right\}$ is formed by copying the larger weight particles and discarding the smaller weight particles, each particle weights $1/N$ in the particle set. Before resampling, the effective particle number N_{eff} is calculated to determine whether it reaches the set threshold of N_s (generally set as 2/3 of the selected particle). If not, resampling should be performed, as shown in Equation (1.21).

$$\begin{cases} N_{\mathrm{eff}} = \dfrac{1}{\sum_{i=1}^{N} \left(\omega_k^i\right)^2} \\ \left\{\mathrm{SOC}_{0;k}^{i^*},\ i = 1,2,\ldots,N\right\} \end{cases} \tag{1.21}$$

5. State estimation: The weight of N particles is obtained after sampling. Since the state estimation object is SOC and the value range is (0, 1), which is the same as the weight obtained after normalization, the average value of the weight after resampling can be directly calculated to obtain the SOC at time point k, as shown in Equation (1.22).

$$\mathrm{SOC}_k = \frac{\sum_{i=1}^{N} \mathrm{SOC}_{0;K}^{i^*}}{N} \tag{1.22}$$

6. Prediction of capacity and SOH: Based on the battery capacity estimation method of Equation (1.21), the variation of battery capacity of the process can be calculated directly after the SOC is obtained. i_k is the terminal current that the measured noise is added. The SOH calculation method in Equation (1.22) is incorporated into the algorithm, as shown in Equation (1.23).

$$\begin{cases} Q_k = \dfrac{i_k T}{3600\left(\mathrm{SOC}_k - \mathrm{SOC}_{k-1}\right)} \\ \mathrm{SOH}_k = \dfrac{Q_k}{Q_{\mathrm{rate}}} * 100\% \end{cases} \tag{1.23}$$

The following is the online parameter identification process, which can obtain the change process of the five parameters, including ohmic resistance R_0, polarization resistance R_1 and R_2, and polarization capacitance C_1 and C_2 during the whole experiment. The transfer function is shown in Equation (1.24).

$$\begin{cases} V(s) = U_{OC}(s) - U(s) \\ G(s) = \dfrac{V(s)}{I(s)} = R_0 + \dfrac{R_1}{1 + R_1 C_1 s} + \dfrac{R_2}{1 + R_2 C_2 s} \end{cases} \tag{1.24}$$

The transfer function in Equation (1.25) only establishes the mathematical relationship between the parameters. Also, the Laplace z-transformation in the equation is utilized to obtain the standard differential equation by substitution of s and z, that is, the bilinear substitution. T is the sampling time, and the result is shown in Equation (1.26).

$$\begin{cases} s = \dfrac{2}{T}\dfrac{1 - z^{-1}}{1 + z^{-1}} \\ G\left(z^{-1}\right) = \dfrac{k_3 + k_4 z^{-1} + k_5 z^{-2}}{1 + k_1 z^{-1} + k_2 z^{-2}} \end{cases} \tag{1.26}$$

In Equation (1.26), k_1, k_2, k_3, k_4, and k_5 are the conversion coefficients, which have a direct mathematical relationship to the parameters to be identified and T, and these coefficients are taken advantage of to obtain R_0, R_1, R_2, C_1, and C_2, as shown in Equation (1.27).

$$\begin{cases} R_0 = \dfrac{-k_3 + k_4 - k_5}{1 - k_1 + k_2}; \; R_0 + R_1 + R_2 = \dfrac{k_3 + k_4 + k_5}{4T^2} \\ R_1 R_2 C_1 C_2 = \dfrac{1 + k_2 - k_1}{16}; \; R_1 C_1 + R_2 C_2 = \dfrac{1 - k_2}{4T} \\ R_1 R_2 C_1 + R_1 R_2 C_2 + R_0 R_2 C_2 + R_0 R_1 C_1 = \dfrac{k_3 - k_5}{4T} \end{cases} \tag{1.27}$$

In Equation (1.27), to obtain model parameters R_0, R_1, R_2, C_1, and C_2, the sampling period T is known, and k_1, k_2, k_3, k_4, and k_5 should be calculated first. To facilitate further parameter identification, Equation (1.27) is conducted parameter decoupling, and the general form of the model state difference equation is obtained, as shown in Equation (1.28).

$$V(k) = -k_1 V(j-1) - k_2 V(j-2) + k_3 I(j) + k_4 I(j-1) + k_5 I(j-2), \quad j = 3,4,\ldots,n \tag{1.28}$$

In Equation (1.28), j is the serial number of test data V and I, and n is a natural number. However, in practical application, there are certain functional relationships between the open-circuit voltage U_{OC} and SOC, the battery operating temperature T_{em}, and the life S, and the time domain integration is carried out. The result is shown in Equation (1.29).

$$\begin{cases} U_{OC} = f\left(\text{SOC}(t), T_{em}(t), S(t)\right) \\ \dfrac{dU_{OC}}{dt} = \dfrac{\partial U_{OC}}{\partial \text{SOC}}\dfrac{\partial \text{SOC}}{\partial t} + \dfrac{\partial U_{OC}}{\partial T_{em}}\dfrac{\partial T_{em}}{\partial t} + \dfrac{\partial U_{OC}}{\partial S}\dfrac{\partial S}{\partial t} \end{cases} \tag{1.29}$$

The above equation is simplified to facilitate the identification of model parameters. Since the sampling time is 0.1 seconds, the variation in SOC and battery life S during the sampling period is almost negligible. The constant discharge test of the battery is carried out in an incubator; hence, the temperature of the battery changes slowly with no variation, as shown in Equation (1.30).

$$\begin{cases} \dfrac{\partial \text{SOC}}{\partial t} = \dfrac{\partial S}{\partial t} = 0 \\ \dfrac{\partial T_{em}}{\partial t} = 0 \end{cases} \tag{1.30}$$

Therefore, through the above simplification, it can be considered that the open-circuit voltage U_{OC} of the battery in the process of discharge remains unchanged and is equal to the terminal voltage at the time point before the battery begins to discharge. The value can be obtained from Equation (1.30). In the actual iteration process, there exists a 'data saturation phenomenon', which may lead to large changes in errors, resulting in abnormal variations in battery parameters, and the accuracy of the model can be affected. A forgetting factor is introduced on the basis of recursive least squares (RLS) method, and the forgetting factor recursive least squares (FFRLS) method is established for parameter identification, which can reduce the phenomenon of 'data saturation', as shown in Equation (1.31).

$$\begin{cases} e(i) = V(i) - x^T(i)\hat{\theta}(i-1) \\ K(i) = \dfrac{P(i-1)x(i)}{z + x^T(i)P(i)x(i)} \\ P(i) = \left[I - K(i)x^T(i)\right]P(i-1)/z \\ \hat{\theta}(i) = \hat{\theta}(i-1) + K(i)e(i) \end{cases} \tag{1.31}$$

In Equation (1.31), i is the number of iterations, $e(i)$ is the estimation error of $V(i)$, $\hat{\theta}(i)$ is an expression of the column matrix of the calculated values of coefficients $k_1 \sim k_2$ during the ith iteration, $x(i)$ is the input value column matrix of the experimental matrix, $P(i-1)$ is the covariance matrix, and $K(i)$ is the gain matrix. It can be observed from Equation (1.31) that the first iteration started with the input of the third group of test data, and the $\hat{\theta}$ and P in the first iteration should be set according to actual needs. $e(i)$, $K(i)$, and $P(i)$ are obtained by calculation, respectively, and then, the $\hat{\theta}(i)$ is obtained. In the above equation, the smaller the z is, the faster the iteration convergence speed is, and the greater the influence of new sampling data on the identification parameters is, but the variation of iteration error also increases. z is generally within the range of (0.95, 1), which is 0.998 in the research. Then, the first iteration is completed. The fourth group of experimental data can be introduced into x^T to start the second iteration. By replacing a set of data, $e(i)$ goes to zero. At this point, the iteration is complete, and then, the $\hat{\theta}(i)$ is the values from k_1 to k_5.

2. Experimental results
    1. Estimation of SOC and SOH under constant current condition

       The experimental results are conducted with the SOC and SOH estimation under the constant current condition. The constant current condition is that under 1 C (40.00 A) constant current condition, the battery is discharged from 4.20 V at full voltage to 2.75 V at cutoff voltage. The current and voltage are shown in Figure 1.6.

       From Figure 1.6, although the battery current is theoretically constant, the actual value cannot be kept constant but fluctuates sharply within a small range. The experimental data were imported into the algorithm to obtain the state estimation results of the lithium-ion battery under constant current conditions, as shown in Figure 1.7.

       In Figure 1.7, the error of the forgetting factor dual-particle filter (FFDPF) algorithm for SOC estimation is less than that of other algorithms. The nonlinear error in the extended Kalman filter (EKF) algorithm leads to the accumulation of estimation error, resulting in the divergence of estimation results. The estimation result of the unscented

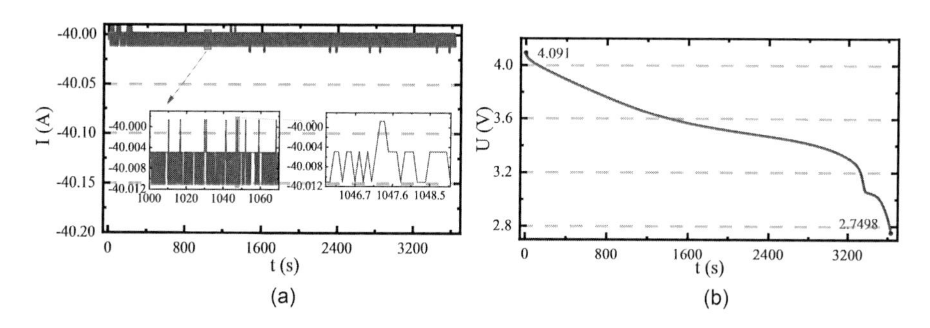

FIGURE 1.6 Variation of current and voltage under constant current conditions. (a) Variation of current. (b) Variation of voltage.

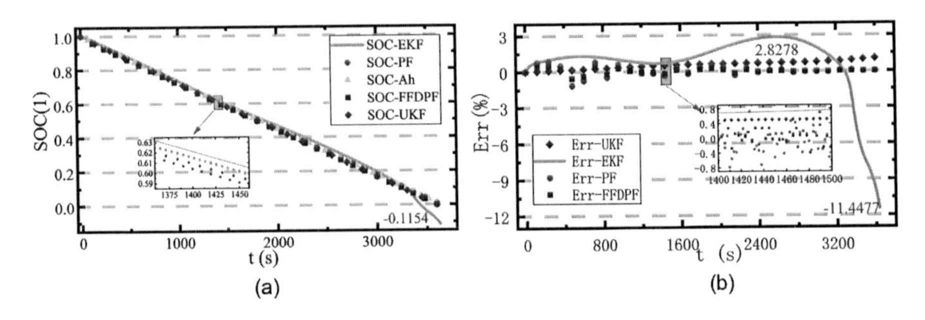

FIGURE 1.7 Battery SOC estimation results under constant current discharge conditions. (a) Result of SOC estimation. (b) Comparison of estimation errors.

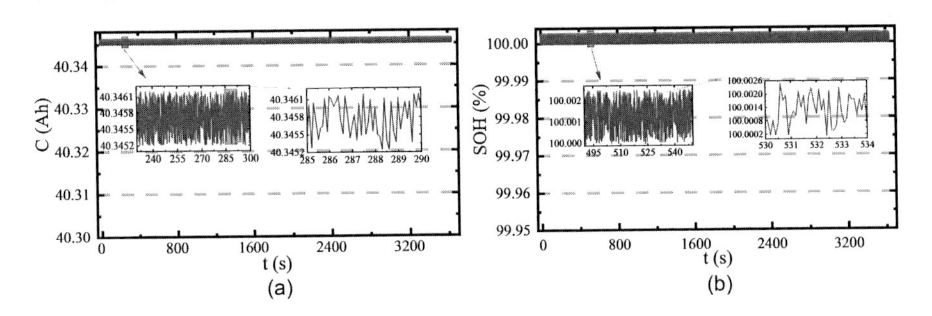

FIGURE 1.8 Battery state of health estimation results under constant current discharge conditions. (a) Result of capacity estimation. (b) Result of SOH estimation.

Kalman filter (UKF) algorithm also diverges in the later period. Compared with the particle filter (PF) algorithm, the FFDPF algorithm also tends to converge in the later stage, but the amplitude of error waviness of the algorithm is smaller. Due to the higher battery current under constant current conditions, the internal reaction of the battery is more intense, which makes the capacity increase slightly larger, as shown in Figure 1.8.

In Figure 1.8, the maximum estimation error of SOH is within 0.87%. The estimated battery capacity can effectively predict the SOH with an accuracy of 99.7%, which lays the foundation for the safety management of the BMS.

2. SOC and SOH estimation in dynamic stress test (DST) condition
Under DST conditions, the battery is charged and discharged in a constant current cycle with a charging current of 0.5 C and a discharging current of 1 C until the battery voltage reaches 2.50 V cutoff voltage. The variations of current and voltage are shown in Figure 1.9.

In Figure 1.9, the current, voltage, and time are imported into the forgetting factor dual particle filtering algorithm together to obtain the SOC value in the whole experiment process. By comparing the

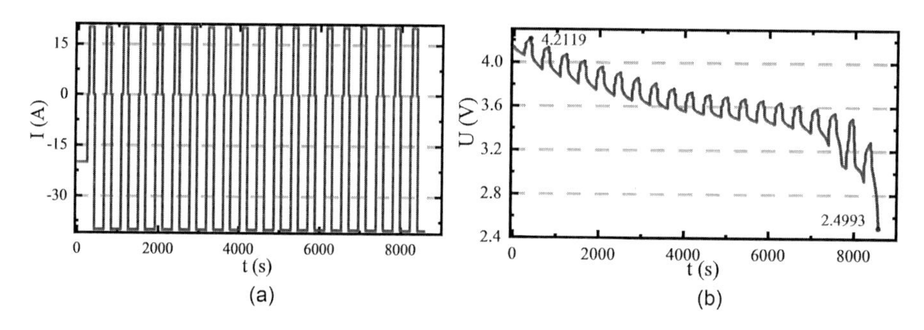

FIGURE 1.9 Variations of current and voltage under DST conditions. (a) The variation of current. (b) The variation of voltage.

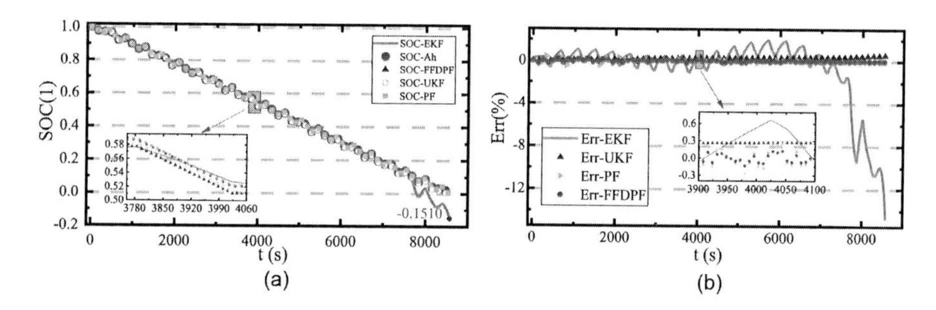

FIGURE 1.10 Battery SOC estimation results under DST conditions. (a) Result of SOC estimation. (b) Comparison of estimation errors.

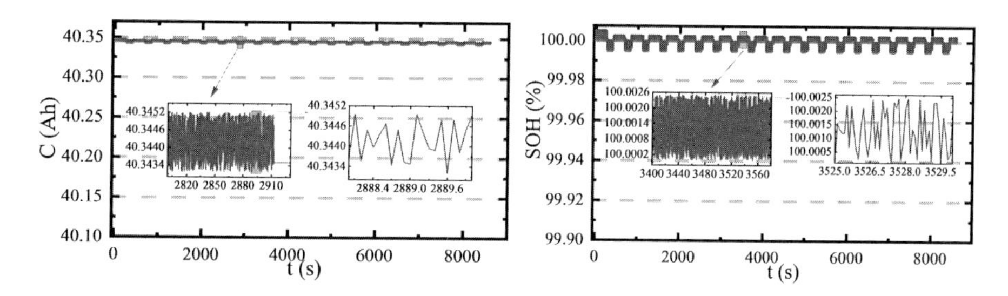

FIGURE 1.11 Battery SOH estimation results under DST conditions. (a) Result of capacity estimation. (b) Result of SOH estimation.

estimated results with the EKF, UKF, and PF algorithms, the errors are obtained, as shown in Figure 1.10.

In Figure1.10a, the FFDPF algorithm has better convergence of SOC estimation results compared with other algorithms. When the battery is close to full discharge, the error of EKF estimation results rapidly accumulates, resulting in a divergence phenomenon. The SOH estimation results under DST conditions are shown in Figure 1.11.

The capacity estimation result is described in Figure 1.11. During the whole implementation process, the battery capacity does not change significantly, which is determined by the stability of the capacity of the lithium-ion battery itself. The capacity of the new battery is 40.3432 Ah. At this time, the battery is still in a relatively active state. During the experiment, the capacity increases slightly, which is the expression of the activation of active substances inside the batteries. In experimental test, the capacity and SOH of the battery change significantly only after hundreds of full charge–discharge cycles.

In addition, SOH is also evaluated from the perspective of other performance parameters of the batteries. This is because the performance of some battery parameters changes after a period of the charging and discharging cycles. By looking for the relationship between the changed performance parameters and the initial value, the state of the battery after the performance decline can be obtained, which is used to characterize the health status of the batteries. SOH is defined from the perspective of battery internal resistance, as shown in Equation (1.32).

$$SOH = \frac{R_{EOL} - R}{R_{EOL} - R_N} \tag{1.32}$$

In Equation (1.32), R_{EOL} is the internal resistance of the battery at the end of its life, R_N is the resistance of the battery when it leaves the factory, and R is the internal resistance of the battery in its current state. This estimation method does not need to consider the change in battery capacity. As long as the internal resistance of the battery is obtained, the change of SOH can be estimated according to the change of internal resistance. The internal resistance of the battery changes with the service life of the batteries. The longer the service life of the battery is, the greater the internal resistance of the battery is. SOH is defined from the perspective of battery cycle times, as shown in Equation (1.33).

$$SOH = \frac{N_r}{N_t} \times 100\% \tag{1.33}$$

In Equation (1.33), N_r is the number of cycles remaining and N_t is the total number of cycles. It can be observed from Equation (1.33) that the total number of cycles of the battery is certain. With the use of the battery, the number of cycles of the battery increases, the number of remaining cycles of the battery is less, and the SOH value of the battery is smaller. Therefore, the number of cycles of the battery can also reflect the SOH of the batteries. Considering that in the actual use of the vehicle, users generally do not charge–discharge completely, it is difficult to obtain the effective number of cycles. Thus, using the number of cycles to estimate the SOH does not have much practical meaning. At present, the definition of lithium-ion battery SOH is mainly based on the above three battery performance parameters.

1.4.3 State of Power

The state of power (SOP) is the power output capability of the battery when discharging, and SOP is often used as its constant index [114–117]. The SOP of the battery refers to the maximum power that can be continuously discharged for t time based on the current state without violating the preset use restrictions of the batteries. Therefore, SOP is one of the measurement standards for the power performance of electric vehicles. The SOP can evaluate the charge–discharge power capabilities of the power battery in different health states and states of charge. In addition, it can optimize the matching relationship between the power battery and the vehicle dynamics, as well as the regenerative braking energy recovery performance of electric vehicles.

For vehicle battery pack power supply design, when a conservative battery power design is adopted to meet the power requirements, it is easy to cause or reduce excessive battery cost and weight, which affects the performance of the vehicles. However, when the design of battery power is relatively small, it may cause overdischarge damage to the battery during usage. The SOP of the power battery changes with the discharge temperature, SOC, and SOH. Especially, the SOP drops significantly under low temperature, low SOC, and low SOH conditions. Besides, if constant peak power is used as a limiting condition, it may cause harm [118]. Therefore, it is essential to estimate the peak power in real time with high accuracy.

1.4.4 State of Energy

The state of energy (SOE) is a direct reflection of the remaining available energy of the battery, and it is also the key to energy optimization and management. At present, there are few real-time estimation types of research on SOE in the application market, and battery energy is managed only by estimating the SOC value [119,120]. However, SOC describes the remaining percentage of active materials in batteries and cannot provide more accurate energy information, which is essentially different from SOE. With more application scenarios of lithium-ion batteries, the battery application conditions are extremely complex, and the efficiency and adaptability of management need to be improved. For more detailed battery management, it is necessary to know the remaining battery capacity and power loss of the batteries. Therefore, the proposal and estimation of SOE have practical engineering significance.

There is a positive correlation and nonlinear functional relationship between SOE and SOC values. The positive correlation between SOE and SOC comes from the fact that SOE is the product of open circuit voltage (OCV) and residual capacity, while SOC is the percentage of residual capacity to the rated capacity. The nonlinear relationship between SOE and SOC is due to the decreasing trend of OCV with the development of the battery discharge process. Also, the heat generated by the internal resistance and electrochemical reaction consumes the energy stored in the battery. During the use of the lithium-ion battery, the discharge current changes dynamically with the load, the battery temperature changes significantly, and the performance of the lithium-ion battery also changes. Therefore, under the same SOC, when the battery temperature and discharge rate change, the discharge efficiency of the battery changes and affects the SOE value of the battery. In the actual operating conditions, the discharge current and temperature changes caused by the load result in a great interference to the accurate SOE estimation of the batteries.

SOC is usually used to characterize the battery SOE, but SOE is not the same as the SOC value [121]. The main differences include: (1) SOC describes the percentage of active materials in the battery, and the factors that affect battery SOE are except SOC value that it also contains different application conditions of the batteries. SOC provides the percentage of the battery's capacity, which cannot provide more accurate energy information. (2) There is a positive correlation and nonlinear functional relationship between SOE and SOC values. The SOC decreases linearly with discharge current, but SOE is the product of OCV and capacity that as the battery's discharge process progresses, the OCV shows a decreasing trend. The battery's heat caused by the internal resistance and electrochemical reaction also consumes the battery's energy. The relationship between SOE and current is not linear, resulting in a difference between SOC and SOE. (3) Discharge current and temperature variation cause changes in battery discharge efficiency, which in turn affects the performance of lithium-ion batteries. Therefore, under the same SOC, when the battery temperature and discharge rate change, the SOE is not at a constant value. The SOE directly reflects the energy that the battery can release, which is a key parameter for the management and optimization of the power battery pack. The definition of SOE is shown in Equation (1.34).

$$\mathrm{SOE}(t)=\left(E_c-E_d\left(t\right)\right)/E_c*100\% \tag{1.34}$$

In Equation (1.34), SOE(t) is the remaining energy of the battery at time point t, is E_c the total energy of the battery, and $E_d(t)$ is the accumulated energy released by the battery at time point t. Normally, the SOE reaches the maximum when the battery is fully charged, and the SOE is zero when the battery is discharged to the cutoff voltage.

1.4.5 Remaining Useful Life

1. Cycle life

   Cycle life is an important parameter to evaluate the technical economy of lithium-ion batteries. The battery undergoes a complete charge–discharge process, which is called a cycle [122–126]. Under a certain discharge system, the number of cycles that the battery can withstand before the capacity of a lithium-ion battery drops to a certain value (usually 80% of the rated capacity) is called the cycle life or service period of the lithium-ion batteries. The cycle life is affected by the depth of discharge (DOD), so the cycle life should also be indicated at the same time as the DOD. As the number of charge–discharge cycles increases, the capacity degradation of lithium-ion batteries is an inevitable process. It is because some irreversible electrochemical processes occur inside the battery during the charge–discharge cycle, causing the battery's discharge capacity to decay. These irreversible factors are mainly as follows.

1. During the charge–discharge cycle, the active surface area of the electrode decreases continuously, which increases the working current density and polarization.
2. The active materials on the electrode fall off or transfer.
3. During the working process of the battery, some electrode materials corrode.
4. Dendrites are formed on the electrodes during the cycle, causing micro-short circuits inside the battery.
5. Aging and loss of diaphragm.
6. The irreversible crystal form of the active material changes during the charge–discharge cycle, thus reducing the activity.
7. Temperature is undoubtedly one of the key factors affecting the life of lithium-ion batteries. Too high or too low temperature leads to a decrease in active lithium-ion content, thus reducing the life of lithium-ion batteries.

2. Storage life

The capacity of the battery changes after long-term storage, which is called storage performance. Although the battery does not release electric energy during storage, there is always a self-discharge phenomenon inside the battery. Even in dry storage, moisture, air, carbon dioxide, and other substances lead to the spontaneous oxidation–reduction reaction of the positive and negative electrodes, which consumes active substances of the battery resulting in a decrease in battery capacity. In addition, if stored in a wet area, this situation will be more serious. The magnitude of the self-discharge is expressed by the time that the battery capacity decreases to a specified capacity, that is, storage life or shelf life.

1.4.6 Temperature Performance

The battery temperature is the battery surface heating phenomenon of the battery that is due to chemical, electrochemical changes, electronic migration, material transfer, and other reasons in the internal structure, which is a normal phenomenon. If the heat generated by the battery cannot be completely dissipated to the environment, it causes the continuous accumulation of heat inside the batteries. Once the accumulation of heat causes the high-temperature point inside the battery, it is possible to cause the heat out of control of the batteries. Lithium-ion batteries have an optimum operating temperature range and need to avoid thermal runaway, thus requiring thermal management [127–129].

The temperature characteristics of the battery indicate the performance of the power battery due to the temperature changes. The operating temperature of conventional lithium-ion batteries ranges from –20°C to +60°C, and low-temperature lithium-ion batteries made of special materials can be discharged in the −40°C alpine environment. However, the nominal voltage and maximum capacity of the battery will vary with temperature, such as the voltage is 4.20 V, put it in the –40°C environment, and the voltage quickly drops below 3.40 V. Due to the material characteristics of the lithium-ion battery, charging in a low-temperature environment can cause serious damage to the batteries.

The battery temperature conversion is the process of converting the parameters of the power battery capacity and the electrolyte-specific gravity at different temperatures into the values at the standard temperature. The temperature coefficient of the cell is the ratio of the capacity available to the power cell relative to the available capacity at the standard temperature. The temperature factor, a battery characteristic, is very important because the battery temperature is a major factor affecting the battery power output. Due to the different temperatures of the working environment and different voltage, current, and power, the operating state of the battery at the extreme temperature needs to be budgeted in the circuit design.

Therefore, when using lithium-ion batteries as energy supply equipment, the temperature should be considered. To obtain the most effective and efficient charging method, a satisfactory temperature should be selected, because once the temperature is too high, the battery releases a lot of heat. If the heat rises suddenly, it damages the battery easily, and under severe conditions, it catches fire. In addition to affecting the normal operation of the battery, the temperature also affects the key parameters in the battery, such as internal resistance. The lower the ambient temperature of a lithium-ion battery, the greater the internal resistance of the battery. Due to the increase in the internal resistance of the battery, the power consumption of the battery is very fast, which leads to the rapid capacity consumption of the battery. The relationship between the heat generation of lithium-ion battery and the current flowing through the battery and internal resistance is shown in Equation (1.35).

$$Q = I^2 R_{in} \tag{1.35}$$

In Equation (1.35), Q is the heat generation rate of lithium-ion batteries. I is the current flowing through the lithium-ion battery. R_{in} is the internal resistance of the lithium-ion batteries. Under normal room-temperature operating conditions, the capacity changes in direct proportion to the increase in temperature.

1.5 CONCLUSION

This chapter mainly starts with the working mechanism of the lithium-ion battery and discusses the basic concept, development status, and development trends of lithium-ion batteries. The research on lithium-ion batteries has broad prospects for development and is widely used in all walks of life. The industrial chain formed by the lithium-ion battery industry occupies a large share in the world market, which plays an important role in promoting the emerging intelligent industry. Also, as one of the new energy sources, it adjusts the energy supply structure of the mainstream market. It can be observed that the accurate measurement of the voltage, capacity, internal resistance, energy, and power density affects the accurate estimation of the relevant state of the lithium-ion batteries, thus affecting the application of lithium-ion batteries in practice. The SOC, SOH, SOP, SOE, and service life are described from different aspects, which makes the research of lithium-ion batteries more standardized and comprehensive. Lithium-ion batteries serve as an essential energy source in today's world due to technological advancement and usage in every aspect of human lives. Therefore, it is of great practical significance to study the reliability, safety, and service life of the lithium-ion battery. Lithium-ion battery systems are complex systems that combine chemical, electrical, and mechanical properties, so they must be designed with a wide range of characteristics in mind. Based on an understanding of the basic principles of lithium-ion batteries, this chapter analyzes the basic parameters of lithium-ion batteries, such as terminal voltage, electromotive force, capacity, internal resistance, and the core elements of battery state, such as SOC, DOD, cycle life, and self-discharge rate. The lithium-ion battery industry is an important direction of global high-tech development now. Because the lithium-ion battery has good electrochemical stability, large energy density, long battery life, and maintenance-free characteristics, it is widely used in electric vehicles and a variety of energy storage equipment. With the electric vehicles, lithium-ion batteries are gradually moving toward the market, so the usage and consumption of lithium resources in the world have increased significantly, resulting in the huge and broad development prospects of the industrial chain.

2 Equivalent Modeling and Parameter Identification

Chunmei Yu, Haotian Shi, Jie Cao, Weijia Xiao, and Wenhua Xu
Southwest University of Science and Technology

Shunli Wang
Sichuan University
Southwest University of Science and Technology

Xiao Yang, Xin Xiong, Yongcun Fan, and Yuhong Jin
Southwest University of Science and Technology

CONTENTS

DOI: 10.1201/9781003333791-2

Introduction: This chapter introduces the modeling knowledge, common working conditions, and parameter identification methods of lithium-ion batteries. Battery modeling serves as the basis for the research of lithium-ion batteries [130–133]. The selection of model type and the accuracy of the model are very important for the research of lithium-ion batteries. The accurate estimation of lithium-ion battery state of charge (SOC) depends on the equivalent model to a great extent, especially its characterization of battery dynamic characteristics. Therefore, understanding the establishment of the lithium-ion battery model lays the foundation for the next chapter of lithium-ion battery SOC estimation and subsequent multistate monitoring of the batteries. After the model is established, it is necessary to select the appropriate identification method to obtain the lithium-ion battery parameters. Then, the system model parameters are identified and optimized according to the simulation results, which are used to observe whether the model fitting effect is consistent with the real-time working conditions.

2.1 OVERVIEW OF BATTERY EQUIVALENT CIRCUIT MODELING

The equivalent circuit models, parameter identification [134,135], and state-space equation play a very important role in the SOC estimation process, which involves multidisciplinary integration. The establishment of the battery model mainly includes the following steps.

1. Selecting the appropriate model: The technical route of battery modeling is determined according to the performance characteristics, application, and model accuracy. The appropriate model is selected from the appropriate model, including the model structure with unknown parameters obtained according to the mechanism. The electrochemical model is established that is based on the internal electrochemical reaction of the battery, or a black-box model.
2. Designing experimental schemes and identifying model parameters: The identification of battery model parameters should be based on the actual measurement data. The identification of model parameters aims to minimize the error between the fitted curve and the measured curve. The commonly used identification methods of model parameters include the recursive least-squares method, genetic algorithm, neural network method, Kalman filtering (KF) algorithm, etc.

3. Verification of model accuracy: There are mainly four ways of verification. (a) Using prior knowledge verification to judge whether the model is practical or not according to the existing knowledge of the system. (b) After a model is identified by one set of data, the applicability is verified by another set of data that are not involved in the identification. (c) Analyzing the actual response and impulsion response by comparison. (d) The autocorrelation function of the excitation signal is used for verification.

Three types of commonly used models of lithium-ion batteries are electrochemical models based on chemical reactions within the battery, neural network models that simulate the work of the human brain, and equivalent circuit models built using electronic components. Electrochemical models, including the pseudo-two-dimension (P2D) model and single-particle model (SPM), are rarely used in engineering applications because of their complex structure and high computational cost. The neural network model has a large estimation error in the case of insufficient sample data [136,137]. The convergence and stability of the algorithm are temporal, which cannot guarantee the timeliness of the model, so it is currently in the simulation experiment stage. The equivalent circuit model is currently the most widely used battery model, and its estimation accuracy and robustness are better than the other two types of models. Since the battery is a highly nonlinear system, it is difficult to accurately estimate the SOC and SOH of the battery due to factors such as temperature, working conditions, and aging. In order to improve the estimation accuracy of SOC and SOH, the influence of environmental factors should be fully considered in battery modeling. The key points of the model establishment are described as follows.

1. The logic structure of the circuit model is shown in Figure 2.1.
2. Principle description
   The battery simulation model is to verify the accuracy of parameters set in the model, so the input is current, and the output is terminal voltage. In the actual battery management system, the current and terminal voltage are both inputs.
3. Modeling steps
   (a) Selecting the appropriate equivalent circuit model. (b) Determining the input, output, and state variables. (c) Writing the state-space equation. (d) Simulation software modeling. (e) Parameter identification by

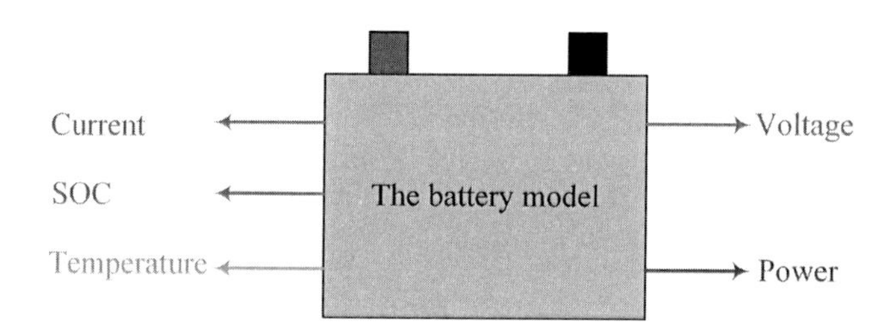

FIGURE 2.1 Logic structure diagram of the circuit model.

experimental data. (f) Modifying model parameters based on variables, such as temperature and SOC. (g) Simulate the circuit and verify its characterization effect.

4. Model selection
   Different battery equivalent models have a great influence on the accuracy of the battery simulation. In theory, the more complex the structure, the higher the computational complexity, and the hardware requirements are increased accordingly. In the equivalent model, the accuracy of the circuit containing the RC loop is high. Generally, the accuracy is higher when the RC order is higher within a certain range.
5. Parameter identification
   According to the experimental results of the hybrid pulse power characterization (HPPC) test, curve fitting is performed by combining with the least-square method to identify relevant parameters.

2.2 MODELING TYPES AND CONCEPTS

The techniques of battery modeling are classified into mechanism modeling [138,139], experimental modeling [140,141], and hybrid modeling methods [142]. Based on scientific equations such as physical equations and concepts of chemical reactions, the mathematical expression of the modeling framework is constructed. The experimental object is regarded as a black box. The model is constructed by experimenting with the variant law on the dynamic characteristic parameters of the target object. The battery is a nonlinear electrochemical energy storage system, and its internal interactions and reactions are difficult to be explained by precise equations. Experimental data models, such as neural networks, involve a great deal of feedback and learning, so hybrid modeling is more prevalent. The electrical characteristic model is usually divided into three parts: the black-box model, electrochemical model, and an equivalent circuit model. The relationship of the modeling process is described in Figure 2.2.

It is necessary to select the correct battery model to get a suitable scheme for obtaining the battery state conveniently and accurately; it is necessary to select the correct battery model to get a suitable scheme. The commonly used equivalent circuit models are given below.

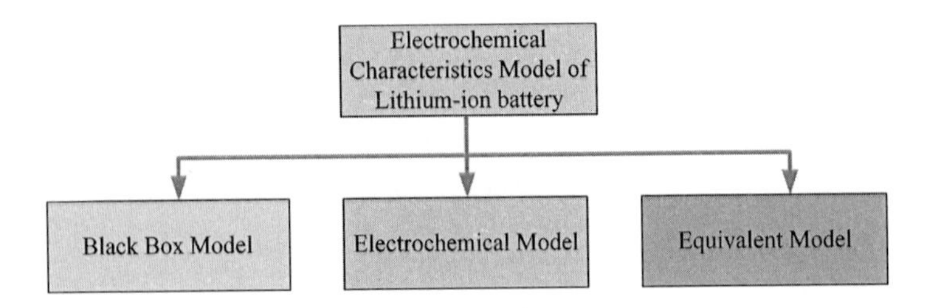

FIGURE 2.2 Connection process of the modeling concept.

1. When the equivalent model of internal resistance is used, the ideal voltage source is implemented to reflect the open-circuit voltage. The internal resistance represented by R_0 and the open-circuit voltage represented by U_{OC} U_{OC} are both related to the SOC levels. The position of the load-discharge is measured only by the positive and the negative of the current pulse.
2. The resistor–capacitor equivalent model is achieved by using circuit resistance and capacitance, which can accurately explain the surface effect. There are three resistors and two capacitors in the model, where C_a is the battery capacity and C_b is the small capacitance generated from the surface effect of the lithium-ion batteries.
3. The Thevenin equivalent model can also be accomplished by the combination of the internal resistance model and the analog circuit of the resistance capacity model, considering the polarization effects.
4. The battery test manual proposes an equivalent model of the PNGV. The PNGV model adds a capacitor Cb to the Thevenin model, which is used to represent the change of the open circuit voltage caused by the accumulation of the load current.

Among the four standard models, the internal resistance model does not consider the transient characteristics of the battery in the electrochemical response stage, and its conversion process cannot be defined in terms of convenience and efficiency in calculation. The resistance–capacitance equivalent model can better simulate the dynamical properties of the battery and increase the definition of the surface impact, temperature, and cell polarization effects that are ignored. Because considering the current accumulation effect, the PNGV model has high accuracy, but the introduction of the series-connected capacitor C_b makes the system vulnerable to cumulative errors in long-term simulation. The error in the polarization effect can be resolved by the Thevenin model [143]. It is very appropriate for the battery charging and discharging study of transient energy. The Thevenin model is a simple structure compared to the PNGV model, the GNL model, and other models that are a nonlinear lower-order model and requires fewer parameters, so it can fulfill engineering application requirements in terms of precision.

2.3 EQUIVALENT CIRCUIT MODELING

According to different modeling principles, battery models are divided into the mechanism and empirical modeling types. The mechanism model is based on the electrochemical theory and uses mathematical methods to describe the reaction process inside the batteries. This model can describe the basic law of battery decay. Based on the mechanism of capacity decay, the electrochemical dynamics and charge transfer processes occurring in the battery are considered. The empirical model is a kind of model that analyzes a lot of actual experimental data and predicts the behavior of the battery without considering the physical and chemical reaction process in the batteries. Polynomial, exponential, power function, logarithm, and trigonometric functions are usually used to express the empirical model.

At present, the most commonly used empirical model is the equivalent circuit model, which is a model based on circuit components such as the power supply, inductance, capacitance, and resistance without considering the chemical composition inside the battery and its corresponding reactions. The equivalent circuit model contains many kinds of model structures. The simple and practical model structure shortens the development cycle and reduces the development cost, but the model accuracy is often lower. Meanwhile, the complex model improves the battery characterization accuracy, but the difficulty of identifying model parameters increases accordingly. Therefore, in practical applications, several factors should be weighed, and the actual accuracy and calculation complexity requirements should be considered comprehensively to establish a suitable battery model. Then, an appropriate battery model should be selected to describe the cycle of the charge–discharge. The following is an analysis of several commonly used equivalent circuit models.

2.3.1 Rint Model

The Rint model is also called the internal resistance model. In the Rint model, the battery is equivalent to the series connection between the ideal voltage source U_{OC} and the resistance R_0. It is a relatively rough battery model, which consists of only a constant voltage source and a resistor in series [144]. However, the polarization characteristics are not considered in this model, so the accuracy of the model is low, and can only approximately represent the ohmic resistance of lithium-ion batteries, as shown in Figure 2.3.

In Figure 2.3, the Rint model regards the lithium-ion battery pack as an ideal power source. U_{OC} is a constant value that represents the open-circuit voltage of the battery, R_0 is the ohmic resistance of the battery, and U_L is the terminal voltage between the positive and negative ends of the batteries. In the application process, the battery operating characteristics are characterized by equating the battery as a series connection of a voltage source and a resistor. Due to the existence of the internal resistance, during the charge–discharge process, some of the energy converted into electricity is consumed by the internal resistance in the form of heat generation. Then, the charge–discharge efficiency is analyzed and carried out to establish the mathematical expression of the working characteristics [145]. To simplify the calculation, U_{OC} and R_0 are set as constant values in the model, ignoring the characteristic that the internal resistance of the battery varies with electrolyte concentration and SOC value, which is inconsistent with the actual parameters of the batteries. The Rint model has low

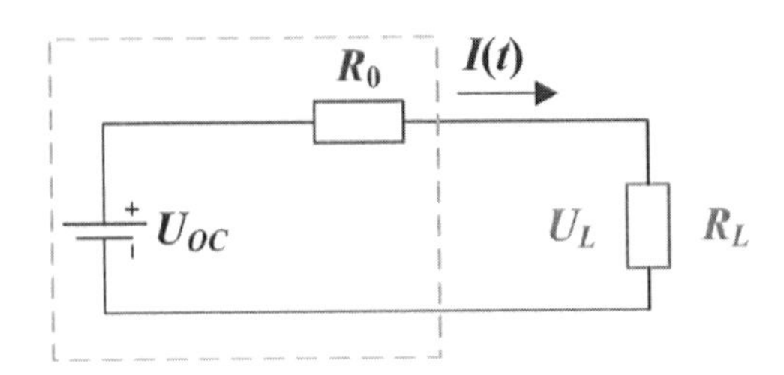

FIGURE 2.3 Internal resistance equivalent structure diagram.

accuracy and is unsuitable for the actual SOC estimation of lithium-ion batteries. The state-space description of the Rint model is expressed by Equation (2.1).

$$U_L = U_{OC} - I(t) * R_0 \tag{2.1}$$

From the analysis of the state space shown in Equation (2.1), the actual output voltage of the battery pack is quite different from the equivalent simulation process of the internal resistance model of the lithium-ion battery pack. Also, the transient characteristics of the electrochemical reaction process are not considered in the model, which cannot accurately characterize the changes in the working process.

2.3.2 RC Model

The RC model has two capacitors and three resistors where the high capacity C_b demonstrates its energy storage potential. The C_s capacitance defines the surface potential and diffusion effect. R_t is the battery terminal resistance, R_s is the surface resistance, and R_e is the end resistance. The RC model structure is shown in Figure 2.4.

When the model is used in an operating environment to simulate the battery behavior at a specific SOC value, the parameters of the model can be considered constants. However, when the influence of these parameters varies along with the temperature changes and needs to be considered, a large SOC working range must be simulated, which makes the process of arranging and processing model parameterization complicated. The battery terminal voltage is shown in Equation (2.2).

$$\begin{cases} \begin{bmatrix} V_{cb}^* \\ V_{cs}^* \end{bmatrix} = \begin{bmatrix} -\dfrac{1}{(R_e+R_s)C_b} & \dfrac{1}{(R_e+R_s)C_b} \\ \dfrac{1}{(R_e+R_s)C_s} & -\dfrac{1}{(R_e+R_s)C_s} \end{bmatrix} \begin{bmatrix} V_{C_b} \\ V_{C_s} \end{bmatrix} + \begin{bmatrix} \dfrac{R_s}{(R_e+R_s)C_b} \\ \dfrac{R_e}{(R_e+R_s)C_s} \end{bmatrix} I_L \\ V_L = \begin{bmatrix} \dfrac{R_s}{R_e+R_s} & \dfrac{R_e}{R_e+R_s} \end{bmatrix} \begin{bmatrix} V_{C_b} \\ V_{C_s} \end{bmatrix} + \begin{bmatrix} \dfrac{R_s+R_e}{R_e+R_s} + R_t \end{bmatrix} I_L \end{cases} \tag{2.2}$$

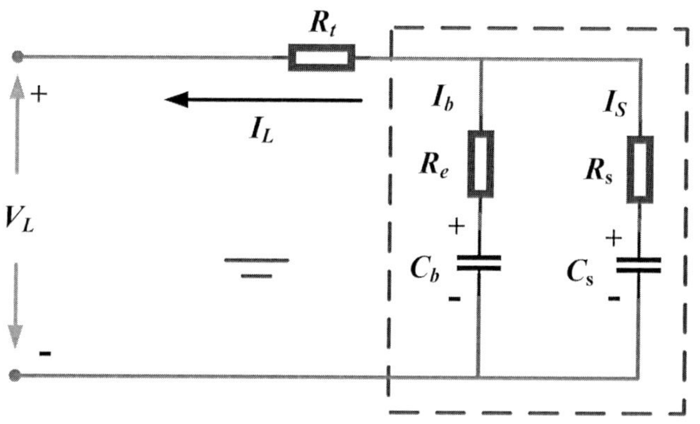

FIGURE 2.4 Schematic diagram of the RC model.

2.3.3 Thevenin Model

Based on the internal resistance model, the Thevenin model is constructed by considering the internal polarization effect during the charge–discharge process of the lithium-ion battery pack. Thevenin model considers that the characteristics of batteries are similar to those of capacitors [146,147]. In this model, the ideal voltage source U_{OC} describes the open-circuit voltage of batteries, the resistance R_0 is the internal resistance of batteries, and the capacitor C_p and resistor R_p are in the parallel connection type to describe the overpotential of batteries. Compared with the Rint model, it has the advantage of adding an RC circuit, which can fully reflect the battery voltage lag phenomenon during the charge–discharge process. The model enhances the nonlinear characterization effect of the model by characterizing the obstruction of polarization effect to current. It realizes the effective characterization of electrochemical polarization and concentration difference polarization effect of the battery through the parallel loop of R_p and C_p. Since the Thevenin model is a first-order model, there are some errors in the dynamic output voltage curve of the model compared with the actual battery output in practical applications. The structure of the Thevenin model is shown in Figure 2.5.

In Figure 2.5, U_{OC} is the open-circuit voltage, R_0 is the ohmic resistance, R_p is the polarization resistance, and C_p is the polarization capacitance. The parallel circuit of R_p and C_p describes the polarization process. U_L is the closed-circuit voltage after the battery is connected to the external circuit. The model takes into account the polarization process of the battery, and the first-order model constructed is relatively simple to calculate. Firstly, the state-space equation should be obtained according to the basic operating characteristics of the capacitor components. The relationship between the current flowing through the battery polarization capacitor C_p and its two closed-circuit voltages is shown in Equation (2.3).

$$I_{Cp}(t) = C_p \frac{dU_{Cp}(t)}{dt} \tag{2.3}$$

In Equation (2.3), based on the analysis of the equivalent circuit composition, through the application of the knowledge of circuit science and Kirchhoff's voltage law (KVL), the voltage relationship in the equivalent circuit of Figure 2.5 can be obtained, as shown in Equation (2.4).

$$R_0 * I_{Cp}(t) + R_0 * \frac{U_{Cp}(t)}{R_P} + U_{Cp}(t) = U_{OC} - U_L \tag{2.4}$$

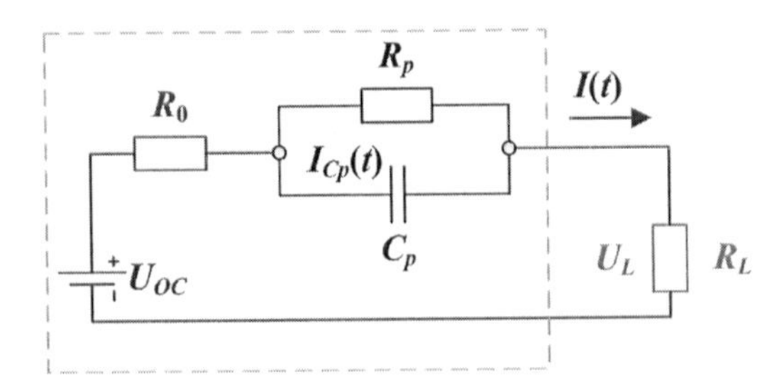

FIGURE 2.5 Thevenin equivalent model structure diagram.

In Equation (2.4), taking $U_{Cp}(t)$ as the state variable, the above two expressions are combined to describe the comprehensive working state. The voltage $U_{Cp}(t)$ at both ends of the equivalent circuit and the capacitor is set as the state variable, according to which the calculation process is analyzed to obtain the state-space expression of the equivalent model, as shown in Equation (2.5).

$$R_0 * C_P \frac{dU_{Cp}(t)}{dt} + \left(1 + \frac{R_0}{R_p}\right) U_{Cp}(t) = U_{OC} - U_L \tag{2.5}$$

In Equation (2.5), U_{OC} is the open-circuit voltage, R_0 is the ohmic resistance, R_p is the polarization resistance, C_p is the polarization capacitance, and U_L is the closed-circuit voltage after being connected to the external circuit. The parameters in the equivalent model are fixed values and cannot be modified and adjusted. After the internal resistance is equivalently processed, the equivalent model takes into account the influence of the polarization effect. The RC parallel circuit of the capacitive device C_p is added with the dynamic characteristics, so that the equivalent circuit model can have better dynamic simulation characteristics of the battery.

2.3.4 PNGV Model

The PNGV model is the standard battery model in the PNGV Battery Test Manual, which contains two resistors and two capacitors, one less than the RC circuit model. The characteristic of this model is that it takes into account the error caused by the current accumulation during the charge–discharge process of the battery OCV characteristics. Compared with the Thevenin model, the PNGV model adds a large capacitance C_b to describe the OCV change [148]. Because it has the characteristics of battery polarization and ohmic resistance, the circuit model is more accurate. The model structure is shown in Figure 2.6.

The PNGV model is a typical nonlinear lumped parameter circuit, which can be used to predict the terminal voltage variation of the battery under the condition of hybrid power pulse characterization pulse load. In Figure 2.6, U_{OC} is the ideal voltage source, R_0 is the ohmic resistance, R_p is the polarization resistance, and C_p

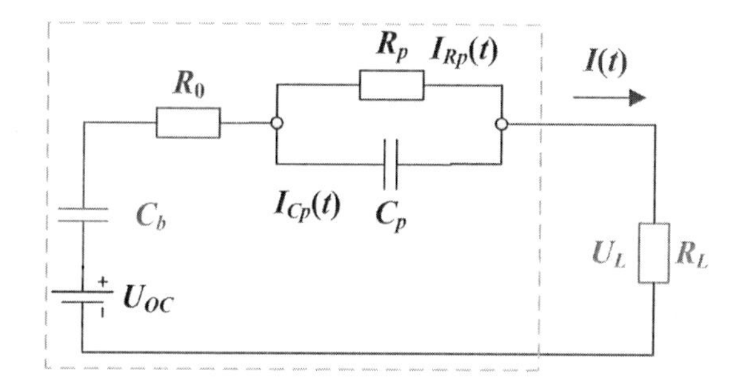

FIGURE 2.6 Lithium-ion battery PNGV equivalent model.

is the polarization capacitance. The parallel circuit of R_p and C_p is used to reflect the production of the battery polarization process to greatly reduce the differences in the working process and the impact on the working environment. C_b is the self-discharge capacitance, which is used to describe the influence of the current accumulation effect on battery open-circuit voltage. Both U_{OC} and C_b indicate the differences in the open-circuit voltage, such as voltage, temperature, and capacity, between the cells, and the differences increase with the aging of the battery. The model does not consider the effect of self-discharge and temperature. By analyzing the illustrated equivalent circuit model, the state-space equation of each parameter can be obtained, as shown in Equation (2.6).

$$U_L = U_{OC} - C_b\left(\int i(t)\,dt\right) - R_0 I_L - R_p I_p \tag{2.6}$$

In Equation (2.6), the parameter identification of the PNGV model is combined with the HPPC test. In the HPPC experiment for the parameter identification, four different sample times of a, b, c, and d are selected for the mathematical state-space description. After the matrix description is performed, the relationship between different sample times can be obtained, as shown in Equation (2.7).

$$\begin{bmatrix} U_{La} \\ U_{Lb} \\ U_{Lc} \\ U_{Ld} \end{bmatrix} = \begin{bmatrix} 1 & \int_0^a I_a(t)\,dt & I_a(t) & I_a(t) \\ 1 & \int_0^b I_b(t)\,dt & I_b(t) & I_b(t) \\ 1 & \int_0^c I_c(t)\,dt & I_c(t) & I_c(t) \\ 1 & \int_0^d I_d(t)\,dt & I_d(t) & I_d(t) \end{bmatrix} \begin{bmatrix} U_{OC} \\ 1/C_b \\ R_0 \\ R_p \end{bmatrix} \tag{2.7}$$

In Equation (2.7), U_{OC} is the open-circuit voltage, R_0 is the ohmic resistance, R_p is the polarization resistance, C_p is the polarization capacitance, I_L is the current, I_p is the current through polarization internal resistance, and U_L is external load closed-circuit voltage. The subscript parameters a, b, c, and d, respectively, represent four different sample times, which are used to characterize the state values of the parameters at different times.

2.3.5 GNL Equivalent Model

The general nonlinear (GNL) model is an improvement and extension of the PNGV model. In the GNL model, the constant voltage source U_{OC} is used as the open-circuit voltage of the battery, C_{cap} is a large energy storage capacitor to describe the capacity of the battery, two parallel RC networks are used to describe the change of battery terminal voltage caused by the polarization effect, R_e is the ohmic resistance of the

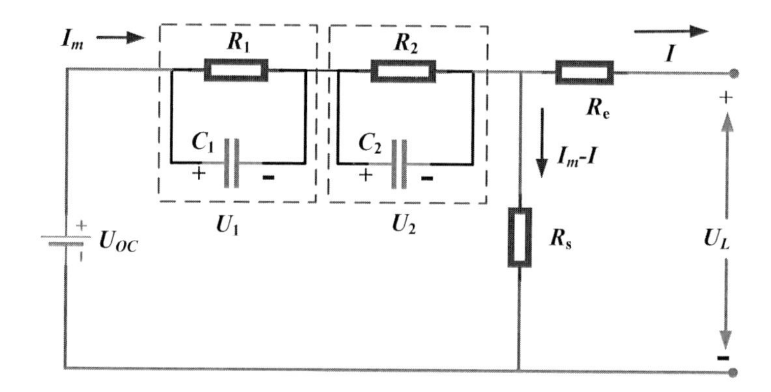

FIGURE 2.7 Lithium-ion battery GNL equivalent model.

battery, and R_s is the self-discharge resistance of the batteries. The equivalent circuit structure of the GNL model is shown in Figure 2.7.

With the introduction of the RC series network and a second-order RC parallel network in the GNL model, it is more accurate than the Thevenin model and PNGV model; in addition, the model can simulate the ohmic polarization, electrochemical polarization, concentration polarization, and self-discharge phenomena of the batteries. GNL model takes into account the relationship between the battery parameters, the temperature, and the SOC value [149]. The model performs fitting experiments on model parameters at specific SOC points, which discretizes the model parameters into functions corresponding to SOC. It fuses battery experimental data at different temperatures to establish the functional relationship between temperature and model parameters. Also, the complexity of the GNL model is getting higher and higher. It is used for PC calculations, which is still a big challenge for embedded processors.

By analyzing the GNL model, the voltage–current relationship in the circuit can be obtained, as shown in Equation (2.8).

$$\begin{cases} I_m = \dfrac{U_{1.t}}{R_1} + C_1 \dfrac{dU_{1.t}}{dt} \\ I_m = \dfrac{U_{2.t}}{R_2} + C_2 \dfrac{dU_{2.t}}{dt} \\ U_L = U_{\mathrm{OC}} - IR_e - U_{1.t} - U_{2.t} \\ U_L = -IR_e + (I_m - I) \cdot R_s \end{cases} \tag{2.8}$$

In Equation (2.8), in contrast to the Thevenin and PNGV models, the GNL series model adds an RC series network and a second-order RC parallel network, which increases the model accuracy. The ohmic polarization, electrochemical polarization, and concentration can all be simulated using this model. In the GNL model, the relation between battery parameters of temperature and SOC is taken into consideration.

2.3.6 Second-Order Equivalent Model

The second-order RC model is formed by adding an RC branch to the Thevenin model, so it is called the improved Thevenin model. Compared with the Thevenin model, these two RC circuits can fully reflect the battery impedance caused by two different polarization effects: electrochemical polarization and concentration difference polarization. Through the exploration of the above improvement ideas, the second-order model is obtained, as shown in Figure 2.8.

In Figure 2.8, U_{OC} is the open-circuit voltage, R_0 is the ohmic resistance, R_1 is the polarization resistance, C_1 is the polarization capacitance, R_2 is the surface effect resistance, C_2 is the surface effect capacitance of the lithium-ion battery pack, $I(t)$ is the load current, and U_L is the closed-circuit voltage when the battery pack is connected to a load. By analyzing the circuit structure of the above equivalent model, the knowledge of circuit analysis is applied to obtain its state and observation. According to Kirchhoff's circuit law, the circuit dynamic model of the second-order model is shown in Equation (2.9).

$$\begin{cases} \left[\frac{dU_1}{dt} = -\frac{U_1}{R_1C_1} + \frac{I_L}{C_1}\right] \Leftarrow \left\langle I_L = C_1\frac{dU_1}{dt} + \frac{U_1}{R_1}\right\rangle \\ \left[\frac{dU_2}{dt} = -\frac{U_2}{R_2C_2} + \frac{I_L}{C_2}\right] \Leftarrow \left\langle I_L = C_2\frac{dU_2}{dt} + \frac{U_2}{R_2}\right\rangle \\ U_L = U_{OC} - IR_0 - U_1 - U_2 \end{cases} \tag{2.9}$$

In Equation (2.9), by taking the parameters SOC, U_1, and U_2 into a parameter matrix $\begin{bmatrix} \text{SOC} & U_1 & U_2 \end{bmatrix}^T$ as state variables, the state space of the lithium-ion battery pack is obtained. The calculation process is shown in Equation (2.10).

$$\begin{bmatrix} \text{SOC}(k+1) \\ U_1(k+1) \\ U_2(k+1) \end{bmatrix} = \begin{bmatrix} 1 & 0 & 0 \\ 0 & e^{-\frac{\Delta t}{R_1C_1}} & 0 \\ 0 & 0 & e^{-\frac{\Delta t}{R_2C_2}} \end{bmatrix} \times \begin{bmatrix} \text{SOC}(k) \\ U_1 \\ U_2 \end{bmatrix} + \begin{bmatrix} R_1\left(1 - e^{-\frac{\Delta t}{R_1C_1}}\right) \\ R_2\left(1 - e^{-\frac{\Delta t}{R_2C_2}}\right) \end{bmatrix} [I_L] \tag{2.10}$$

For the equivalent circuit model, the closed-circuit voltage U_L of the lithium-ion battery pack is used as the output of the nonlinear SOC estimation model. Furthermore, the output current I_L is used as the input of the nonlinear system, and the observation equation $H(*)$ to obtain the parameter U_L is shown in Equation (2.11).

$$U_L = H\begin{pmatrix} f(\text{SOC}) & U_1 & U_2 \end{pmatrix} = f(\text{SOC}) - U_1 - U_2 - I(t)R_0 \tag{2.11}$$

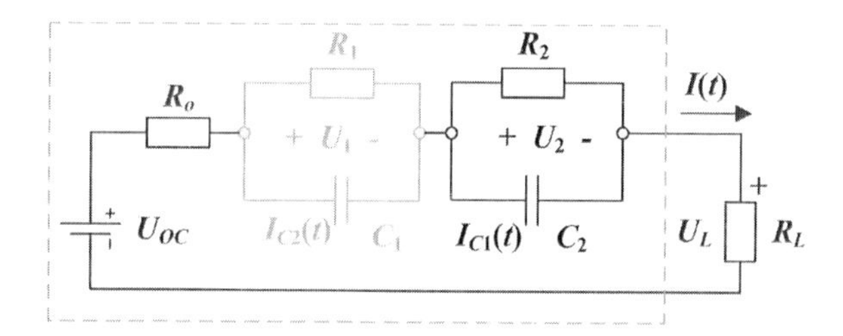

FIGURE 2.8 Second-order equivalent circuit model of lithium-ion battery.

2.3.7 Compound Equivalent Model

The equivalent circuit modeling process is based on experimental data and uses the method of establishing an electrical model to simulate the working process of the battery. Regardless of internal chemical reactions, it has strong applicability. The equivalent model is often used to simulate the dynamic characteristics of the battery. However, a simple model cannot reflect the dynamic changes of the battery because it may lead to inaccurate estimation results [150]. Complex models need to determine too many parameters, and the amount of calculation is highly increased, which may cause the problem of parameter divergence. The battery pack is constructed by cascading multiple battery cells to achieve high-power applications and provide the required energy. A model construction method based on circuit equivalence is usually selected to construct a battery pack composite equivalent circuit model to construct a high-fidelity equivalent circuit model with the greatest extent and meet the requirements of the most complex working conditions. This book uses lithium-ion battery packs as an example, which briefly explains the construction of the compound equivalent model.

It is designed to combine the advantages of different equivalent circuit models to achieve an accurate description of the working state of the lithium-ion battery packs, comprehensive characterization accuracy, and computational complexity. The composite equivalent circuit of the lithium-ion battery pack is proposed and constructed with circuit equivalent. The proposed S-ECM model realizes the accurate mathematical expression of the operating conditions and the working process by simulating different effects inside the cascaded lithium-ion battery packs. Also, the model adds parallel resistance based on the first-order RC equivalent circuit to characterize the self-discharge effect. Based on the PNGV equivalent, this model introduces a resistance parallel circuit with a series of reverse diodes to characterize the difference in internal resistance during charge–discharge. Based on the Thevenin equivalent, this model adds a series of power supplies and resistors at both ends of the electromotive force to characterize the influence of the equilibrium state and accurately describe the working process of the lithium-ion battery pack. Based on fully considering the working characteristics of the lithium-ion battery pack, the framework of its equivalent model is realized. Through the experimental analysis of working characteristics and the identification of the state parameters, the effective mathematical state-space description of the model is carried out.

The equivalent mechanism of each part of the lithium-ion battery pack S-ECM model is described as follows. (1) The electromotive force in the model comes from a constant voltage source U_{OC}, and a large resistance R_s is added in parallel at both ends to characterize the self-discharge effect. Then, through series connection, the internal resistance R_0 is the ohmic effect. (2) Using the first-order RC parallel circuit to characterize the polarization effect, the resistance R_d and R_c parallel circuits with reverse diodes are improved in series to characterize the difference in charge–discharge internal resistance, which further improves the accuracy of its working state description. (3) Considering the consistency difference between the monomers in the group equivalence process, it carries out an equivalent description of the influence of the equilibrium state on the description of the working state.

The phenomenon leads to the composition of the output voltage U_L to change and shorten the operating voltage range. Therefore, a time-varying voltage source U_δ connected in reverse series with the open-circuit voltage source U_{OC} is used for characterization. This phenomenon causes the additional accumulation of the ohmic resistance R_0 to increase, which gradually increases the heating phenomenon. Consequently, the time-varying resistance parameter R_δ is used to describe the effect. The existing SOC estimation methods have not yet fully considered the influence of the above factors. If the comprehensive influence of these parameters can be considered in the SOC estimation process, it provides an effective solution to the existing problems of lithium-ion battery packs. The S-ECM equivalent model of the lithium-ion battery pack is shown in Figure 2.9.

In the equivalent model shown in Figure 2.9, the meaning of each parameter is described as follows. U_{OC} is the open-circuit voltage of the battery pack, which is a value changing with the SOC value. R_s is a large resistance used to characterize the self-discharge effect of the lithium-ion battery packs. R_0 is the ohmic resistance, which is used to characterize the voltage drop between the positive and negative electrodes of the lithium-ion battery pack caused by the ohmic effect during charge–discharge.

By using a first-order RC parallel circuit, the relaxation effect during the characterization process is simulated and characterized, and then, the transient response of

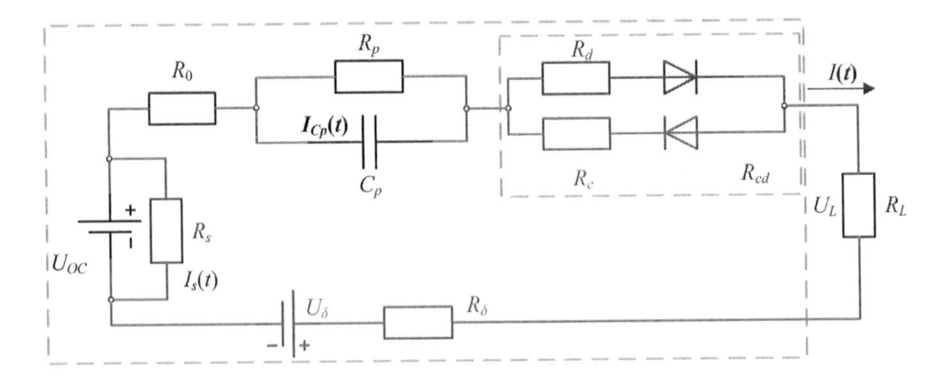

FIGURE 2.9 Improved S-ECM model diagram of the lithium-ion battery pack.

the battery pack is expressed. R_p is the polarization resistance, and C_p is the polarization capacitance of the lithium-ion battery pack. The parallel circuit of R_p and C_p reflects the generation and elimination process of the polarization effect of the lithium-ion battery pack. R_d is the difference in internal resistance exhibited by the lithium-ion battery pack during the discharge period. R_c and R_d are the difference in the internal resistance of the lithium-ion battery pack during the charge and discharge processes, respectively. U_δ and R_δ are used to characterize the influence of the balance state between the internally interconnected monomers. $U_L(t)$ is the closed-circuit voltage at both ends of the positive and negative poles during the charge and discharge processes after the lithium-ion battery pack is connected to the external circuit. $I(t)$ is the value of the current.

2.3.8 Voltage Matching Equivalent Circuit Model

Considering the needs of engineering application, the dynamic characteristics of the lithium-ion battery should be characterized by an uncomplicated equivalent model, so the Thevenin model is selected. The Thevenin model is obtained by connecting the Rint model in series with a resistor–capacitor (RC) loop [151]. The basic idea is to use an RC parallel circuit to characterize the polarization effect of the battery during use, which is to some extent making up for the shortcomings of the Rint model that cannot characterize the dynamic characteristics of lithium-ion batteries. It visualizes the complex and abstract dynamic response problem of batteries through a circuit model and is easy to be implemented in engineering, so it is one of the most commonly used models. The Thevenin equivalent circuit model can only simulate the charging process or discharging process of lithium-ion batteries. The voltage matching model is improved by adding two diodes with different directions to the Thevenin model. The diode has unidirectional conductivity, which can, respectively, characterize the discharge and charge states of the lithium-ion battery and match the corresponding identification parameters to improve the simulation accuracy of the model, as shown in Figure 2.10.

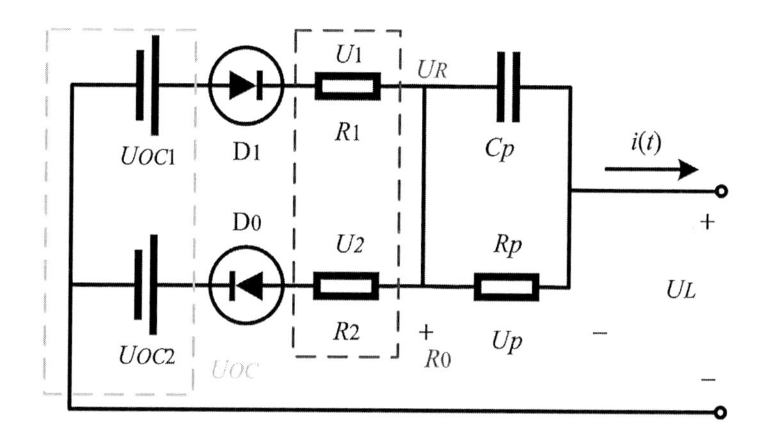

FIGURE 2.10 Voltage matching equivalent circuit equivalent model.

As shown in Figure 2.10, two diodes D_0 and D_1 are added into the voltage matching equivalent circuit model, which is used to distinguish the open-circuit voltage and ohm resistance in the charge–discharge stage. When the current $i > 0$, it is the discharging process of the battery, and when the current $i < 0$, it is the charging process of the battery. Like the Thevenin model, the open-circuit voltage of the voltage matching equivalent circuit model is U_{OC}, but it is divided into two stages: discharge open-circuit voltage U_{OC1} and discharge open-circuit voltage U_{OC2}. Similarly, the ohmic resistance of the voltage matching equivalent circuit model is R_0, which is divided into two stages: discharging resistance R_1 and charging resistance R_2. When $i > 0$, it means that the battery is being discharged. When $i < 0$, it means that the battery is charging. According to Kirchhoff's law, the equivalent circuit expressions for the charge–discharge phases of the battery can be obtained, as shown in Equation (2.12).

$$\begin{cases} U_L = U_{OC1} - U_{R1} - U_P (U_L > 0) \\ U_L = U_{OC2} - U_{R2} - U_P (U_L < 0) \\ i(t) = C\dfrac{dU_P}{dt} + \dfrac{U_1}{R_1}(i > 0) \\ i(t) = C\dfrac{dU_P}{dt} + \dfrac{U_2}{R_2}(i < 0) \end{cases} \tag{2.12}$$

In Equation (2.12), the open-circuit voltage can be characterized by the state varible SOC, and a nonlinear function relationship can be obtained. Combined with the definition of SOC, the calculation expression of SOC can be obtained, as shown in Equation (2.13).

$$\text{SOC}_t = \text{SOC}_{t0} - \frac{\int_{t0}^{t} i(t)\eta\, dt}{Q_0} \tag{2.13}$$

In Equation (2.13), SOC_{t0} is the initial power of the battery, SOC_t is the power of the battery at the time point k, Q_0 is the rated capacity, η is the Coulomb efficiency coefficient, and $i(t)$ is the charge–discharge current. The state-space variable $x_k = \left[\text{SOC}_k, U_{P,k}\right]^T$, the input variable $U_k = \left[i_k\right]$, and the output variable $y_k = \left[U_{L,k}\right]$ are selected. A discrete state-space equation can be obtained, as shown in Equation (2.14).

$$\begin{cases} \begin{bmatrix} \text{SOC}_{k+1} \\ U_{P,k+1} \end{bmatrix} = \begin{bmatrix} 1 & 0 \\ 0 & e^{-\frac{\Delta t}{\tau}} \end{bmatrix} \begin{bmatrix} \text{SOC}_k \\ U_{L,k+1} \end{bmatrix} + \begin{bmatrix} -\dfrac{\Delta t}{Q_N} \\ R_p\left(1 - e^{-\frac{T}{\tau}}\right) \end{bmatrix} i_k + \begin{bmatrix} w_{1,k} \\ w_{2,k} \end{bmatrix} \\ U_{L,k} = U_{OC,k} - R_{0,k} I_{k+} \begin{bmatrix} 0 \\ -1 \end{bmatrix}^T \begin{bmatrix} \text{SOC}_k \\ U_{P,k} \end{bmatrix} + v_k \end{cases} \tag{2.14}$$

In Equation (2.14), Δt is the sampling time interval, $\tau = R_P C_P$, w_k is the state error, and v_k is the measurement error, which is the zero-mean white noise of the covariance matrices Q and R, respectively. The first sub-equation is the state equation for the SOC estimation, and the second sub-equation is the systematic observation equation.

2.3.9 Improved Second-Order RC Equivalent Model

Compared with other equivalent circuit models, the second-order RC model reflects the changing law of the batteries more accurately. It has two RC loop circuits, which can fully represent the battery impedance caused by two different polarization effects: electrochemical polarization and concentration difference polarization. However, the widely used second-order RC equivalent circuit model does not consider the self-discharge effect of lithium-ion batteries. When the lithium-ion battery is not connected to a load, the power storage capacity also decreases over time, which leads to the change in the open-circuit voltage. The equivalent circuit model does not consider the self-discharge effect on the corresponding error when selecting the open-circuit voltage. Thereby, the parameter identification accuracy of the lithium-ion battery equivalent circuit model is reduced, which affects the accuracy of the SOC estimation. Therefore, an improved second-order equivalent circuit model is established that considers the self-discharge effect, as shown in Figure 2.11.

In Figure 2.11, U_{OC} is the open-circuit voltage of the lithium-ion battery pack, R_0 is the ohmic resistance, R_a is a self-discharge resistance used to represent the self-discharge process, R_{p1} is the polarization resistance of the battery, C_{p1} is the polarization capacitance, R_{p2} is the surface effect resistance of the battery, C_{p2} is the surface effect capacitance of the lithium-ion battery, $I(t)$ is the line current, and U_L is the closed-circuit voltage when the battery pack is connected to a load. By analyzing the circuit structure of the above equivalent model, the knowledge of circuit analysis is applied to obtain its state and observation. According to Kirchhoff's circuit law, the electrical characteristics of the second-order equivalent circuit model are shown in Equation (2.15).

$$\begin{cases} U_L = U_{OC} - U_a - U_R - U_{p1} - U_{p2} \\ i_L = C_{p1}\dfrac{dU_{p1}}{dt} + \dfrac{U_{p1}}{R_{p1}} = C_{p2}\dfrac{dU_{p2}}{dt} + \dfrac{U_{p2}}{R_{p2}} \end{cases} \tag{2.15}$$

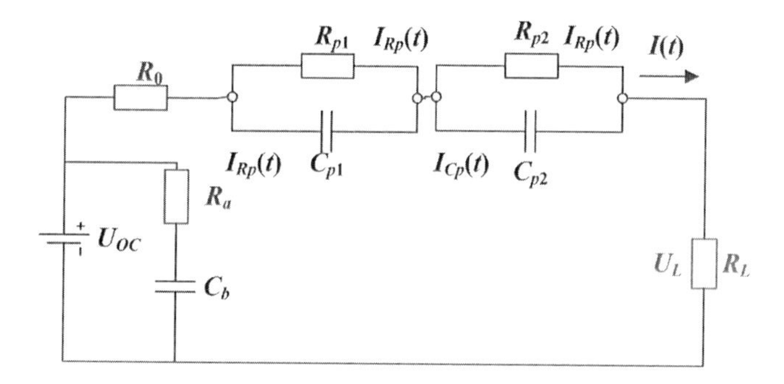

FIGURE 2.11 Improved second-order equivalent circuit model.

The state-space variables are initialed as $x_k = \begin{bmatrix} SOC_k, & U_{p1,k}, & U_{p2,k}, & U_a \end{bmatrix}^T$, the input variable is i_k, and the output variable is obtained as shown in $y_k = U_{L,k}$. The state-space expression equation of the equivalent model can be obtained, as shown in Equation (2.16).

$$\begin{cases} x_{k+1} = \begin{bmatrix} 1 & 0 & 0 & 0 \\ 0 & e^{-\frac{\Delta t}{\tau_1}} & 0 & 0 \\ 0 & 0 & e^{-\frac{\Delta t}{\tau_2}} & 0 \\ 0 & 0 & 0 & 1 \end{bmatrix} x_k + \begin{bmatrix} -\frac{\Delta t}{Q_N} \\ R_1\left(1-e^{-\frac{\Delta t}{\tau_1}}\right) \\ R_2\left(1-e^{-\frac{\Delta t}{\tau_2}}\right) \\ \frac{1}{C_b} \end{bmatrix} i_k + w_k \\ y_k = U_{OC,k} - R_0 i_k + \begin{bmatrix} 0 \\ -1 \\ -1 \\ -1 \end{bmatrix} x_k + v_k \end{cases} \tag{2.16}$$

In Equation (2.16), Δt is the sampling interval, $\tau_1 = R_{p1}C_{p1}$, $\tau_2 = R_{p2}C_{p2}$, w_k is system noise, and v_k is the measurement error. The measurement method of R_a is described as follows: The fully charged lithium-ion battery is placed in a constant temperature environment for 30 days, and the voltage change is monitored for identification. It can be observed from the KVL equation that the voltage–current relationship in the resting state is shown in Equation (2.17).

$$U_b = \Delta U_{OC} = U_{OC0}\left(1-e^{-\frac{t}{R_aC_b}}\right) \tag{2.17}$$

In Equation (2.17), after measuring C_b, the self-discharge resistance R_a can be calculated by Equation (2.18).

$$R_a = \frac{-t}{C_b \times Ln\left(1-\frac{\Delta U_{OC}}{U_{OC0}}\right)} \tag{2.18}$$

A simulation model is built in Simulink, which can be used to calculate the accuracy of parameter identification, as shown in Figures 2.12 and 2.13.

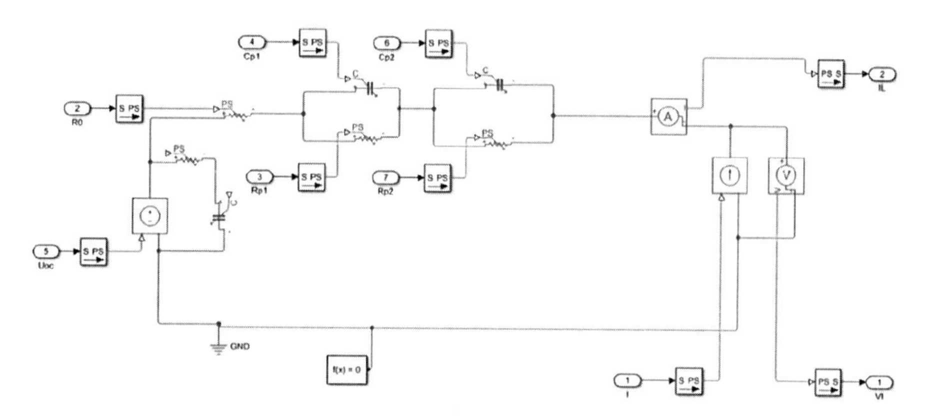

FIGURE 2.12 Improved second-order equivalent circuit model simulation.

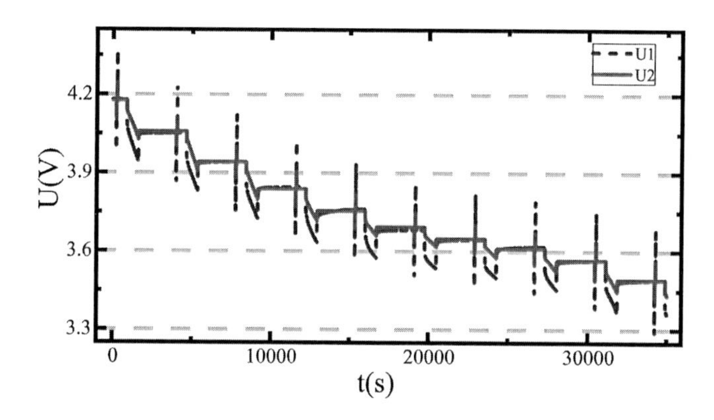

FIGURE 2.13 Second-order RC equivalent model simulation results.

Figure 2.13 shows the parameter identification accuracy verification when the second-order RC equivalent circuit model does not consider the self-discharge effect, and there is always a large error in the discharge process The simulation results of the model after considering the effect of self-discharge are obtained as shown in Figure 2.14.

In Figure 2.14, after considering the self-discharge resistance, the result of parameter identification accuracy verification is obtained. It can be observed that after adding the self-discharge process, the model can simulate the actual voltage output well and achieve high accuracy.

2.4 INTRODUCTION OF COMMON WORKING CONDITIONS

2.4.1 Hybrid Pulse Power Characterization Test

The HPPC test is a battery performance test method proposed in the 'Freedom CAR Battery Test Manual for Power-Assist Hybrid Electric Vehicles manual', which is used to reflect the power battery pulse charge–discharge performance of a feature

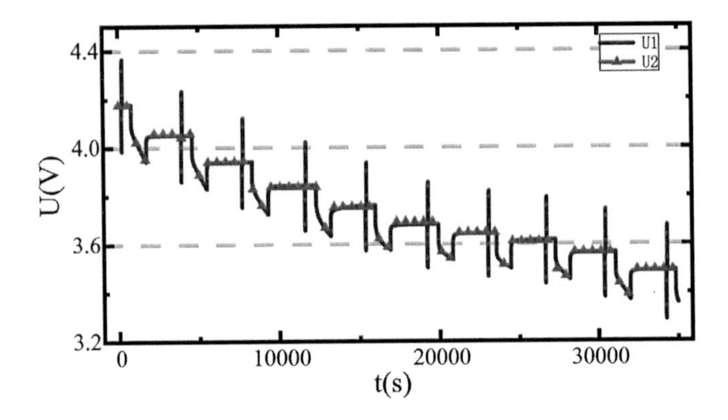

FIGURE 2.14 Improved second-order RC equivalent model simulation results.

[152]. After the HPPC experimental test, the model parameters are identified. HPPC tests are performed by using dedicated battery testing equipment, which is used to test the internal resistance of the battery. The HPPC experiment process is described as follows: In the cyclic test, the lithium-ion battery is subjected to a power pulse test at every SOC point with equal interval, and SOC levels are selected from 0 to 1. The process of the pulse test is to discharge the lithium-ion battery at a 1 C rate for 10 seconds and then charge it at a 0.75 C rate for 10 seconds after resting for 40 seconds. The battery needs to be placed between adjacent pulse tests for a long time to restore electrochemical and thermal equilibrium. The whole experiment is mainly composed of a single repeated charge and discharge pulse test [153].

All the parameters of the Thevenin circuit model are obtained through the HPPC experiment. The disadvantage of this experiment is that it takes a long time, usually about 12 hours, and most of the time is consumed in the repeated shelving stage, which causes great losses to the experimental equipment and operators. Therefore, two test channels are used to solve the time-consuming problems. In the same batch, two lithium-ion batteries with similar capacities are selected to conduct the HPPC test and also to ensure the consistency of other influence quantities except for the experimental quantity. Each channel only cycles the power pulse test five times. The first channel is conducted with the pulse test when SOC is equal to 1, 0.8, 0.6, 0.4, and 0.2, and the second channel is conducted with the pulse test when SOC is equal to 0.9, 0.7, 0.5, 0.3, and 0.1. With this improved scheme, the experimental time is reduced to about 7 hours. When the difference between the two groups of batteries is small, the experimental results are consistent with the original HPPC experimental results. Therefore, the improved experimental procedure is accurate and feasible.

2.4.2 Beijing Bus Dynamic Stress Test

The Beijing bus dynamic stress test (BBDST) working condition is obtained through the actual data collection of the Beijing buses. The rated parameters of the lithium-ion battery pack on the vehicle are 380 V and 360 Ah. The electric bus is used for

BBDST operating condition testing. The working condition steps mainly include starting, acceleration, sliding, braking, constant speed, rapid acceleration, and braking. The counting of the running time and power for specific each step is summarized, as shown in Table 2.1.

In Table 2.1, P (kW) is the output power of the battery under the real bus starting, accelerating, and taxing conditions. As can be seen from the table, the time required to the complete BBDST condition was 300 seconds. Considering that the battery parameters of different test objects are different, the test steps need to be changed according to certain conversion rules. The transformation process is carried out as follows: The test conditions are applied to battery cells or battery modules, as shown in Equation (2.19).

$$P_c = \frac{V_c Q_c}{V_b Q_b} P_b = \lambda_b P_b \tag{2.19}$$

In Equation (2.19), P_b, V_b, and Q_b are the power, voltage, and capacity of the reference battery, respectively. P_c, V_c, and Q_c are the power, voltage, and capacity respectively of the tested cell or module. b is the battery equivalent coefficient. It is necessary to increase the battery test power to test the battery overload capacity, as shown in Equation (2.20).

TABLE 2.1
BBDST Working Conditions

	P(kW)		*t*(seconds)		
Step	Air Conditioner On	Air Conditioner Off	Complete Steps	Total	Status Description
1	45	37.5	21	21	Start
2	80	72.5	12	33	Accelerate
3	12	4.5	16	49	Slide
4	−15	-15	6	55	Brake
5	45	37.5	21	76	Accelerate
6	12	4.5	16	92	Slide
7	−15	-15	6	98	Brake
8	80	72.5	9	107	Accelerate
9	100	92.5	6	113	Rapid acceleration
10	45	37.5	21	134	Constant speed
11	12	4.5	16	150	Slide
12	−15	−15	6	156	Brake
13	80	72.5	9	165	Accelerate
14	100	92.5	6	171	Rapid acceleration
15	45	37.5	21	192	Constant speed
16	12	4.5	16	208	Slide
17	−35	−35	9	217	Brake
18	−15	−15	12	229	Brake
19	12	4.5	71	300	Stop

$$P_c' = \alpha P_c \tag{2.20}$$

In Equation (2.20), α is the overload stress coefficient. The test conditions are applied to two different vehicles. The above working condition is derived from vehicle a, the power at time point t is P_a, and the application object of the working condition is vehicle b. The rated voltage and capacity of the power battery pack of vehicle a are V_a and Q_a, respectively, and the mass of the tested vehicle is m_a. Also, the rated voltage and capacity of the battery pack of vehicle b are V_b and Q_b, and the mass of the test vehicle is m_b. Then, the working condition is applied when the vehicle is b and the power is calculated, as shown in Equation (2.21).

$$P_b = \frac{V_a Q_a}{V_b Q_b} \frac{m_b}{m_a} P_a = \lambda_v P_b \tag{2.21}$$

In Equation (2.21), λ_v is the vehicle equivalent coefficient.

2.4.3 Dynamic Stress Test

In practical application, the real-time current of lithium-ion batteries is complex and changeable. In different working conditions, there are often sudden switching and stopping of current, so it is necessary to strictly require the model dynamic simulation performance of the battery. The self-defined dynamic stress test (DST) working condition experimental data are used to simulate the model to further verify the SOC estimation of the lithium-ion batteries under more complex working conditions [154]. The experimental procedures of the DST are described as follows.

1. The battery is charged to a maximum terminal voltage of 4.20 V with a constant current of 1 C. And then, it is charged with a constant voltage until the charging current rate drops to 0.05 C.
2. After charging is completed, the battery is set aside to stabilize the battery voltage. Since the selected lithium-ion battery has a small capacity, the residence time is selected as 30 minutes.
3. The constant current discharge is carried out at a rate of 0.5 C for 4 minutes and then let the battery rest for 30 seconds after stopping the discharge.
4. The lithium-ion battery is charged with a constant current rate of 0.5 C for 2 minutes. After stopping the charging, the battery rests for 30 seconds.
5. The constant current discharge is performed at a rate of 1 C for 4 minutes.

Steps (4) and (5) are repeated until the battery discharge is over.

2.5 OFFLINE PARAMETER IDENTIFICATION

In the process of data analysis and processing, the effective data segment is extracted firstly from the original experimental data [155]. Then, the extracted data segment is analyzed and effective processing methods are adopted to obtain the relationship between the internal parameters of the equivalent circuit model and the SOC value.

2.5.1 Point Calculation

The equivalent circuit model is only a rough description of the characteristics of lithium-ion batteries. It is necessary to identify the parameters of the equivalent model based on the experimental data, which is used to describe the working characteristics of batteries accurately and establish a state-space model. The experimental process is shown in Figure 2.15.

First, all the voltage data are extracted from the original data, and the differences in the terminal voltage of the lithium-ion battery are depicted during the HPPC experimental test, as shown in Figure 2.16.

From the analysis of the graph obtained from the experimental results, it can be observed that the voltage changes significantly after 1 hour of constant current discharge. After the discharge was completed, the battery voltage gradually stabilizes after prolonged storage, indicating that internal chemical reactions and thermal effects had reached equilibrium. The battery voltage at the time is its open-circuit voltage. Therefore, the relationship curve between the open-circuit voltage and SOC is obtained. A comparison of the two curves has shown to have the same fitting. Useful data segments are extracted from the data of the overall process, and then, parameter identification is carried out. For the Thevenin model, the parameters that need to be identified are ohmic resistance R_0, polarization internal resistance R_p, and polarization capacitance C_p. These parameters can be identified according to the power pulse test phase of the HPPC test experiment. Consequently, the data of the

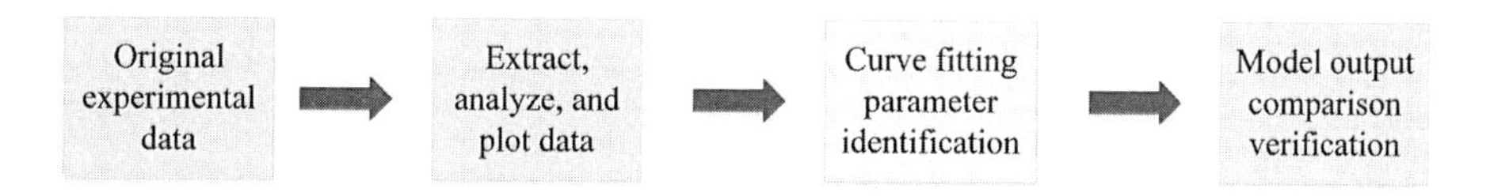

FIGURE 2.15 Data processing and model validation.

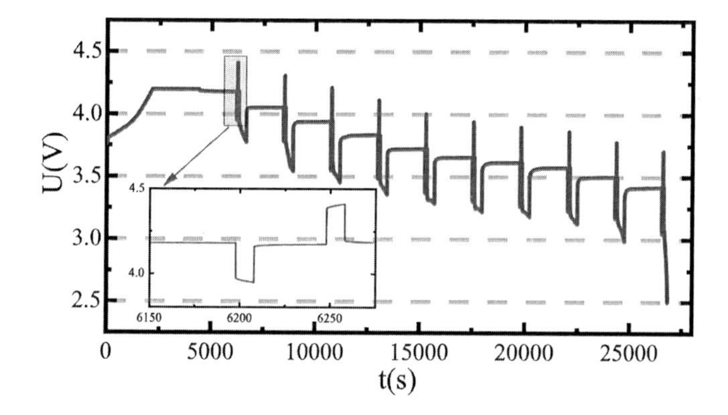

FIGURE 2.16 HPPC test experiment voltage response curve.

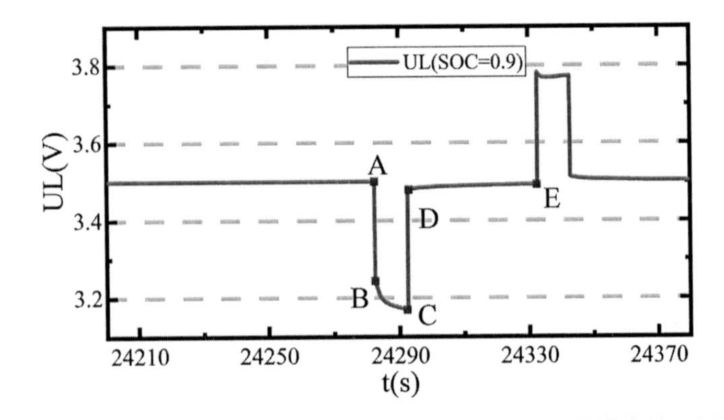

FIGURE 2.17 Pulse test chart when SOC = 0.9.

pulse test in each cycle are extracted first. The impulse response curve of the lithium-ion battery is obtained when the SOC is 0.9, as shown in Figure 2.17.

In Figure 2.17, the transient and steady-state characteristics of lithium-ion batteries are also reflected. When the pulse discharge starts, the battery voltage drops instantaneously, and then, the voltage slowly drops with time during the discharge. When the discharge ends, the battery voltage rebounds immediately. During the resting period, the voltage gradually rises and stabilizes. Meanwhile, the voltage response during charge–discharge is the opposite. There are two methods for parameter identification using pulse test data: One is the calculation method by taking points, and the other is the curve fitting method. First, extract all the voltage data from the original data, and depict the changes in the terminal voltage of the lithium-ion battery during the HPPC test experiment, as shown in Figure 2.18.

All the current data are extracted from the original data to describe the current changes in the lithium-ion battery during the HPPC test process, as shown in Figure 2.19.

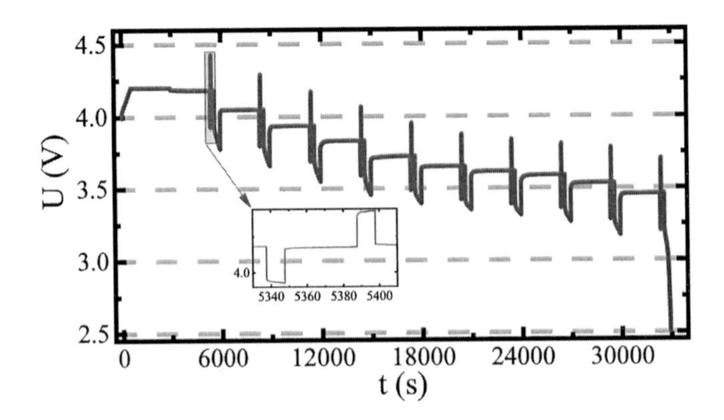

FIGURE 2.18 HPPC test experiment voltage response curve.

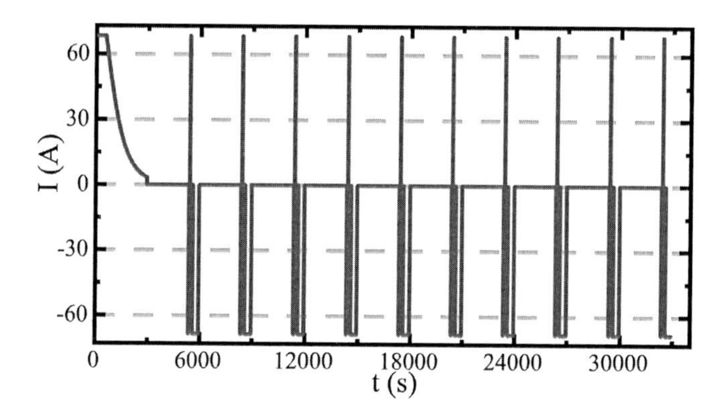

FIGURE 2.19 HPPC test experiment response curve.

The relevant data are extracted from the experimental results of the overall process, and then, parameter identification is carried out. For the Thevenin model, the parameters that need to be identified are ohmic resistance R_0, polarization internal resistance R_p, and polarization capacitance C_p. These parameters are identified from the power pulse test phase of the HPPC test experiment, so the data segment of the pulse test in each cycle must be extracted first. When the SOC is 0.7, the impulse response curve of the lithium-ion battery is shown in Figure 2.20.

In Figure 2.20, the identification process of the point-taking calculation method is described as follows. First, five points A, B, C, D, and E are obtained. From the Thevenin equivalent model, the rapid change of voltage at the beginning of discharge and charging is due to the instantaneous response caused by the current flowing through R_0. The RC circuits can be used in the Thevenin model to explain the gradual drop of voltage during discharge. The value of the ohmic resistance parameter is obtained from the sections AB and CD. Also, the value of the internal polarization resistance and the polarization capacitance of the RC circuit is obtained from the sections of BC and DE. From this, the calculation equation of the parameter is obtained, in which the calculation expression of R_0 is shown in Equation (2.22).

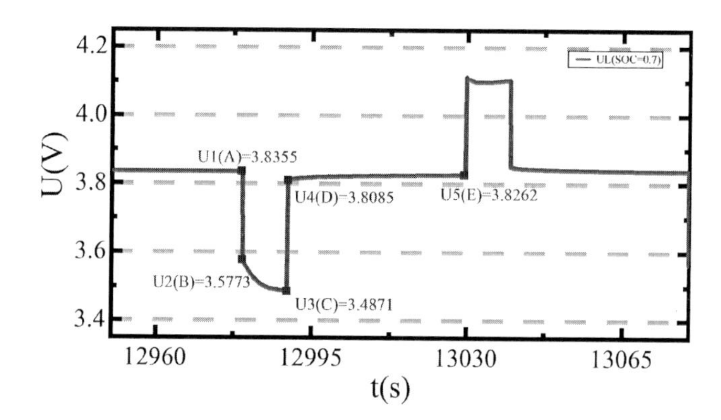

FIGURE 2.20 Pulse test chart when SOC = 0.7.

$$R_0 = \frac{|\Delta U_{AB}| + |\Delta U_{CD}|}{2I} \tag{2.22}$$

In Equation (2.22), ΔU_{AB} is the voltage difference of the AB section, ΔU_{CD} is the voltage difference of the CD section, and the discharge current is characterized by I. The voltage response of the DE segment is analyzed corresponding to the zero-input response in the Thevenin model, and the mathematical relationship is obtained by the KVL of the equivalent circuit model, as shown in Equation (2.23).

$$\begin{cases} U_D = U_A - U_p e^{-\frac{t_D - t_A}{\tau}} \\ U_E = U_A - U_p e^{-\frac{t_E - t_D}{\tau}} \end{cases} \tag{2.23}$$

In Equation (2.23), U_A is the voltage at point A, which is the open-circuit voltage at the current stage. Taking point D as the zero-time points, the time from point D to E is $t_E - t_D$, and U_D and U_E represent the voltage of points D and E. The calculation equation of the time constant is constructed for τ, as shown in Equation (2.24).

$$\tau = -\frac{t_E - t_D}{\ln\left(\frac{U_A - U_E}{U_A - U_D}\right)} \tag{2.24}$$

After the time constant is obtained, only one parameter R_p or C_p needs to be calculated, and the other can be derived from Equation (2.25).

$$\tau = R_p C_p \tag{2.25}$$

The voltage response of the BC section corresponds to the zero-state response of the Thevenin model. According to the KVL relationship in the model, the set is shown in Equation (2.26).

$$\begin{cases} U_C = U_A - IR_0 - IR_p\left(1 - e^{-\frac{t_C - t_B}{\tau}}\right) \\ U_B = U_A - IR_0 \end{cases} \tag{2.26}$$

In Equation (2.26), U_C is the voltage at point C. Taking point B as the zero-time point, the time from point B to point C is $t_B - t_C$. Solving it to obtain the calculation equation of polarization internal resistance R_p is shown in Equation (2.27).

$$R_p = \frac{U_B - U_C}{I\left(1 - e^{-\frac{t_C - t_B}{\tau}}\right)} \tag{2.27}$$

TABLE 2.2
Parameter Fitting Result

τ	$R_0(\Omega)$	$R_p(\Omega)$	C_p(F)
0.36737624	0.004326	0.000527964	6958.360967
0.312561072	0.004835	0.000140234	2228.850513
0.319941747	0.007286	0.000935037	3421.703185
0.369667569	0.005796	0.000265601	1391.814065
0.401429972	0.006542	0.00090463	4437.505023
0.333883416	0.005912	0.000255268	1307.972699
0.348955741	0.005943	0.000235589	1481.206277
0.375727008	0.005905	0.000224778	1671.547113
0.378136003	0.005637	0.000176885	2137.753394
0.332731927	0.006058	0.000279054	1192.355491

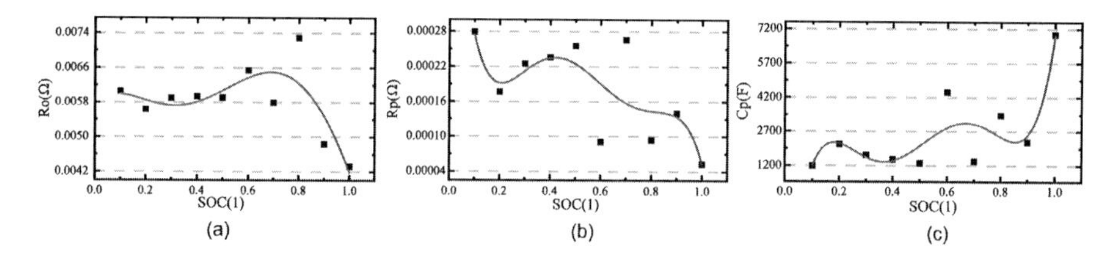

FIGURE 2.21 Thevenin model parameter identification results. (a) Ohm internal resistance R_0. (b) Polarized internal resistance R_p. (c) Polarization capacitance C_p.

In Equation (2.27), using the point calculation method to identify the parameters of each SOC point stage, the ohmic resistance, polarization internal resistance, and polarization capacitance data from SOC from 0.1 to 1 are shown in Table 2.2.

The identified parameters are obtained and drawn in the coordinate system. Taking SOC as the independent variable and the parameters as the dependent variable, a scatter plot of the relationship between R_0, R_p, C_p, and SOC is obtained, and the fitting curve is obtained by the least-square polynomial fitting method, as shown in Figure 2.21.

The results of parameter identification are verified by HPPC working conditions, and the accuracy of the model is determined by the error between the simulated voltage and the actual voltage. The verification results are shown in Figure 2.22.

In Figure 2.22a, U_2 is the output terminal voltage of the model and U_1 is the voltage at the actual terminal of the battery. It can be observed that the voltage at the output terminal of the model is in good agreement with the actual value, which illustrates the rationality of the Thevenin equivalent circuit model and also proves the feasibility and reliability of the parameter identification method. The error curve is shown in Figure 2.22b.

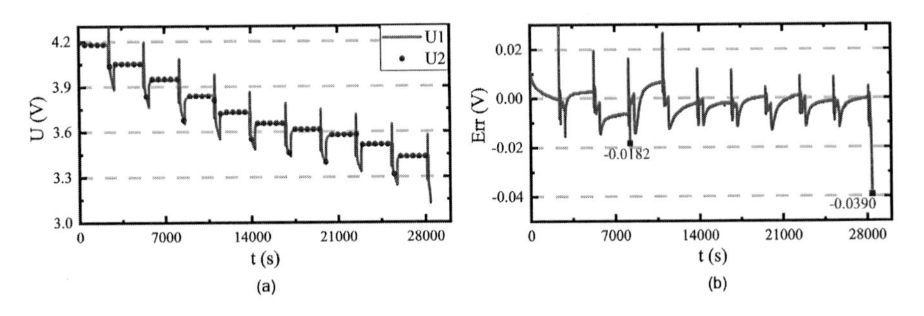

FIGURE 2.22 Parameter identification result graph. (a) Voltage output comparison. (b) Voltage error.

2.5.2 Curve Fitting

Curve fitting is a method in which model parameters are obtained by fitting the entire pulse discharge curve through a formal function with parameters. Its adoption rate of data is much higher than that of taking the point equation method [156]. For curve fitting, only a fitting function expression needs to be given, which contains the parameters to be obtained. The curve fitting is mainly for the relatively complicated calculation parameters, such as R_p and C_p in the Thevenin model, while the ohmic resistance R_0 does not require curve fitting and can be obtained directly according to Equation (2.22). The fitting curve segment selected here is the *BC* segment. The zero-state response curve and the circuit expression are obtained from the KVL relationship of the Thevenin model, as shown in Equation (2.28).

$$U_L = U_{OC} - IR_0 - IR_p\left(1 - e^{\frac{-t}{\tau}}\right) \tag{2.28}$$

In Equation (2.28), the parameter relationship is subtracted, and a parameterized expression form is obtained, as shown in Equation (2.29).

$$y = a - b\left(1 - e^{\frac{-x}{c}}\right) \tag{2.29}$$

In Equation (2.29), y is the ordinate variable that is the terminal voltage U_L, x is the abscissa variable at the time point t, and a, b, and c are coefficients of the parameters, corresponding to $U_{OC} - IR_0$, IR_p, and τ. According to the known conditions of open-circuit voltage, current and ohmic resistance, the range of parameters is set to make the fitting curve better express the real voltage change. The voltage simulation curve when the SOC value is 0.9 is shown in Figure 2.23.

By using the curve fitting method to identify the parameters of each SOC point stage, the ohmic resistance, polarization internal resistance, and polarization capacitance data for the SOC variation from 0.1 to 1 are shown in Table 2.3.

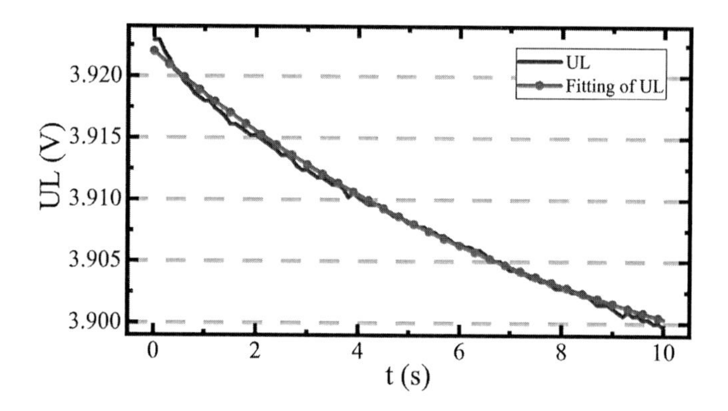

FIGURE 2.23 Curve fitting method to identify parameters.

TABLE 2.3
Parameter Identification Result

SOC	a	b	c	$R_0(\Omega)$	$R_p(\Omega)$	$C_p(\Omega)$
1.0	4.103	0.02552	7.505	0.001878994	0.000587646	12771.29261
0.9	3.979	0.02757	7.409	0.001885902	0.000634851	11670.45149
0.8	3.870	0.02963	7.797	0.001904323	0.000682287	11427.74949
0.7	3.771	0.03102	7.902	0.001878994	0.000714294	11062.67263
0.6	3.653	0.02848	7.534	0.002060906	0.000655806	11488.15959
0.5	3.573	0.02223	7.422	0.002047090	0.000511888	14499.27598
0.4	3.532	0.02135	7.650	0.002063209	0.000491624	15560.67330
0.3	3.501	0.02248	7.978	0.002056301	0.000517644	15412.12611
0.2	3.444	0.02491	8.126	0.002083933	0.000573600	14166.67463
0.1	3.364	0.03091	7.889	0.002141500	0.000711761	11083.77701

The identified parameters are obtained and drawn in the coordinate system. Taking SOC as the independent variable and the parameters as the dependent variable, a scatter plot of the relationship between R_0, R_p, C_p, and SOC is obtained. The fitting curve is obtained by the least-square polynomial fitting method, as shown in Figure 2.24.

Figure 2.24a–c shows the changing trend of R_0, R_p, and C_p with SOC, respectively. It can be observed from the figure that the internal parameters of the Thevenin model fluctuate within a certain range with SOC changes. Therefore, in the case of low accuracy requirements for the model, to minimize the amount of calculation, these parameters are replaced by their corresponding values or by the look-up table method. To get more accurate simulation data, the polynomial fitting equation of the curve is obtained. The parameter fitting polynomial functions corresponding to the graph are shown in Equations (2.30)–(2.32).

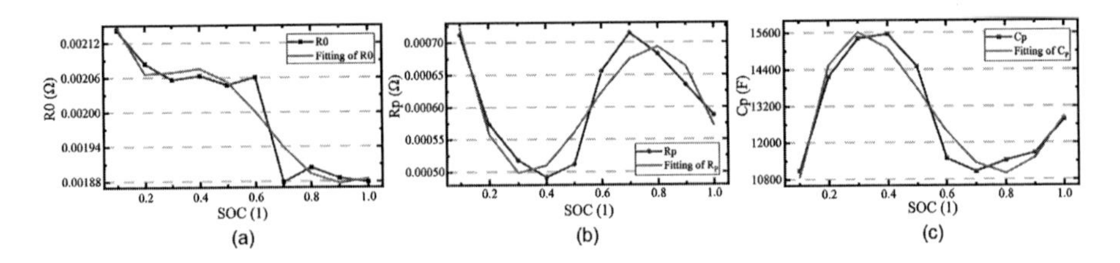

FIGURE 2.24 Thevenin model parameter identification result. (a) Ohm internal resistance R_0. (b) Polarized internal resistance R_p. (c) Polarization capacitance C_p.

$$R_0 = -0.005179 * u^5 + 0.01863 * u^4 - 0.02471 * u^3 + 0.01545 * u^2 - 0.004758 * u + 0.002464 \quad (2.30)$$

$$R_p = 0.03256 * u^5 - 0.0831 * u^4 + 0.07091 * u^3 - 0.02125 * u^2 + 0.0005915 * u + 0.0008495 \quad (2.31)$$

$$C_p = (-4.596e+05) * u^5 + (1.148e+06) * u^4 + (-8.816e+05) * u^3 + (1.401e+05) * u^2 + (6.67e+04) * u + 2105 \quad (2.32)$$

As can be observed from Figure 2.24 and Equations (2.30)–(2.32), the parameters within the Thevenin model fluctuate within a certain range when the SOC changes. Therefore, when the accuracy of the model is not required, the average value of each parameter can be used as the final result, and the table look-up method is also used to obtain the result. If more accurate simulation data are needed, polynomial fitting equations for the curves need to be obtained. The parameter fitting polynomials corresponding to the figure are shown in Table 2.4.

TABLE 2.4
Parameter Identification Result

R_0: $y = 0.00245 - 0.00479 * u + 0.0215 * u^2 - 0.04248 * u^3 + 0.03672 * u^4 - 0.01151 * u^5$
C_p: $y = 8336.79306 + 17063.97174 * u + 167066.86614 * u^2 - 703922.67073 * u^3 + 863340.03368 * u^4 - 339190.57413 * u^5$
R_p: $y = 0.000686263 + 0.00173 * u - 0.02048 * u^2 + 0.05921 * u^3 - 0.0646 * u^4 + 0.02404 * u^5$
U_{OC}: $y = 3.42621 - 0.54017 * u + 13.07197 * u^2 - 55.87842 * u^3 + 105.54386 * u^4 - 90.28558 * u^5 + 28.84722 * u^6$

2.5.3 Equivalent Circuit Model Parameter Validation

After obtaining the mathematical description of each parameter of the model, it is necessary to verify the accuracy of the model parameters. The identified parameters and current values of the HPPC test are input into the Thevenin equivalent circuit model. The output voltage response data from the model are compared with the actual voltage data to verify the model and improve the model based on the verified results [157]. The inputs of the Thevenin model are I, R_0, R_p, C_p, and U_{OC} from top to bottom. They have a functional relationship with SOC as an independent variable and change along with the SOC and current variation. The inputs are the terminal voltage and the load current. The Ampere-hour integration module obtains the real-time changes of the SOC. Then, it is connected to the input and end of the parameter function, to obtain the model parameters corresponding to the constant changes of the current input. Therefore, the control quantity of the model is the input current value, and the response is the terminal voltage of the model. Based on this model, the working condition of the lithium-ion battery is analyzed. The internal structure of the Thevenin model is shown in Figure 2.25.

In Figure 2.25, the circuit diagram of the Thevenin model is designed, where each circuit element is a time-varying controllable parameter. The input current I acts on a controllable current source as a load, and the output voltage U is obtained through a voltmeter. As shown on the right side of Figure 2.25, the effectiveness and accuracy of model parameter identification are analyzed and compared with the actual terminal voltage data of the lithium-ion battery under the same input current. If the deviation is large, the parameter identification is not accurate enough, or the model itself has defects.

As shown in Figure 2.26a, the real U is the model output voltage, and the model U is the actual output voltage of the battery. It can be observed that the model output voltage is in good agreement with the actual value, indicating the rationality of the Thevenin equivalent circuit model and the feasibility and reliability of the parameter identification method.

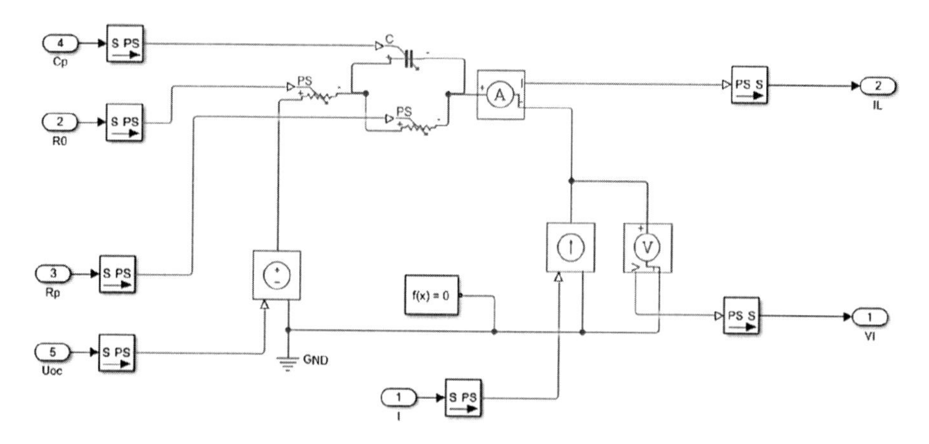

FIGURE 2.25 Model verification Simulink simulation.

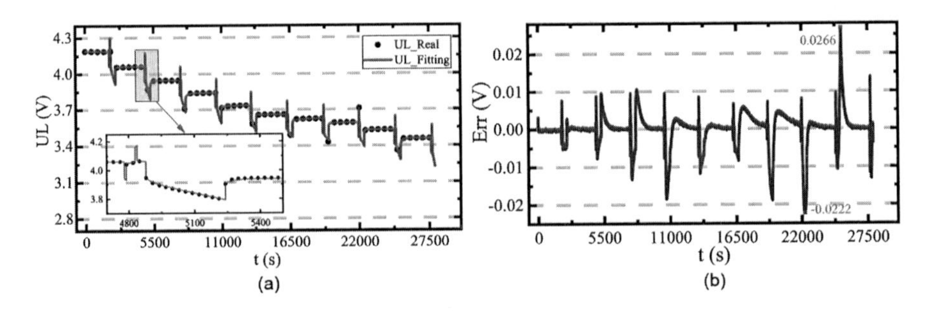

FIGURE 2.26 Terminal voltage output curve comparison. (a) Output comparison. (b) Output error.

It can be observed from the error curve in Figure 2.26b that there is no divergence of the model in the overall process, and a large error occurs in the power pulse test phase. The reason is that the sudden change of the input current is caused by the internal chemical reaction of the battery, leading to rapid changes in terminal voltage. Also, the Thevenin model does not perfectly show the response effect of lithium-ion batteries, but its error fluctuates within an acceptable and reasonable range. Even in the final stage when the SOC value is very low, the model cannot be very accurate in the case of a good reaction at the present point, the maximum error does not exceed 0.11 V, and the maximum voltage of the lithium-ion battery is 4.40 V. Consequently, the model can reach an accuracy of more than 96%.

2.5.4 Model Parameter Identification

The improved PNGV equivalent circuit model is used for parameter identification. According to the 'U.S. Freedom CAR Battery Experiment Manual', standard HPPC current pulse tests are carried out on lithium-ion batteries. The SOC is 0.1, 0.2, ..., 0.9, 1.0, and the charge–discharge rates are set to 1 C. The step content of a single cycle is the 10 seconds constant current discharge pulse, the 40 seconds shelving, and the 10 seconds constant current charging pulse. The interval of each pulse cycle period is 1 hour. In the HPPC experiment, the current pulse curve and the corresponding voltage change curve are obtained. The voltage output waveform of the HPPC experiment terminal is shown in Figures 2.27 and 2.28.

From the HPPC test battery voltage response curve, features are extracted and calculated to obtain the model parameters for ten SOC points. The features are described as follows.

1. The vertical change of the voltage at the time point t_1 of the discharge start and the time point t_2 of the discharge, in which the voltage transient is stopped by the existence of the ohmic resistance of the battery.
2. The battery terminal voltage drops slowly during $t_1 \sim t_2$, which is the process of charging the polarized capacitor by the discharge current, which is the zero-state response of the double RC series circuit.

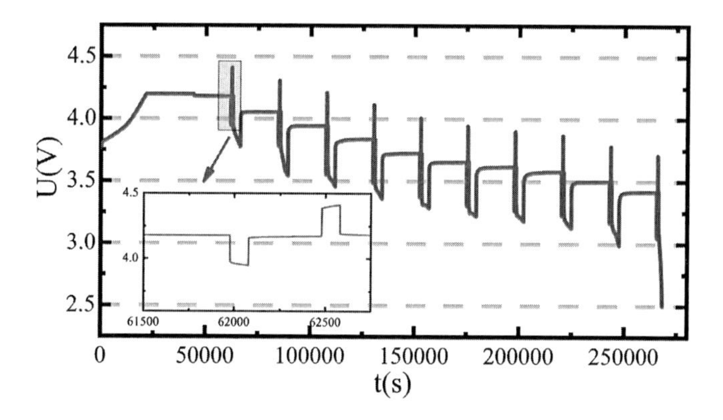

FIGURE 2.27 Schematic voltage curve diagram of HPPC test.

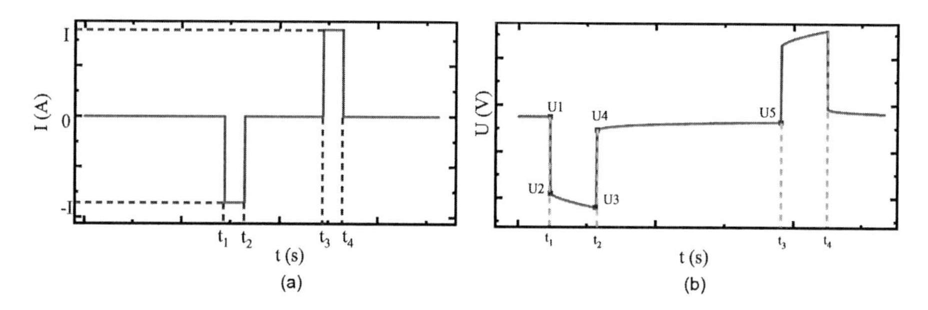

FIGURE 2.28 Schematic current and voltage curve in HPPC test. (a) Current curve. (b) Voltage curve.

3. The battery terminal voltage slowly rises during $t_2 \sim t_3$, which is the process of discharging the polarized capacitor to the polarized resistor, which is the zero-input response of the double RC series circuit.
4. The battery terminal voltage U_1 before the discharge pulse is slightly higher than the terminal voltage U_5 at the end of the discharge to reach a steady state, which is the voltage difference caused by the integration of the discharge current on the energy storage capacitor.

It can be observed from the above characteristics that the values of R_0 and C_b in the improved PNGV model are obtained from the characteristics (1) and (4). Meanwhile, the values of polarization resistance and capacitance in the dual RC circuit can be obtained by the characteristics of (2) and (3) in the battery terminal electrical curve identification process.

1. Open-circuit voltage

   The open-circuit voltage U_{OC} is the stable voltage at the positive and negative ends of the battery when the battery is left resting for a long time. Experiments show that the terminal voltage after the battery is left resting for 1 hour is equal to the open-circuit voltage of the batteries. The battery is discharged at a constant current rate of 1 C for 6 minutes after measuring the open-circuit voltage of the battery at SOC = 1 to reduce the SOC by 0.1 to obtain the open-circuit voltage of the lithium-ion battery corresponding to different SOC states. After resting for 1 hour, when the battery voltage reaches a steady-state condition, the open-circuit voltage of the battery is measured again, and cycle the test until the SOC is reduced to 0.

2. Ohmic resistance

   It can be observed from mathematical and mechanism analysis that the sudden change in the terminal voltage at the start and stop time points of discharge is caused by the ohmic resistance. The value of the ohmic resistance is calculated by Ohm's law, as shown in Equation (2.33).

$$R_0 = \frac{(U_1 - U_2) + (U_4 - U_3)}{2I} \tag{2.33}$$

3. Double RC circuit

   The improved PNGV model is a typical dual RC circuit model. The HPPC experiment is performed on lithium-ion batteries. When the battery is in the pulse discharge phase, the current direction of the improved PNGV model is described as follows: As expressed by I_b in the figure, the current direction is supposed to be positive at this time. For the reference direction of voltage and current, the KVL and KCL equations are obtained, as shown in Equations (2.34)–(2.36).

$$U_L = U_{OC}(\text{SOC}) - i(t)R_0 - U_S - U_L \tag{2.34}$$

$$\frac{dU_S}{dt} = \frac{i(t)}{C_S} - \frac{U_S}{R_S C_S} \tag{2.35}$$

$$\frac{dU_L}{dt} = \frac{i(t)}{C_L} - \frac{U_L}{R_L C_L} \tag{2.36}$$

In Equations (2.34)–(2.36), U_S is the terminal voltage of the parallel circuit composed of R_S and C_S, and U_L is the terminal voltage of the parallel circuit composed of R_L and C_L. During the battery discharge, the polarized capacitors C_S and C_L are in a charged state, and the voltage of the R_c parallel circuit rises exponentially. After the battery enters a rest from the discharged state, the capacitors C_S and C_L are discharged to their respective parallel resistors, and the voltage drops exponentially. The values of R and C in the model are related to the SOC estimation of the batteries. Using Simulink software to curve-fit the data obtained from the HPPC experiment and then use the undetermined parameter method to find the values of R_S, R_L, C_S, and C_L in the improved PNGV model, the specific method is described as follows:

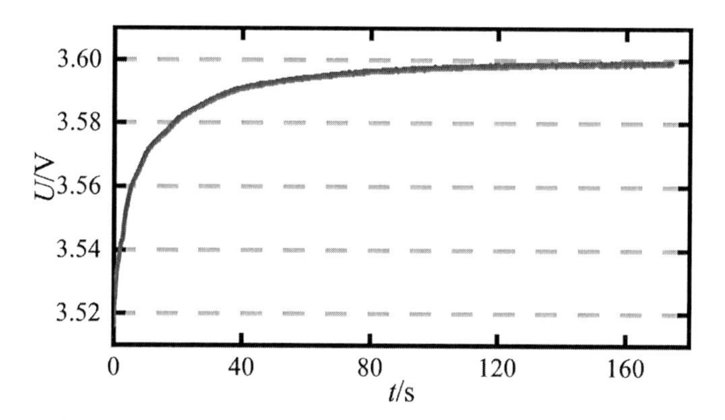

FIGURE 2.29 Terminal voltage variation of when polarization effect disappears.

First, the battery terminal voltages are obtained at different SOC values and at the same stage as $U_4 \sim U_5$ from the HPPC experimental data. This research selects the case where the SOC value is 0.4, and the terminal voltage curve is shown in Figure 2.29.

In Figure 2.29, the terminal voltage change at this stage is the process of discharging the polarized capacitor to the polarized resistor, which is the zero-input response of the double RC circuit. At this level, the value of I_b in the circuit is 0, and the output equation of battery voltage is obtained, as shown in Equation (2.37).

$$U_L = U_{OC} - IR_s e^{-\frac{t}{\tau_s}} - IR_L e^{-\frac{t}{\tau_L}} \tag{2.37}$$

The coefficients of the equation shown in Equation (2.37) are replaced by Equation (2.38).

$$U_L = f - ae^{-ct} - be^{-dt} \tag{2.38}$$

Second, using f, a, b, c, and d as undetermined parameters, using experimental data and Equation (2.38) as the target equation for single-exponential and double-exponential curve fitting, the results are obtained accordingly, as shown in Figure 2.30.

Finally, the results of double-exponential curve fitting are obtained. Comparing Equations (2.37) and (2.38), the mathematical relationship functions are obtained, as shown in Equation (2.39).

$$\begin{cases} U_{OC} = f, R_S = \dfrac{a}{I}, R_L = b/I \\ C_S = \dfrac{1}{R_S c}, C_L = 1/(R_L d) \end{cases} \tag{2.39}$$

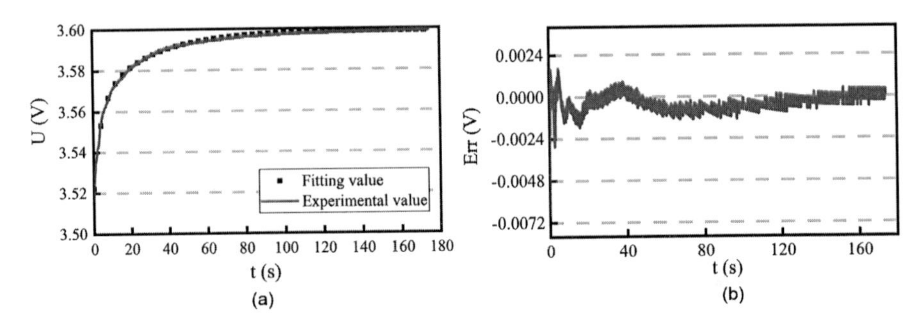

FIGURE 2.30 Terminal voltage fitting of polarization reaction disappearance process. (a) Double-exponential fitting curve. (b) Double-exponential fitting error.

TABLE 2.5
Double-Exponential Fitting Results

SOC	R_0(mΩ)	R_s(mΩ)	R_s(kF)	R_L(mΩ)	C_L(kF)	U_{OC}(V)
0.9	2.2867	0.9734	3.879633291	0.22620	108.7544032	3.984
0.8	2.2967	0.5754	2.187991245	0.88600	21.0023850	3.880
0.7	2.4033	0.8038	3.640885484	0.74980	106.7805431	3.786
0.6	2.3833	0.7968	5.548276217	0.19364	147.0450532	3.682
0.5	2.4467	0.7704	5.106321790	0.39240	60.5181662	3.632
0.4	2.4300	0.8178	4.814145114	0.71880	37.8972369	3.599
0.3	2.4467	1.1822	5.598150640	0.37340	175.2678794	3.551
0.2	2.6667	0.8738	3.075588934	0.41000	60.3121758	3.462
0.1	2.9933	1.6028	2.015858355	0.49600	40.9283197	3.417

2.5.5 Double-Exponential Fitting Results

Double-exponential fitting is conducted, according to which the fitting results are shown in Table 2.5.

Analyzing the parameter identification results, the time constant τ_s of the parallel circuit composed of R_S and C_S is small, which is used to simulate the rapid voltage change process of the battery when the current changes suddenly. The time constant of the parallel circuit is composed of R_L and C_L is larger than τ_S, which is used to simulate the process of slow voltage stabilization. The establishment of the improved PNGV model and the SOC estimation of lithium-ion batteries need to obtain the characteristic curves of R_S, R_L, C_S, C_L, R_0, and U_{OC} for the various SOC levels. U_{OC} varies significantly from different SOC values. A seventh-order polynomial is used as the fitting function of the U_{OC} value for different SOC levels to obtain the characteristic curve of U_{OC} with SOC. The U_{OC}-SOC fitting polynomial is shown in Equation (2.40).

$$f(x) = B_1x^7 + B_2x^6 + B_3x^5 + B_4x^4 + B_5x^3 + B_6x^2 + B_7x + B_8 \tag{2.40}$$

In Equation (2.40), the independent variable x is the SOC value; the dependent variable is the parameter curve to be fitted; $B_1 \sim B_8$ are the polynomial undetermined coefficients. Using simulation to establish the polynomial curve fitting on the experimental data, the results are obtained, as shown in Table 2.6.

In addition, the characteristic curves of R_S, R_L, C_S, C_L, and R_0 can be obtained by polynomial fitting to the data identified. When using the least-squares method to fit the data onto a polynomial, if the parameter change law is not obvious, the higher the polynomial order (sixth-order and above), the more likely it is to produce oscillations. Therefore, this chapter compares the data in Table 2.6. When performing polynomial fitting, it is found that the fitting effect of fourth-order polynomial and fifth-order polynomial is similar, so the fourth-order polynomial is selected for fitting, and the fitting polynomial is described as shown in Equation (2.41).

$$f(x) = P_1x^4 + P_2x^3 + P_3x^2 + P_4x + P_5 \tag{2.41}$$

The meaning of each parameter in Equation (2.41) is similar to that in Equation (2.40), and the fitting results are shown in Table 2.7.

2.5.6 Experimental Verification

The improved simulation system model is established in the simulation software for parameter verification. The specific structure is shown in Figure 2.31.

In Figure 2.31, the parameter values of the improved PNGV model under different SOC states obtained by the above identification are verified. First, import the charge–discharge current values and the battery terminal voltage value obtained

TABLE 2.6
Open-Circuit Voltage Fitting Results

Parameter	B_1	B_2	B_3	B_4
U_{OC}(V)	−23.60229	141.34077	−314.9228	345.34531
Parameter	B_5	B_6	B_7	B_8
U_{OC}(V)	−200.1546	60.21383	−7.88447	3.77173

TABLE 2.7
Polynomial Fitting Results

Parameter	P_1	P_2	P_3	P_4	P_5
R_0(Ω)	0.02176	−0.0529	0.04554	−0.01654	0.004568
R_S(Ω)	0.03249	−0.0669	0.04916	−0.0158	0.002803
C_S(F)	1771000	−3139000	1604000	−180400	26910
R_L(Ω)	−0.06047	0.1192	−0.07955	0.02085	−0.0013
C_L(F)	6417000	−4156000	−6328000	5408000	−14110

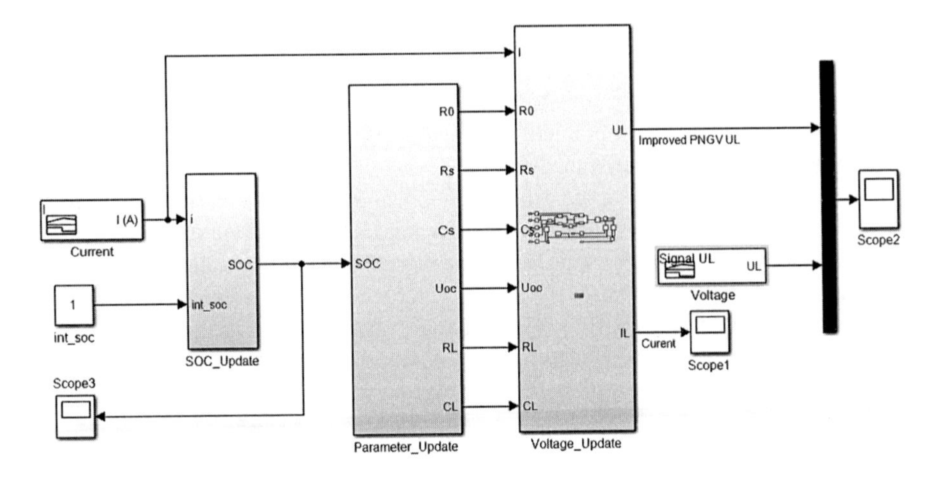

FIGURE 2.31 Improved PNGV simulation system model.

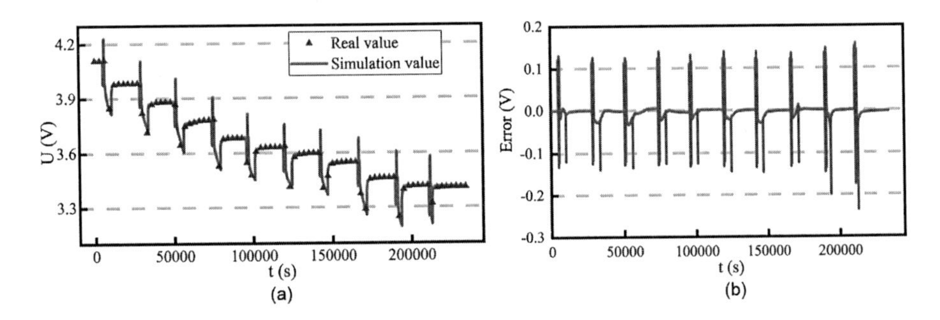

FIGURE 2.32 Parameter identification verification results. (a) Simulated and real voltage curve. (b) Error of simulated and real voltage.

in the HPPC experiment into the current and voltage modules, respectively. Then, respectively fit the R_0 that is obtained by the Simulink software and curve fitting, six polynomial functions of SOC, R_S(SOC), C_S(SOC), R_L(SOC), C_L(SOC), and OCV-SOC are input to the model parameter update subsystem for simulation. The obtained simulation and measured terminal voltage are compared. The estimation curve and the error curve of the two battery models are obtained, as shown in Figure 2.32.

In Figure 2.32, the error curve of the simulated and real voltage is described under the periodic current discharge.

1. When the current pulse applies, the battery terminal voltage suddenly changes, causing the error between the experimental and simulated values to increase to 0.1–0.2 V. After several HPPC experimental results' analysis under different discharge rates, it is believed that at the time point when the current pulse applies, the large error of the battery terminal voltage is caused by environmental factors and measurement deviations.

2. When the battery charges within the range of SOC value varying from 0% to 10%, the error between the experimental value and the simulated value of the terminal voltage is slightly larger than that of the SOC value changes into varying from 10% to 100%, mainly because when the SOC value is less than 10%, the internal electrochemical reaction is changed drastically, and the parameter values in the battery model are changed highly.
3. Except for the above individual time points, when the SOC value of the battery changes from 10% to 100%, the simulated value of the terminal voltage is consistent with the experimental value, and the average error is less than 0.035 V, which accounts for 0.9% of the nominal voltage.

2.6 ONLINE PARAMETER IDENTIFICATION

Battery parameters can be obtained by the online and offline identification methods. Most offline parameter identification methods are based on the obtained experimental data. These data can only reflect the changes in battery parameters under certain conditions, but the operating conditions of lithium-ion batteries vary greatly in different operating conditions and working environments. In addition, it is impossible to do this work for all batteries, and when the identified battery parameters are applied to the same type of battery, the simulation accuracy is low [158]. Therefore, the model cannot rely on offline parameter identification under these conditions. In addition, the aging of the battery is an important reason that affects the battery parameters. In the use of batteries, the parameters of batteries change slowly with the aging process, and the parameters need to be refitted. The results of offline parameter identification are relatively fixed. In many cases, it cannot meet actual engineering requirements. Therefore, the equivalent model based on the offline method does not have high adaptability and cannot accurately estimate SOC and SOH values. The online identification method uses the input and output data onto the battery system obtained in real time and in the past period to approximate the real value of the system parameters by a dynamic programming method.

2.6.1 Recursive Least-Square Method

At present, recursive least-square (RLS), dual-extended Kalman filter (DEKF), and genetic algorithm (GA) are widely used in parameter identification, which is a recursive online identification method. It is not necessary to store and calculate all the previous input and output data repeatedly in the computer, so they can greatly reduce the amount of data storage and calculation. It is especially suitable for online real-time identification calculation. The least-square identification is a process in which the model parameters are constantly corrected by the error between the model output and the actual output under a given structural framework, and the optimal model is finally obtained. To realize real-time control, the least-square method must be optimized into the recursive least-square (RLS) method, which is mainly used for online identification [159,160].

However, in practical applications, only one online parameter identification is not good, and the main reasons are obtained as follows. (1) The internal resistance of the battery is a slow variable. When identifying such parameters, there is the so-called data saturation phenomenon, and the data correction ability is slowly lost. (2) The phenomenon of asynchronous sampling of voltage and current makes the data samples have a time delay. When the sampling step is relatively small, the calculated voltage and current no longer maintain the due change relationship. Some online identification methods adjust the internal resistance value at the time delay, resulting in large fluctuations in the identification results. (3) When the model has colored noise, some online estimation is not optimal.

Using the online parameter identification method to identify the battery model parameters can better observe the changes in the model parameters. In addition, the online parameter identification method can also reduce manual calculations, saving time and avoiding calculation errors. At present, the commonly used online parameter identification methods mainly include the recursive least-square, multi-innovation least-square, and extended Kalman filtering (EKF) algorithm. The chapter takes the Thevenin equivalent circuit model as an example to briefly introduce these three methods. The Thevenin model is shown in Figure 2.33.

In Figure 2.33, U_{OC} is the open-circuit voltage, U_O is the terminal voltage, R_0 is the ohmic resistance, and U_R is the ohmic voltage, which is the transient voltage drop effect of the battery during charge–discharge. The RC parallel loop is composed of a polarization resistance R_1 and a polarization capacitor C_1 to characterize the polarization effect of the lithium-ion battery, where U_1 is the polarization voltage. According to Kirchhoff's law, the Thevenin equivalent circuit model is analyzed, and the voltage and current expressions of the equivalent circuit are shown in Equation (2.42).

$$\begin{cases} U_O = U_{OC} - U_R - U_1 \\ I(t) = C_1 \dfrac{dU_1}{dt} + \dfrac{U_1}{R_1} \end{cases} \tag{2.42}$$

Among them, the open-circuit voltage U_{OC} is characterized by the state variable SOC, and the nonlinear function relationship can be obtained. The least-square (LS) method is a classic data processing method, in which the approximation principle is used to minimize the sum of the distance squares between the estimated and actual

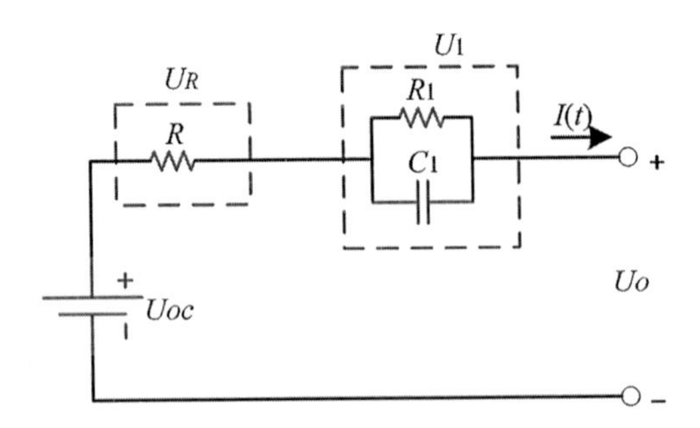

FIGURE 2.33 Thevenin equivalent circuit diagram.

data to find the best function match of the data. The model parameters are identified, as shown in Equation (2.43).

$$y = a_1x_1 + a_2x_2 + \cdots + a_nx_n + b + e \tag{2.43}$$

In Equation (2.43), there are N groups of input and output data, as shown in Equation (2.44).

$$\begin{cases} Y = \begin{bmatrix} y(1) \\ y(2) \\ \vdots \\ y(N) \end{bmatrix}, \theta = \begin{bmatrix} a_1 \\ a_2 \\ \vdots \\ a_n \\ b \end{bmatrix}, E = \begin{bmatrix} e(1) \\ e(2) \\ \vdots \\ e(N) \end{bmatrix} \\ X = \begin{bmatrix} x_1(1) & x_2(1) & \cdots & x_n(1) \\ x_1(2) & x_2(2) & \cdots & x_n(2) \\ \vdots & \vdots & \vdots & \vdots \\ x_1(N) & x_2(N) & \cdots & x_n(N) \end{bmatrix} \end{cases} \tag{2.44}$$

The identification model can be changed, as shown in Equation (2.45).

$$Y = X\theta + \xi \tag{2.45}$$

The criterion function is obtained, as shown in Equation (2.46).

$$J = \sum_{i=1}^{N} e(i)^2 = \xi^T \xi = (Y - X\theta)^T (Y - X\theta) \tag{2.46}$$

To make J the smallest, the mathematical treatment is conducted, as shown in Equation (2.47).

$$\begin{cases} \dfrac{dJ}{d\hat{\theta}} = -2X^T\left(Y - X\hat{\theta}\right) = 0 \\ \hat{\theta}_{LS} = \left(X^T X\right)^{-1} X^T Y \end{cases} \tag{2.47}$$

In Equation (2.47), θ is the identification result. Recursive least-square (RLS) is a derivation of the LS algorithm to solve the disadvantage that the RLS algorithm cannot be updated in time-varying systems. RLS is a recursive estimation algorithm. The estimation result at the time point $t+1$ is modified by the estimation result at time point t and the 'innovation' superimposed, which is calculated according to the input and output at time point $t+1$. The recursive least-square identification algorithm can reduce the amounts of calculations and the storage of data in the computer.

Also, it can identify the characteristics of the system in real time. Based on the LS method, the calculation equations can be obtained, as shown in Equations (2.48)–(2.51).

$$\hat{\theta}_N = \left(X_N^T X_N\right)^{-1} X_N^T Y_N \tag{2.48}$$

$$\hat{\theta}_{N+1} = \left(X_{N+1}^T X_{N+1}\right)^{-1} X_{N+1}^T Y_{N+1} \tag{2.49}$$

$$Y_{N+1} = \begin{bmatrix} y(1) \\ y(2) \\ \vdots \\ y(N) \\ y(N+1) \end{bmatrix} = \begin{bmatrix} Y_N \\ y(N+1) \end{bmatrix} \tag{2.50}$$

$$X_{N+1} = \begin{bmatrix} X_N \\ x^T(N+1) \end{bmatrix} \tag{2.51}$$

The parameter P_N is initialed, as shown in Equation (2.52).

$$P_N = \left(X_N^T X_N\right)^{-1} \tag{2.52}$$

According to Equation (2.52), P_{N+1} can be obtained, as shown in Equation (2.53).

$$\begin{aligned} P_{N+1} = \left(X_{N+1}^T X_{N+1}\right)^{-1} &= \left\{ \begin{bmatrix} X_N \\ x^T(N+1) \end{bmatrix}^T \begin{bmatrix} X_N \\ x^T(N+1) \end{bmatrix} \right\}^{-1} \\ &= \left[P_N^{-1} + x(N+1)x^T(N+1) \right]^{-1} \end{aligned} \tag{2.53}$$

The deduced RLS method identification equation is shown in Equation (2.54).

$$\begin{cases} \hat{\theta}_{N+1} = \hat{\theta}_N + \gamma(N+1) P_N x(N+1) \left[y(N+1) - x^T(N+1)\hat{\theta}_N \right] \\ P_{N+1} = P_N - \gamma(N+1) P_N x(N+1) x^T(N+1) P_N \\ \gamma(N+1) = 1 / \left[1 + x^T(N+1) P_N x(N+1) \right] \end{cases} \tag{2.54}$$

When applying the RLS method of the battery equivalent model, according to the equivalent circuit model in Figure 2.33, the frequency domain expression of the circuit differential equation can be obtained, as shown in Equation (2.55).

$$U_O(s) = U_{OC}(s) - I(s)\left(R_0 + \frac{R_1}{1 + R_1C_1s}\right) \tag{2.55}$$

In practical application, the input and output data of batteries are obtained by discrete sampling, so it is necessary to change S-domain to Z-domain for calculation. Substituting $s = 2\left(1 - z^{-1}\right)/T\left(1 + z^{-1}\right)$ into Equation (2.55), discretizing Equation (2.55) to obtain Equation (2.56).

$$\frac{E(s)}{I(s)} = \frac{\frac{R_0T + R_1T + 2R_1C_1R_0}{T + 2R_1C_1} + \frac{R_0T + R_1T - 2R_1C_1R_0}{T + 2R_1C_1}z^{-1}}{1 + \frac{T - 2R_1C_1}{T + 2R_1C_1}z^{-1}} \tag{2.56}$$

In Equation (2.56), $E(s) = U_O(s) - U_{OC}(s)$. According to the principle of the RLS algorithm, the parameters can be calculated, as shown in Equation (2.57).

$$\begin{cases} \theta_1 = \dfrac{T - 2R_1C_1}{T + 2R_1C_1} \\ \theta_2 = \dfrac{R_0T + R_1T + 2R_1C_1R_0}{T + 2R_1C_1} \\ \theta_3 = \dfrac{R_oT + R_1T - 2R_1C_1R_0}{T + 2R_1C_1} \end{cases} \tag{2.57}$$

Combining Equations (2.56) and (2.57) can get the parameters that need to be identified, as shown in Equation (2.58).

$$E(k) = -\theta_1E(k-1) + \theta_2I(k) + \theta_3I(k-1) \tag{2.58}$$

The RLS method can be used to estimate the values of θ_1, θ_2, and θ_3, and then, according to Equation (2.57), the various parameters can be calculated, as shown in Equation (2.59).

$$\begin{cases} R_0 = \dfrac{(\theta_2 - \theta_3)}{(1 - \theta_1)} \\ \tau = R_1C_1 = \dfrac{(1 - \theta_1)}{(2\theta_1 + 2)} \\ R_1 = (1 + 2\tau)\theta_2 - 2R_0\tau - R_0 \\ C_1 = \dfrac{\tau}{R_1} \end{cases} \tag{2.59}$$

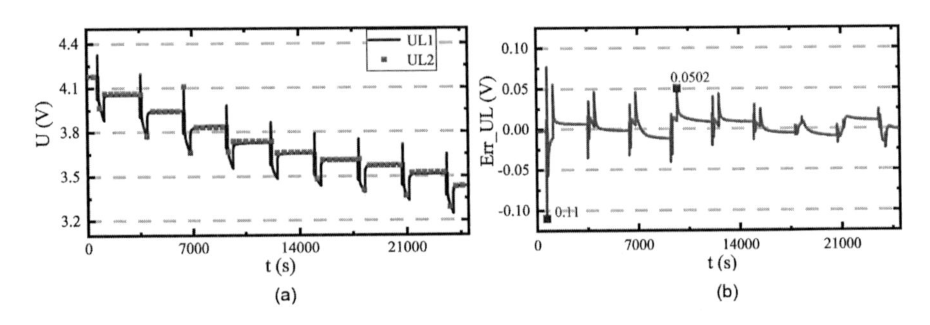

FIGURE 2.34 Parameter identification result graph. (a) Voltage output comparison. (b) Voltage traction error.

The results of parameter identification are substituted for the Thevenin equivalent circuit simulation model to verify the accuracy of parameter identification under working conditions. The voltage and current data under cyclic charge–discharge conditions are imported into the battery equivalent model, and the online parameter identification results and model are verified. The model output voltage is compared with the actual terminal voltage value as well as the error analysis, as shown in Figure 2.34.

Figure 2.34a shows the comparison result between the actual output voltage curve and the model output voltage curve under the HPPC experimental condition. In Figure 2.34a, U_{L1} is the real battery terminal voltage value and U_{L2} is the voltage simulation value output by the constructed model. Figure 2.34b shows the error curve of the two. It can be observed from the figure that, under the experimental condition of HPPC, the output voltage of the equivalent model has a good tracking effect on the actual value, and the average estimation deviation is about 0.03 V. It indicates that the Thevenin equivalent circuit model can well characterize the characteristics of the battery for practical application. The large initial error is due to the initial value error of the parameters in the online parameter identification, and the error decreases after the parameters converge.

2.6.2 Bias Compensation Method

As a complex nonlinear electrochemical system, the model parameters of lithium-ion battery generally show time-varying characteristics; that is, during the process of battery charge–discharge reaction, the model parameters of the battery, open-circuit voltage U_{OC}, ohmic resistance R_0, polarization internal resistance R_p, and polarization capacitance C_p, all change. Because of the noise signal in the battery measurement data, the traditional least-square method cannot identify the battery model parameters with high accuracy. The accuracy of battery model parameters directly affects the estimation accuracy of battery SOC and SOH. The bias compensation recursive least-square (BCRLS) algorithm introduces noise variance estimation, which can compensate the parameters identified by the general least-square method and realize the accurate identification of battery model parameters. The identification process of the bias compensation recursive least-square algorithm is introduced in detail below.

The recursive process of the BCRLS algorithm consists of eight steps. The first is parameter initialization. The second step is the calculation of model prediction output and estimation error. The third and fourth steps are to calculate the Kalman gain matrix and parameter estimation. The fifth and sixth steps are to calculate the error criterion function and estimate the noise variance. Finally, seventh and eight steps are to update the covariance matrix and calculate the parameters after offset compensation. The specific iteration equation is shown in Equation (2.60).

$$\begin{cases} \hat{y}(k) = \varphi^T(k)\theta(k-1) \\ e(k) = y(k) - \hat{y}(k) \\ K(k) = P(k-1)\varphi(k)\left[1+\varphi^T(k)P(k-1)\varphi(k)\right]^{-1} \\ \theta(k) = \theta(k-1) + K(k)e(k) \\ J(k) = J(k-1) + e^2(k)\left[1+\varphi^T(k)P(k-1)\varphi(k)\right]^{-1} \\ \sigma^2(k) = \dfrac{J(k)}{k\left[1+\theta_c(k-1)\lambda\theta(k-1)\right]} \\ P(k) = \left[I - K(k)\varphi^T(k)\right]P(k-1) \\ \theta_c(k) = \theta(k) + k\sigma^2(k)P(k)\theta_c(k-1) \end{cases} \tag{2.60}$$

In Equation (2.60), δ is generally a large positive number and I_0 is the identity matrix. The overall process and operation principles of BCRLS are shown in Figure 2.35.

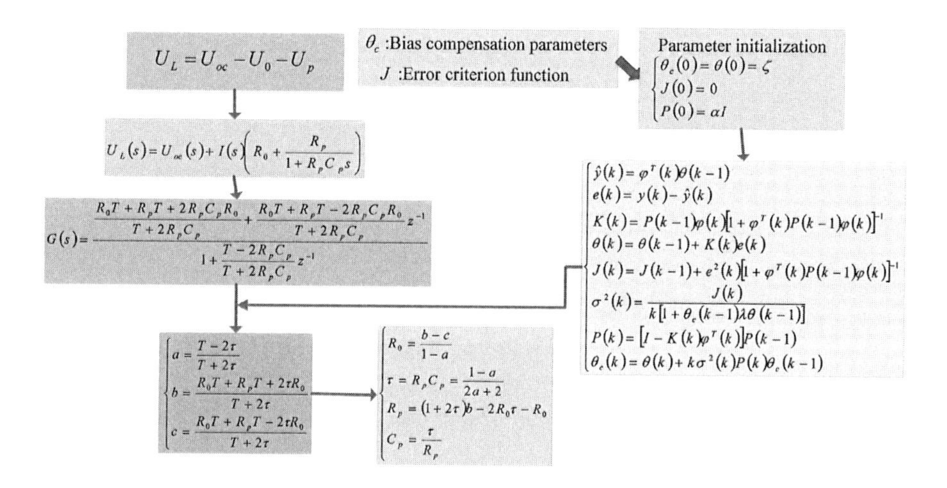

FIGURE 2.35 BCRLS schematic diagram.

2.6.3 Forgetting Factor—RLS Method

The battery has time-varying characteristics in use, and the battery model parameters are constantly changing under different SOC, current, temperature, and aging conditions. When using the battery model for online estimation, if the model parameters can be adaptively adjusted and corrected according to the battery state, the estimation accuracy is improved. Therefore, the recursive least-square method with the forgetting factor can be used to identify the battery model parameters [161,162]. According to the least-square theory, the model expression needs to be discretized first. For the second-order RC equivalent circuit model, the expression of battery terminal voltage is obtained through the Laplace transformation, as shown in Equation (2.61).

$$U_{\text{OCV}}(s) - U_t(s) = \left(R_0 + \frac{R_1}{1+\tau_1 s} + \frac{R_2}{1+\tau_2 s} \right) I(s) \tag{2.61}$$

In Equation (2.61), the mathematical relationship can be expressed with the optimation treatment, and the transfer function is obtained accordingly, as shown in Equation (2.62).

$$G(s) = \frac{U_{\text{OCV}}(s) - U_t(s)}{I(s)} = R_0 + \frac{R_1}{1+\tau_1 s} + \frac{R_2}{1+\tau_2 s} \tag{2.62}$$

By using the bilinear transformation equation, as shown in Equation (2.63).

$$s = \frac{2}{T} * \frac{1-z^{-1}}{1+z^{-1}} \tag{2.63}$$

In Equation (2.63), T is the sampling period. Then, Equation (2.65) is discretized into Equation (2.62), as shown in Equation (2.64).

$$G\left(z^{-1}\right) = \frac{\alpha_3 + \alpha_4 z^{-1} + \alpha_5 z^{-2}}{1 - \alpha_1 z^{-1} + \alpha_2 z^{-2}} \tag{2.64}$$

In Equation (2.64), α_1, α_2, α_3, α_4, and α_5 are corresponding constant coefficients, and then, the discretization equation can be obtained, as shown in Equation (2.65).

$$\begin{aligned} y(k) &= U_{\text{OCV}}(s) - U_t(s) \\ &= \alpha_1 y(k-1) + \alpha_2 y(k-2) + \alpha_3 I(k) + \alpha_4 I(k-1) + \alpha_5 I(k-2) \end{aligned} \tag{2.65}$$

In Equation (2.65), $I(k)$ and $y(k)$ are the input and output of the system, respectively, as shown in Equation (2.66).

$$\begin{cases} \varphi(k) = \left[y(k-1), y(k-2), I(k), I(k-1), I(k-2) \right] \\ \theta(k) = \left[\alpha_1, \alpha_2, \alpha_3, \alpha_4, \alpha_5 \right] \end{cases} \tag{2.66}$$

And then, Equation (2.65) can be expressed, as shown in Equation (2.67).

$$y(k) = \varphi^T(k)\theta(k) \tag{2.67}$$

In the recursive least-square identification method, $\theta(k)$ is taken as the coefficient to be identified, which is then used to deduce the parameters of the battery model accordingly. The recursive least-square method has the attitude of infinite memory, that is, with the increase of recursion times k, old data accumulate continuously, which makes it difficult for new data to play the role of correction, which affects the effect of parameter estimation, especially in time-varying systems. Because of the imbalance between the old and new data, the newly estimated parameters cannot accurately reflect the characteristics of the current system. To avoid the above situation, a forgetting factor of λ is introduced in the RLS algorithm. The function of the forgetting factor is to give a smaller weight to the data with a longer running time in the recognition process, while the latest observation data occupy a larger weight. The genetic factor is introduced to weaken the influence of the old data and enhance the feedback effect of the new data. The steps of the least-square algorithm with the forgetting factor are described as follows.

1. Initializing parameters can be realized, such as $\hat{\theta}(k)$ and covariance matrix $P(k)$.
2. The calculation of $\hat{\theta}(k+1)$ is conducted at the next time point, as shown in Equation (2.68).

$$\hat{\theta}(k+1) = \hat{\theta}(k) + K(k+1)\left[y(k+1) - \varphi^T(k+1)\hat{\theta}(k) \right] \tag{2.68}$$

3. Calculating the posterior error, as shown in Equation (2.69).

$$e(k+1) = y(k+1) - \varphi^T(k+1)\hat{\theta}(k) \tag{2.69}$$

4. Calculating a gain matrix, as shown in Equation (2.70).

$$K(k+1) = P(k+1)\varphi(k+1)\left[\lambda + \varphi^T(k+1)P(k)\varphi(k) + 1 \right]^{-1} \tag{2.70}$$

5. Calculating covariance matrix, as shown in Equation (2.71).

$$P(k+1) = \frac{1}{\lambda}\left[1 - K(k+1) * \varphi^T(k+1)P(k) \right] \tag{2.71}$$

6. Circulating steps (2), (3), (4), and (5).

2.6.4 FFRLS-Based Second-Order RC Model Parameter Identification

To overcome the phenomenon of data saturation, the forgetting factor $\lambda(0<\lambda<1)$ is introduced into the traditional RLS algorithm to weaken the influence of old data and enhance the feedback effect of new data. The basic principle of the forgetting factor recursive least-square (FFRLS) method is to minimize the error sum of squares. The improved objective function is shown in Equation (2.72).

$$J=\sum_{k=n+1}^{N}\lambda^{N-k}\left[e(k)\right]^2 \tag{2.72}$$

Then, the recursive equation of the FFRLS method is modified, as shown in Equation (2.73).

$$\begin{cases}\hat{\theta}(k+1)=\hat{\theta}(k)+K(k+1)\left[y(k+1)-\phi^T(k+1)\hat{\theta}(k)\right]\\ K(k+1)=P(k+1)\phi(k+1)\left[\phi^T(k+1)P(k)\phi(k+1)+\lambda\right]^{-1}\\ P(k+1)=\lambda^{-1}\left[I-K(k+1)\phi^T(k+1)\right]P(k)\end{cases} \tag{2.73}$$

In Equation (2.73), the closer λ is to 1, the better the simulation result. Generally, the value is greater than 0.9. The smaller the value of λ, the smaller the tracking effect of the algorithm, but it causes fluctuations in the algorithm. When $\lambda=1$, it is the standard least-square recursive algorithm. The equivalent circuit model and the battery charge–discharge experiment are inseparable. Through the charge–discharge characteristic experiment and the HPPC experiment, the performance parameters of the lithium-ion battery during operation are studied. Considering the accuracy of the equivalent circuit model and the difficulty of identifying and estimating parameters such as SOC, resistance, and capacitance, the second-order RC equivalent circuit model is finally selected as the lithium-ion battery equivalent circuit model. This model consists of a static ohmic resistance R_i and two RC circuits. The dynamic response is composed in series, as shown in Figure 2.36.

According to Kirchhoff's voltage law, the KVL equation of the circuit can be obtained, as shown in Equation (2.74).

$$U_b=U+U_i+U_1+U_2 \tag{2.74}$$

In Equation (2.74), U_b is the open-circuit voltage of the battery, R_i is the ohmic resistance, and both R_1 and R_2 represent the polarization internal resistance of the batteries. The least-square parameter identification method with the forgetting factor is applied to the identification of the equivalent model parameters of the lithium-ion battery, and the parameters in the second-order RC equivalent circuit model are identified. The battery model converts into a mathematical form of least square, as shown in Equation (2.75).

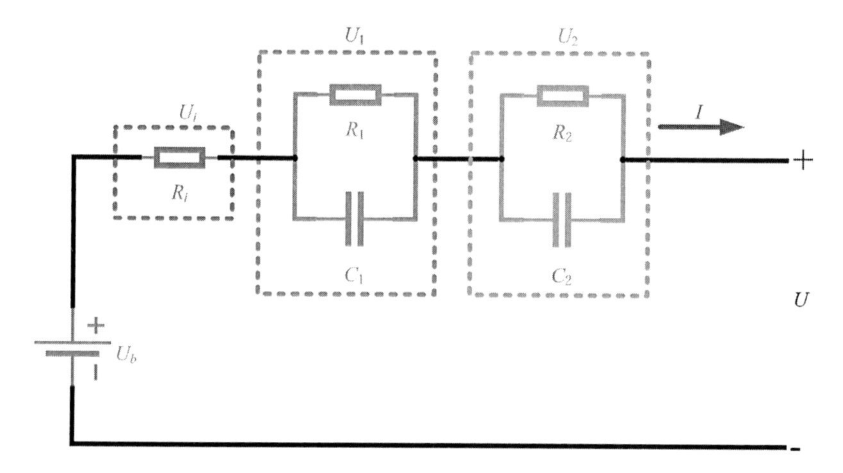

FIGURE 2.36 Second-order RC equivalent circuit model.

$$U_b = \left(\frac{R_1}{R_1 C_1 s + 1} + \frac{R_2}{R_2 C_2 s + 1} + R_i \right) I + U \tag{2.75}$$

Considering the time constant $\tau_1 = R_1 C_1$, $\tau_2 = R_2 C_2$, then (2.75) can be rewritten, as shown in Equation (2.76).

$$\begin{aligned} \tau_1 \tau_2 U_b s^2 + (\tau_1 + \tau_2) U_b s + U_b = \tau_1 \tau_2 R_i I_s{}^2 + \left[R_1 \tau_2 + R_2 \tau_1 + R_i (\tau_1 + \tau_2) \right] I_s \\ + (R_1 + R_2 + R_i) I + \tau_1 \tau_2 U_s^2 + (\tau_1 + \tau_2) U_s + U \end{aligned} \tag{2.76}$$

Supposing $a = \tau_1 \tau_2$, $b = \tau_1 + \tau_2$, $c = R_1 + R_2 + R_i$, $d = R_1 \tau_2 + R_2 \tau_1 + R_i (\tau_1 + \tau_2)$, (2.76) can be simplified, as shown in Equation (2.77).

$$a U_b s^2 + b U_b s + U_b = a R_i I_s{}^2 + d I_s + c I + a U_s^2 + b U_s + U \tag{2.77}$$

Substituting $s = [x(k) - x(k-1)]/T$ and $s^2 = [x(k) - 2x(k-1)]/T^2$ into equation (2.77) for discretization, where T is the sampling time, as shown in Equation (2.78).

$$\begin{aligned} U_b(k) - U(k) = & \frac{-bT - 2a}{T^2 + bT + a} [U(k-1) - U_b(k-1)] \\ & + \frac{a}{T^2 + bT + a} [U(k-2) - U_b(k-2)] \\ & + \frac{cT^2 + dT + aR_i}{T^2 + bT + a} I(k) + \frac{-dT - 2aR_i}{T^2 + bT + a} I(k-1) \\ & + \frac{aR_i}{T^2 + bT + a} I(k-2) \end{aligned} \tag{2.78}$$

And then, the calculation relationship can be obtained, as shown in Equation (2.79).

$$U_b(k) - U(k) = k_1\left[U(k-1) - U_b(k-1)\right] + k_2\left[U(k-2) - U_b(k-2)\right] + k_3 I(k) + k_4 I(k-1) + k_5 I(k-2) \tag{2.79}$$

In Equation (2.79), the parameters are described as shown in Equation (2.80).

$$k_1 = \frac{-bT - 2a}{T^2 + bT + a},\ k_2 = \frac{a}{T^2 + bT + a},\ k_3 = \frac{cT^2 + dT + aR_i}{T^2 + bT + a},\quad k_4 = \frac{-dT - 2aR_i}{T^2 + bT + a},\ k_5 = \frac{aR_i}{T^2 + bT + a} \tag{2.80}$$

Equation (2.80) is substituted into the recursive least-square method, taking $\theta = \left[k_1 \quad k_2 \quad k_3 \quad k_4 \quad k_5\right]^T$ as the direct identification parameter and then deriving the circuit model parameters R_i, R_1, C_1, R_2, and C_2 from the identification results of these parameters, and deriving as follows. Setting $k_0 = T^2 + bT + a$, the parameter values can be obtained, as shown in Equations (2.81)–(2.86).

$$k_0 = \frac{T^2}{k_1 + k_2 + 1} \tag{2.81}$$

$$a = k_0 k_2 \tag{2.82}$$

$$b = \frac{-k_0(k_1 + 2k_2)}{T} \tag{2.83}$$

$$c = \frac{k_0(k_3 + k_4 + k_5)}{T^2} \tag{2.84}$$

$$d = \frac{-k_0(k_4 + 2k_5)}{T} \tag{2.85}$$

$$R_i = \frac{k_5}{k_2} \tag{2.86}$$

In Equations (2.81)–(2.86), initialing $a = \tau_1\tau_2$, $b = \tau_1 + \tau_2$, and $\tau^2 - b\tau + a = 0$, the parameters values of t_1 and t_2 can be obtained, as shown in Equation (2.87).

$$\tau_1 = \frac{b + \sqrt{b^2 - 4a}}{2}, \tau_2 = \frac{b - \sqrt{b^2 - 4a}}{2} \tag{2.87}$$

In Equation (2.87), from $c = R_1 + R_2 + R_i$ and $d = R_1\tau_2 + R_2\tau_1 + R_i(\tau_1 + \tau_2)$, combining equations from Equations (2.81) to (2.86), the values of other parameters can be obtained, as shown in equations from Equations (2.88) to (2.91).

$$R_1 = \frac{(\tau_1 c + \tau_2 R_i - d)}{\tau_1 - \tau_2} \tag{2.88}$$

$$R_2 = c - R_1 - R_i \tag{2.89}$$

$$C_1 = \frac{\tau_1}{R_1} \tag{2.90}$$

$$C_2 = \frac{\tau_2}{R_2} \tag{2.91}$$

From Equations (2.88) to (2.91), the values of the various parameters are calculated for the equivalent circuit model.

2.6.5 Multi-Innovation Least-Square Method

Multi-innovation least-square (MILS) identification algorithm extends the scalar innovation to vector innovation and a data vector to an information matrix based on the traditional RLS identification algorithm. It identifies system parameters by online iteration. According to Kirchhoff's law, the Thevenin equivalent circuit model is analyzed, and the voltage and current expressions of the equivalent circuit are shown in Equation (2.92).

$$\begin{cases} U_L(k) = U_{OC} - U_P(k) - I_L(k)R_0 \\ \dfrac{dU_P(k)}{dk} = \dfrac{I_L(k)}{C_P} - \dfrac{U_P(k)}{C_P R_P} \end{cases} \tag{2.92}$$

In Equation (2.93), the state-space equation of the system obtained from the S-ECM model is a continuous-time model, which does not apply to online parameter identification, so Equation (2.92) can be discretized according to the procedures shown in Figure 2.37.

In combination with Figure 2.37, the differential equation of the discretized system can be obtained, as shown in Equation (2.93).

$$U(k) = \frac{R_p C_p}{R_p C_p + T} U(k-1) + \left(R_0 + \frac{R_p}{R_p C_p + T} \right) I(k) - \frac{R_0 R_p C_p}{R_p C_p + T} I(k-1) \tag{2.93}$$

In Equation (2.93), T is the sampling period and k is the discrete-time constant. The matrices of coefficients and parameters to be identified can be obtained by the difference equation, as shown in Equation (2.94).

$$\begin{cases} \theta = [a_1, a_2, a_3] = \left[-\dfrac{R_p C_p}{R_p C_p + T}, R_0 + \dfrac{R_p}{R_p C_p + T}, -\dfrac{R_0 R_p C_p}{R_p C_p + T} \right] \\ \eta = [R_0, R_p, C_p, R_{cd}] = \left[\dfrac{b_1}{a_1}, \dfrac{a_1 b_1 - b_2}{a_1(a_1+1)}, -\dfrac{a_1{}^2 T}{a_1 b_1 - b_2}, R_{\text{nom}} - \dfrac{b_1}{a_1} \right] \end{cases} \tag{2.94}$$

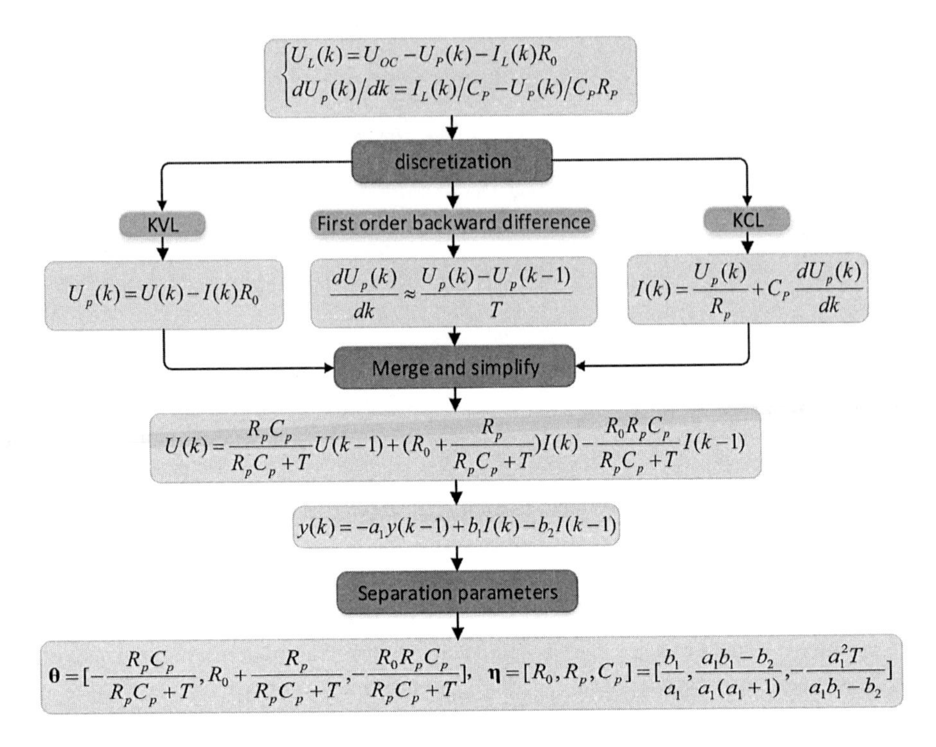

FIGURE 2.37 Discretization process of the state-space equation.

In Equation (2.94), θ and η are the matrices of coefficients and parameters to be identified in the discrete system model, respectively. Then, the autoregressive exogenous (ARX) model for online parameter identification of lithium-ion batteries is achieved, as shown in Equation (2.95).

$$\begin{cases} y(k) = \varphi^T(k)\theta \\ \varphi(k) = \left[U(K-1), I(K), I(K-1)\right] \\ \theta = \left[a_1, a_2, a_3\right] \end{cases} \tag{2.95}$$

In Equation (2.95), $\varphi(k)$, $y(k)$, and $I(k)$ are the data vector, the output vector, and input vector of the discrete system model, respectively. The coefficient matrix θ of the discrete system model is identified by MILS. According to (2.95), all the parameters to be identified in the S-ECM model are acquired by recursion. The MILS algorithm can solve the data saturation problem existing in the traditional RLS algorithm. The whole problem of its algorithm is described as follows. if we assume the ARX model, as shown in Equation (2.96).

$$A\left(z^{-1}\right)y(k) = B\left(z^{-1}\right)u(k) + e(k) \tag{2.96}$$

In Equation (2.96), $y(k)$ is the output sequence of the system, $u(k)$ is the input sequence of the system, $e(k)$ is the Gaussian white noise with zero-means, $A(z^{-1})$ and $B(z^{-1})$ are delay operators polynomials of order n_a and order n_b, respectively, as shown in Equation (2.97).

$$\begin{cases} A(z^{-1}) = 1 + a_1 z^{-1} + a_2 z^{-2} + \cdots + a_{n_a} z^{-n_a} \\ B(z^{-1}) = b_0 + b_1 z^{-1} + b_2 z^{-2} + \cdots + b_{n_b} z^{-n_b} \end{cases} \tag{2.97}$$

And then, the discrete transfer function of the system is shown in Equation (2.98).

$$G(z^{-1}) = \frac{B(z^{-1})}{A(z^{-1})} = \frac{b_0 + b_1 z^{-1} + b_2 z^{-2} + \cdots + b_{n_b} z^{-n_b}}{1 + a_1 z^{-1} + a_2 z^{-2} + \cdots + a_{n_a} z^{-n_a}} \tag{2.98}$$

Then, the discrete transfer function is transformed into a difference equation, as shown in Equation (2.99).

$$y(k) = -\sum_{i=1}^{n_a} a_i y(k-i) + \sum_{i=1}^{n_b} b_i u(k-i) + e(k) \tag{2.99}$$

Taking θ as the coefficient matrix of the model, the expression is obtained, as shown in Equation (2.100).

$$\theta = \left[-a_1, -a_2, \ldots, -a_{n_a}, b_0, b_1, \ldots, b_{n_b}\right] \tag{2.100}$$

Then, the scalar innovation of the system is described, as shown in Equation (2.101).

$$v(k) = y(k) - \varphi^T(k)\hat{\theta}(k-1) \tag{2.101}$$

In Equation (2.101), $\varphi^T(k)$ is the data vector of the current time point, and $\hat{\theta}(k-1)$ is the estimated value of the coefficient matrix at the previous time point. After expanding the scalar innovation, the resulting matrix innovation is expressed by Equation (2.102).

$$V(n_p, k) = \begin{bmatrix} y(k) - \varphi^T(k)\hat{\theta}(k-1) \\ y(k-1) - \varphi^T(k-1)\hat{\theta}(k-2) \\ \vdots \\ y(k-n_p+1) - \varphi^T(k-n_p+1)\hat{\theta}(k-n_p) \end{bmatrix} \tag{2.102}$$

In Equation (2.102), n_p is the length of innovation. Then, the model applicable is obtained to MILS, as represented in Equation (2.103).

$$Y(n_p, k) = \beta^T(n_p, k)\theta(k-1) + V(n_p, k) \tag{2.103}$$

In Equation (2.103), $Y(n_p,k)$ is the system output vector after innovation modification.

According to the description of the MILS algorithm, the recursive process of parameter identification based on this algorithm is described as follows:

Step 1: Initializing the coefficient matrix and covariance matrix.

Step 2: Obtaining the current terminal voltage and operating current measurements.

Step 3: Construct an improved multi-innovation matrix and data matrix.

Step 4: Calculate the gain matrix, as shown in Equation (2.104).

$$K(k)=P(k-1)\beta(n_p,k)\left[I_{n_p}+\beta^T(n_p,k)P(k-1)\beta(n_p,k)\right]^{-1} \quad (2.104)$$

In Equation (2.104), $P(k-1)$ is the covariance matrix, I_{n_p} is the n_p-order identity matrix.

Step 5: Updating the covariance matrix, as shown in Equation (2.105).

$$P(k)=P(k-1)-K(k)\beta^T(n_p,k)P(k-1) \quad (2.105)$$

Step 6: Updating coefficient matrix, as shown in Equation (2.106).

$$\hat{\theta}(k)=\hat{\theta}(k-1)+K(k)\left[Y_{\text{imp}}(n_p,k)-\beta^T(n_p,k)\hat{\theta}(k-1)\right] \quad (2.106)$$

In Equation (2.106), $K(k)$ is the gain matrix.

Step 7: Algorithm iteration cycle until the end.

2.6.6 Extended Kalman Filter and Verification

The extended Kalman filtering (EKF) algorithm is an improvement of the ordinary Kalman filtering (KF) algorithm. The EKF method is an online linearization processing technique, which means linearizing the estimated parameters and then performing a linear KF [163,164]. When the state equation or measurement equation is nonlinear, EKF is usually used. EKF makes a first-order linearization truncation of the Taylor expansion of nonlinear functions, ignoring the other higher-order terms, to transform the nonlinear problem into linearity, and the Kalman linear filtering algorithm can be applied to the nonlinear system. The value of the next time point is estimated through the previous time point, continuously updating the state variables with the observed values of the input and output of the system to realize the optimal estimation [165,166].

When using the EKF method, the process noise and the observation noise are required to be white noise with an approximately Gaussian distribution, which is also a limitation of all KF algorithms. Only under this noise condition can the process noise and observation noise be easily coordinated, but the variance can be controlled. In the estimation process, the Taylor expansion algorithm is used to process the nonlinear state-space equation. After removing the second-order and above high-order terms, a first-order linearized model remains. When the linearized model is obtained, the KF is further used for estimation, and the algorithm flow of the extended Kalman is shown in Figure 2.38.

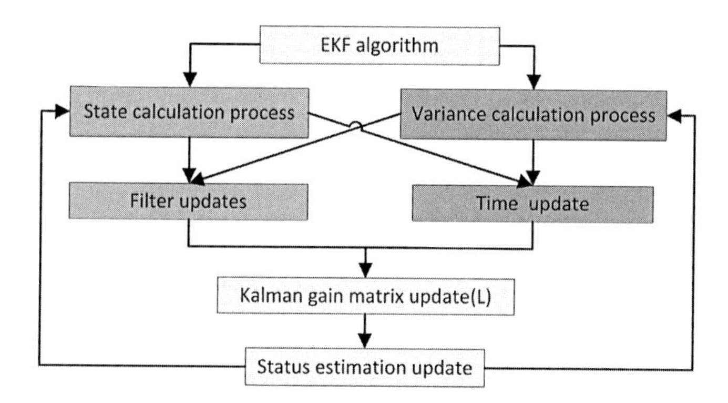

FIGURE 2.38 The flowchart of the EKF algorithm.

At the time point k, it is assumed that the covariance of the state estimation value of $k-1$ and the estimation error at the previous time has been obtained. Now, the time point k is predicted in advance by the state value. In the state equation, the state value at time point k is a function of the state value at time point $k-1$. Therefore, by combining the system state and the error covariance at the previous time point, the prior prediction value for the current time point can be obtained. The two values of the preceding and following time points written in the model are true values, and the true value cannot be obtained in the actual situation [167]. The posterior state estimation is used at the previous time point, so it is also necessary to predict the next step in advance.

For the one-step advance prediction of the observation at time k, according to the observation function in the mathematical model. The observed true value is a function of the estimated true value, but in actual situations, the estimated true value cannot be obtained. In the absence of noise, linearization, and errors, the value is calculated and processed through the mathematical model and directly with the physical method. In the absence of measurement noise, the measured true value is equal because of the linearization error of the mathematical model plus the prior estimation. There is a certain error between the true values, and there is a certain error between all the measured values obtained through the mathematical model and the actual measured values.

KF is an algorithm that recursively estimates the current state of the system according to the input and output data of the system by establishing a linear state-space equation. Because the KF algorithm is completely estimated in the time domain, and there is no mutual conversion between the time domain and the frequency domain, the amount of calculation is small. The expression and observation of the discrete nonlinear system space are obtained, as shown in Equation (2.107).

$$\begin{cases} X_{k+1} = f(X_k, k) + w_k \\ Z_k = h(X_k, k) + v_k \end{cases} \tag{2.107}$$

The first part of Equation (2.107) is the state, and the second part is the observation. k is the discrete-time, X_{k+1} is the n-dimensional state vector, Z_k is the m-dimensional observation vector, and w_k and v_k are independent Gaussian white noises. To apply the KF, the nonlinear functions $f(*)$ and $h(*)$ are expanded around $\hat{X}_k$ by the first-order Taylor expansion, and the expansion result is shown in Equation (2.108).

$$\begin{cases} f(X_k,k) \approx f\left(\hat{X}_k,k\right) + \left.\dfrac{\partial f(X_k,k)}{\partial X_k}\right|_{X_k=\hat{X}_k} \left(X_k - \hat{X}_k\right) \\ h(X_k,k) \approx h\left(\hat{X}_k,k\right) + \left.\dfrac{\partial h(X_k,k)}{\partial X_k}\right|_{X_k=\hat{X}_k} \left(X_k - \hat{X}_k\right) \end{cases} \tag{2.108}$$

Furthermore, Equation (2.108) can be linearized, as shown in Equation (2.109).

$$\begin{cases} X_{k+1} = A_k X_k + B_k + w_k \\ Z_k = C_k X_k + D_k + v_k \end{cases} \tag{2.109}$$

For Equation (2.109), the values of A_k, B_k, C_k, and D_k are calculated, as shown in Equation (2.110).

$$\begin{cases} A_k = \left.\dfrac{\partial f(X_k,k)}{\partial X_k}\right|_{X_k=\hat{X}_k}, B_k = f\left(\hat{X}_k,k\right) - A_k \hat{X}_k \\ C_k = \left.\dfrac{\partial h(X_k,k)}{\partial X_k}\right|_{X_k=\hat{X}_k}, D_k = h\left(\hat{X}_k,k\right) - C_k \hat{X}_k \end{cases} \tag{2.110}$$

Applying the basic KF algorithm to the linearized model Equation (2.110), the recurrence process of the EKF algorithm is shown in Equation (2.111).

$$\begin{cases} \hat{X}_{k+1}^{-} = f\left(\hat{X}_k\right) \\ \hat{P}_{k+1}^{-} = A_k \hat{P}_k A_k^T + Q_{k+1} \\ K_{k+1} = \hat{P}_{k+1}^{-} C_{k+1}^T \left(C_{k+1} \hat{P}_{k+1}^{-} C_{k+1}^T + R_{k+1}\right)^{-1} \\ \hat{X}_{k+1} = X_{k+1}^{-} + K_{k+1}\left[Z_{k+1} - h\left(\hat{X}_{k+1}^{-}\right)\right] \\ \hat{P}_{k+1} = [I - K_{k+1} C_{k+1}] \hat{P}_{k+1}^{-} \end{cases} \tag{2.111}$$

In Equation (2.111), P is the mean square error, K is the Kalman gain, and I is an $n \times m$ unit matrix. Q and R are the variances of process noise w and measurement noise v, respectively, and generally do not change with the system. The initial filter value and the filter variance are $X(0)=E[X(0)], P(0)=\text{var}[X(0)]$. The filtering steps in the $k+1$ cycle are described as follows: first estimate the current state and the mean square error from the state $\hat{X}_k$ and the mean square error $\hat{P}_k$ at time point k to obtain the prior state $\hat{X}_{k+1}$ and prior mean square error $\hat{P}_{k+1}^-$ at the time point $k+1$, and then calculate the Kalman gain K_{k+1} at the current time point according to the observed value at the time point $k+1$. Finally, using K_{k+1} to correct the prior state to get the current state $\hat{X}_{k+1}$, the prior mean square error is corrected to get the current mean square error $\hat{P}_{k+1}$.

When the EKF algorithm is applied to the model parameter identification, assuming that the parameter to be identified is θ, and its value can be obtained from $\theta=\begin{bmatrix} R & R_1 & C_1 \end{bmatrix}^T$. The state observation for θ can be obtained, as shown in Equation (2.112).

$$\begin{cases} \theta_{k+1}=\theta_k+w_k^\theta \\ y_k=h(x_k,u_k,\theta_k)+v_k^\theta \end{cases} \tag{2.112}$$

In Equation (2.112), w_k^θ is the process noise, v_k^θ is the observation noise, $w_k^\theta - N(0, Q^\theta)$, $v_k^\theta - N(0, R^\theta)$, and θ_k is the state variable. Generally, the method is combined with the SOC estimation process. According to Equation (2.112), C_k^θ can be obtained, as shown in Equation (2.113).

$$C_k^\theta=\left.\frac{dh(\hat{x}_k^-,u_k,\theta)}{d\theta}\right|_{\theta=\hat{\theta}_k^-} \tag{2.113}$$

The process steps of using EKF to identify model parameters are described as follows:

1. Initialization θ_0, Q_0^θ, R_0^θ, P^θ
2. Time update, as shown in Equation (2.114).

$$\begin{cases} \hat{\theta}_{k+1}^-=\hat{\theta}_k^+ \\ P_{\theta,k+1}^-=P_{\theta,k}^++Q_k^\theta \end{cases} \tag{2.114}$$

3. Measurement update, as shown in Equation (2.115).

$$\begin{cases} K_{k+1}^\theta=P_{\theta,k+1}^-C_{k+1}^{\theta\ T}\left(C_{k+1}^\theta P_{\theta,k+1}^-C_{k+1}^{\theta\ T}+R^\theta\right)^{-1} \\ \hat{\theta}_{k+1}^+=\hat{\theta}_{k+1}^-+K_{k+1}^\theta\left(y_{k+1}-y_{k+1}^-\right) \\ P_{\theta,k+1}^+=\left(I-K_{k+1}^\theta C_{k+1}^\theta\right)P_{\theta,k+1}^- \end{cases} \tag{2.115}$$

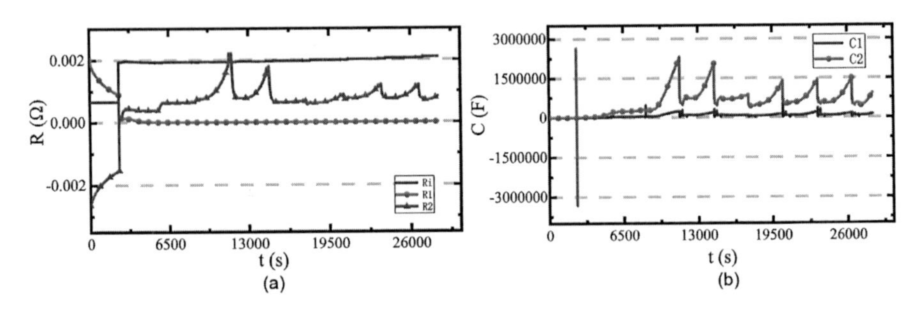

FIGURE 2.39 Online parameter identification results. (a) Resistance identification results. (b) Capacitance identification results.

Through Equations (2.114) and (2.115), the time update and measurement update are continuously performed, and the real-time model parameters can be identified.

The online parameter identification experiment is verified by the FFRLS algorithm. The identified battery model is shown in Figure 2.36, wherein the forgetting factor is 0.998. The identification results are shown in Figure 2.39.

In the initial stage of identification, due to the large difference between the given initial parameter value and the actual parameter value, the identified parameters fluctuate greatly. With the increase of time, the model parameters converge to a stable value.

2.6.7 Comparison of Thevenin and Second-Order RC Modeling

By comparing the Thevenin equivalent model and the second-order RC model, it can be observed that the second-order model is improved from the Thevenin model by adding an RC parallel circuit to the Thevenin model to characterize the phase of slow voltage change during the chemical reaction inside the batteries. Simple models are easy to calculate, but cannot accurately describe the operating characteristics of the batteries. Complex models can better characterize the charge–discharge characteristics of the battery, but the computational effort is greatly increased, reducing the adaptability and generalization of the model. The circuit dynamic model of the second-order differential network is shown in Figure 2.40.

In Figure 2.40, the state-space equation can be obtained from the second-order model, and the results of the model parameter identification are obtained, as shown in Table 2.8.

The parameter fitting polynomials corresponding to the figure are shown in Table 2.9.

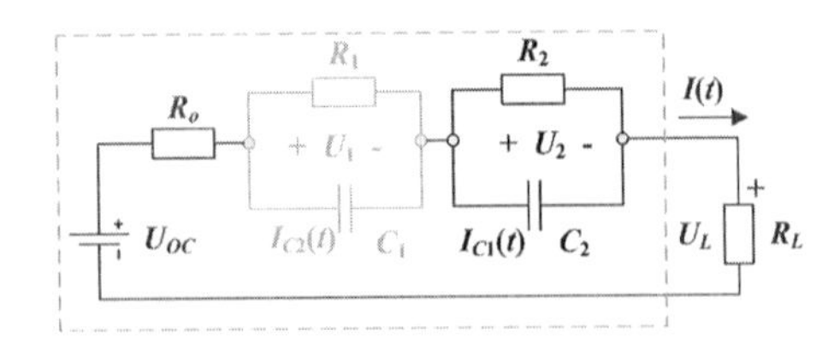

FIGURE 2.40 Second-order equivalent circuit diagram.

TABLE 2.8
Parameter Identification Result

SOC	*a*	*b*	*c*	*d*	*f*
1	0.004982	0.009623	0.8434	0.06552	4.178
0.9	0.007184	0.010750	0.5288	0.04081	4.057
0.8	0.006924	0.011640	0.5880	0.04569	3.948
0.7	0.006316	0.011830	0.7282	0.05564	3.846
0.6	0.005284	0.010710	0.7922	0.05461	3.735
0.5	0.003687	0.008972	0.7725	0.06215	3.657
0.4	0.003432	0.008784	0.8107	0.06008	3.617
0.3	0.003838	0.008912	0.8788	0.05942	3.586
0.2	0.004532	0.009571	0.7056	0.04896	3.530
0.1	0.006485	0.011260	0.8145	0.05779	3.451
SOC	$R_0(\Omega)$	$R_1(\Omega)$	C_1(F)	$R_2(\Omega)$	C_2(F)
1	0.001835	0.00011472	10335.40523	0.000221588	68877.98832
0.9	0.001856	0.000165425	11431.60103	0.000247539	98989.64572
0.8	0.001867	0.000159438	10666.70891	0.000268033	81656.48675
0.7	0.001853	0.000145438	9442.174733	0.000272408	65977.0606
0.6	0.002031	0.000121674	10374.50017	0.000246618	74251.14952
0.5	0.002026	8.49001E-05	15247.30905	0.000206597	77881.52217
0.4	0.002041	7.90283E-05	15608.36372	0.000202268	82289.15029
0.3	0.002038	8.83772E-05	12875.66920	0.000205216	82008.14788
0.2	0.002070	0.000104358	13580.51863	0.000220390	92675.74877
0.1	0.002114	0.000149329	8221.74040	0.000259283	66738.08550

TABLE 2.9
Parameter Fitting Result

R_0: $y=0.0024-0.00439*u^\wedge1+0.02011*u^\wedge2-0.04059*u^\wedge3+0.03569*u^\wedge4-0.01137*u^\wedge5$

R_1: $y=2.07881E\text{-}4-6.7124E\text{-}4*u^\wedge1+7.67998E\text{-}4*u^\wedge2+3.66447E\text{-}4*u^\wedge3-0.0000379487*u^\wedge4-5.1781E\text{-}4*u^\wedge5$

C_1: $y=8399.02507-38777.14471*u^\wedge1+550627.44458*u^\wedge2-1.66094E6*u^\wedge3+1.87552E6*u^\wedge4-724449.44933*u^\wedge5$

R_2: $y=2.53056E\text{-}4+4.51845E\text{-}4*u^\wedge1-0.00544*u^\wedge2+0.01561*u^\wedge3-0.01666*u^\wedge4+0.006*u^\wedge5$

C_2: $y=49855.0442+135771.15179*u^\wedge1+984260.32619*u^\wedge2-5.31421E6*u^\wedge3+7.76762E6*u^\wedge4-3.55285E6*u^\wedge5$

U_{OC}: $y=3.2462+3.0115*u^\wedge1-11.16337*u^\wedge2+20.71096*u^\wedge3-16.53234*u^\wedge4+4.91026*u^\wedge5$

U_{OC1}: $y=3.42621-0.54017*u^\wedge1+13.07197*u^\wedge2-55.87842*u^\wedge3+105.54386*u^\wedge4-90.28558*u^\wedge5+28.84722*u^\wedge6$

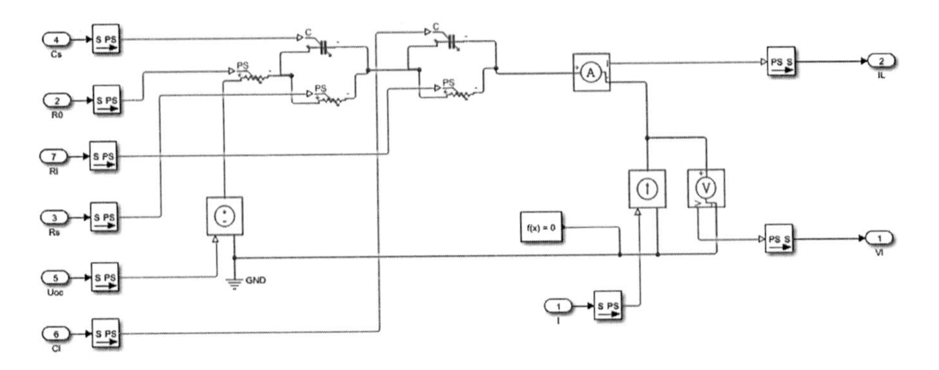

FIGURE 2.41 Second-order equivalent circuit diagram.

According to the structure diagram of the second-order equivalent circuit model, its improved simulation system model is built in Simulink software for the parameter verification, as shown in Figure 2.41.

To better compare the Thevenin model with the second-order model, the outputs of both model circuits are placed in the same simulation, and the two model voltage outputs are displayed inside the same oscilloscope. The second-order model is more reasonable, and the parameter identification accuracy is higher compared with the Thevenin model at 20°C. The second-order circuit model and the parameter identification method and identification results under this model can provide a theoretical basis for the accurate estimation of the internal state variables of lithium-ion batteries.

2.7 CONCLUSION

This chapter mainly introduces four equivalent circuit models and parameter identification methods, and briefly introduces three commonly used battery operating conditions. The Rint model is a relatively simple model, including the ideal battery voltage source U_{OC} and the battery's internal resistance R_0. Since the model does not consider the polarization characteristics of the battery, the accuracy of the model is low, which is rarely seen in practical applications. Compared with the Rint model, the Thevenin model is improved by the addition of an RC circuit. Due to the polarization characteristics of the battery, the RC circuit can reflect the gradual changes in the battery voltage during and after the charge–discharge maintenance. The Thevenin equivalent circuit model is not only simple in structure, but also able to meet the simulation requirements, so the battery Thevenin equivalent circuit model is often used in practical applications. The PNGV equivalent circuit model includes an open-circuit voltage source U_{OC}, polarization resistor R_p, polarization capacitor C_p, an ohmic resistance R_0, and a capacitance C_b. The PNGV model adds the consideration of the impact of the load current on the battery OCV based on the Thevenin model. Since this circuit model covers the characteristics of battery polarization and ohmic resistances, the PNGV equivalent circuit model is more accurate. The second-order equivalent model introduces an RC series network and a second-order RC parallel network. Compared with the Thevenin model and the PNGV model, the accuracy of the model has been improved, which can simulate the ohmic polarization, electrochemical polarization, concentration polarization, and self-discharge effects. At the same time, offline parameter identification and online parameter identification are introduced to provide a basis for follow-up work.

3 Battery State of Charge Estimation

Dan Deng
Guangxi University of Science and Technology

Daniel-Ioan Stroe
Aalborg University

Jialu Qiao, Jingsong Qiu, Peng Yu, and Qingyun Ma
Southwest University of Science and Technology

Shunli Wang
Sichuan University
Southwest University of Science and Technology

Xiao Yang and Yanxin Xie
Southwest University of Science and Technology

CONTENTS

DOI: 10.1201/9781003333791-3

Introduction: For a practical lithium-ion battery pack, safety and stability are the most important. Therefore, the battery management system (BMS) is essential to monitor the operating status of the lithium-ion battery and ensure the safe and suitable operation of the lithium-ion batteries [168,169]. The BMS is a product integrating multiple functions to realize the management of lithium-ion batteries. In addition to monitoring basic parameters such as voltage, current, and temperature, the BMS also uses various strategies to estimate and predict performance parameters such as battery SOC and state of health (SOH). The SOC estimation has always been a hot and difficult issue in this research field [170]. Accurate SOC estimation is a direct representation of the current endurance of the battery, and it is also the cornerstone of estimating the SOH of the batteries.

3.1 OVERVIEW OF SOC ESTIMATION

3.1.1 Definition of SOC

In the design of BMS, there are also several research hotspots, such as estimating the SOC and SOH values of lithium-ion batteries, solving the problem of inconsistent parameters, balancing technology, and performing fault diagnoses on the lithium-ion batteries [171,172]. Since a mature BMS requires not only comprehensive functions but also should have high reliability, it is very difficult to design a perfect lithium-ion BMS.

Although the lithium-ion battery state estimation algorithm is a basic function of the BMS, it has some technical difficulties in its design process. The accuracy of the lithium-ion battery state estimation directly determines whether the lithium-ion BMS can work perfectly, and it is also the most important line of defense to ensure the safety of users. Therefore, it is of great significance to study the state estimation algorithm of lithium-ion batteries. The SOC value of a battery pack is data representing the power of lithium-ion batteries. The data is greater than 0 and less than 1, so it can usually be expressed as a percentage. Its definition is expressed in Equation (3.1).

$$\mathrm{SOC} = \frac{Q_r}{Q_n} \tag{3.1}$$

In Equation (3.1), Q_r is the amount of charge stored in the battery, and Q_n is the rated capacity of the batteries. Both of these data are usually in units of Ah, but SOC is a unitless quantity, just a ratio. The battery SOC reflects the current range of the battery to a certain extent. It plays an important role in improving battery efficiency, slowing down battery aging, and preventing overcharge and overdischarge. The battery SOC cannot be directly measured and must be calculated by estimation methods. At present, many researchers have made great efforts to find real-time, accurate, and reliable SOC estimation methods. A large number of research results have emerged. The common SOC estimation methods include ampere–hour integration, open-circuit voltage (OCV), and model-based methods. The estimation method is based on impedance, static battery characteristics, fuzzy logic, machine learning, and special measurement methods. Taking into account that the lithium-ion battery is affected by other factors during the charge–discharge process, such as the ambient temperature, the size of the experimental current, and the number of cycles, Equation (3.1) cannot accurately estimate the SOC of the battery. Therefore, based on the above, a more accurate calculation equation is proposed as shown in Equation (3.2).

$$\mathrm{SOC} = \left(1 - \frac{\gamma Q_i}{Q_o}\right) \tag{3.2}$$

In Equation (3.2), Q_i is the number of batteries that have been discharged at the time point i, and γ is the battery efficiency parameter, which can be obtained from experimental data. Estimating the remaining battery power has always been the focus and difficulty of BMS. The need for the high-precision remaining power estimation of the lithium-ion battery has always been a prerequisite for prolonging the service life of the lithium-ion battery and exerting discharge efficiency. For example, temperature and the SOC value at the previous time point affect the estimation of the remaining battery power. In addition, polarization effects and battery life also affect the estimation of the remaining battery power, which brings great nonlinear problems to the battery model.

3.1.2 Main Affecting Factor Analysis

In electric vehicles, the battery pack exhibits a high degree of nonlinearity, and the SOC is affected by many factors such as discharge rate, battery temperature, self-discharge rate, and cycle times, which brings great difficulties to the accurate SOC estimation [173,174].

1. Discharge rate
   Under the condition that other influencing factors remain unchanged, the discharge capacity of the battery decrease as the discharge rate increases. This is because the active material inside the battery has a limited depth of action with the thickness of the electrode. When a large current is discharged, the greater the discharge rate, the shallower the depth of action, and the lower the utilization rate, so the battery capacity is smaller, and vice versa, the greater the capacity.
2. Battery temperature
   The battery power and active material utilization rate increase with the rise of battery temperature, which was mainly caused by the change in temperature performance of the electrolyte. When the temperature of the battery rises, the viscosity of the electrolyte decreases and the activity increases, thereby increasing the ion diffusion movement ability, ultimately improving the utilization rate of active materials, and increasing the actual available power of the battery. Conversely, when the battery temperature drops, the active material utilization rate decreases and the actual available power decreases. Therefore, the actual available power of the battery is directly proportional to the battery temperature. In practical applications, the charge–discharge working temperature range of lithium iron phosphate polymer battery is 0°C–45°C.
3. Number of cycles
   After a period of use, the standard total capacity of lithium-ion batteries changes to a certain extent. At first, the power increases, and it remains the same for a period, and then the power gradually decreases. For lithium iron phosphate batteries, the cycle life of the battery is generally expressed by the number of charge–discharge times when the total discharge capacity drops to 80% of the nominal capacity.

4. Self-discharge rate
   Self-discharge, also known as charge retention capacity, refers to the storage capacity and retention capacity of the battery in an open-circuit state under specific environmental conditions. In the case of battery self-discharge, the SOC value decreases with increasing storage time. In general, the self-discharge rate can be expressed as the percentage of capacity reduction per unit of time, and the expression is shown in Equation (3.3).

$$I_{\mathrm{sdr}} = (C_a - C_b)/(C_a * T) * 100\% \tag{3.3}$$

   In Equation (3.3), I_{sdr} is the self-discharge rate, C_a is the battery power before storage, C_b is the battery power after storage, and T is the battery storage time. The self-discharge rate of lithium-ion batteries is related to many factors such as the number of cycles, ambient temperature, and storage time. It is generally calculated and derived by experimental methods.

3.1.3 Traditional Estimation Methods

The discharge of power batteries in electric vehicles is a very complex electrochemical reaction process. Factors such as discharge rate, ambient temperature, battery internal resistance, and self-discharge rate affect the estimation and prediction of the battery SOC value. At the same time, these factors vary from each other. It changes as the number of cycles increases, which increases the difficulty of battery modeling and SOC estimation algorithm [175,176]. At present, the estimation methods of battery SOC mainly include discharge test method, OCV method, ampere–hour measurement method, internal resistance measurement method, load voltage method, and neural network method. These methods are introduced and analyzed in detail below.

1. Discharge test method
   The discharge test method is an experimental method that uses a constant current to continuously discharge the battery until the battery terminal voltage reaches the discharge cutoff voltage. It is the most reliable SOC estimation method and applies to all types of batteries. However, it also has several defects which greatly limit its application. First, it requires a lot of measurement time. The SOC value at each time point can only be calculated after the entire discharge test is over, and real-time SOC estimation cannot be achieved. Second, the battery must be forced to stop its previous work and switch to a constant current discharge state. Therefore, the discharge test method is not suitable for estimating the SOC estimation of power batteries with real-time estimation requirements, but it can be used for the maintenance of power batteries and the determination of battery model parameters.
2. OCV method
   The OCV method estimates SOC according to the approximately linear relationship between SOC and OCV. The actual operation process is described

as follows. First, the battery needs to rest for a long enough time to ensure the stability of its OCV and other parameters. Then, the OCV of the battery is measured to obtain the SOC value according to the known approximate linear function relationship. The advantage of the OCV method is its simplicity and strong operability. However, the OCV can only be obtained after the battery is fully rested until the electrochemical reaction of the battery is stable, but it is impossible to track the SOC value in real time. In addition, although the functional relationship between SOC and OCV is relatively fixed in a short time, the relationship between SOC and OCV changes with the aging of the battery, which also affects the accuracy of SOC estimation. Therefore, the applicability of this method is poor.

3. Ampere–hour integration method
   The ampere–hour integration method estimates the SOC value by calculating the accumulated power of the battery during the charging or discharging process and compensating the estimated SOC value according to the temperature and the charge–discharge rate. It is currently the most commonly used and simplest SOC estimation method, and it has been successfully applied to the power estimation of electronic consumer products. If the initial SOC value is specified as SOC_0, the SOC value of the present state can be calculated by Equation (3.4).

$$SOC = SOC_0 - \frac{1}{Q_n}\int_0^t \eta_i \, dt \tag{3.4}$$

   In Equation (3.4), Q_n is the nominal capacity. i represents the battery current, which is positive when discharging and negative when charging. η is the battery efficiency coefficient, including the temperature influence coefficient η_r and the charge–discharge rate coefficient η_i, where η_i can be obtained by the Peukert. Three issues need to be considered when this method is used for SOC estimation. The method itself cannot provide the initial battery value SOC_0. Inaccurate current measurement increases the SOC estimation error, which becomes larger and larger after a longtime accumulation. The battery efficiency coefficient η must be considered when estimating SOC value, although the accuracy of current measurement can be solved by using high-performance current sensors, which highly increases the cost of the system. At the same time, to solve the problem of battery efficiency coefficient η, it is necessary to establish empirical equations for temperature influence coefficient η_r and charge–discharge rate coefficient η_i through a large amount of experimental data. If the SOC estimation accuracy requirement is not high, the ampere–hour measurement method can be applied to the calculation of the remaining battery power of electric vehicles.

4. Internal resistance measurement method
   The internal resistance measurement method refers to the method of measuring the AC internal resistance by exciting the battery with different frequencies of alternating current and calculating the SOC value using the battery static model of the remaining power and the AC internal resistance.

In the later stage of battery discharge, the internal resistance measurement method has high estimation accuracy and battery adaptability and is generally used in combination with the ampere–hour measurement method. However, the method only considers two aspects of discharge current and internal resistance but does not consider the influence of environmental temperature, battery cycle times, the imbalance between battery cells, and other factors on the internal resistance of the batteries. In addition, the cause of battery internal resistance is very complicated, so the accuracy of the SOC estimation value obtained by the internal resistance measurement method is not high. In addition, there are great differences in AC impedance at different current frequencies, and electric vehicle power batteries do not work at a fixed AC frequency. Consequently, the internal resistance measurement method cannot meet the SOC estimation requirements of lithium-ion batteries for electric vehicles.

5. Neural network method
The neural network is an artificial intelligence technology that has been developed in recent years and uses simulating brain signal processing. A neural network is a complex nonlinear system formed by a large number of simple neurons connected extensively. It can automatically summarize the collected data and obtain the internal laws of the data onto it. It also has a nonlinear mapping capability and is suitable for noncausal relationships with more complex nonlinearities, accuracy identification, judgment, reasoning, and classification issues. The battery is a highly nonlinear system, and it is usually difficult to establish a reasonable and accurate mathematical model for its charge–discharge process. Therefore, given external incentives, the neural network method can use the learning ability of the neural network and the parallel structure to simulate the nonlinear characteristics of the battery to estimate the SOC value.

SOC estimation often uses a typical three-layer neural network, in which the number of neurons in the input and output layers is determined by the actual system needs. Also, the number of neurons in the middle layer depends on the system complexity and analysis accuracy requirements. In the neural network method, the system input includes battery voltage, ambient temperature, charge–discharge current, battery internal resistance, and accumulated discharge power. Whether the input type and quantity are reasonable or not, they directly affect the calculation cost and accuracy of the method model. The disadvantage of the neural network method is that a large amount of experimental data is needed to train the parameters, but the SOC estimation accuracy is greatly affected by the training data and the training method.

6. Kalman filter
Lithium-ion batteries are a typical nonlinear system. With the development of Kalman filter (KF) technologies, improved KF strategies can be used in complex nonlinear systems. The nonlinear KF mainly includes extended KF and unscented Kalman filtering (UKF) algorithms. Among them, the extended Kalman filter (EKF) uses the partial differentiation method for

linearization. It uses the partial differentiation method to generate the Jacobian matrix in each iteration cycle of the algorithm, so the calculation is more complicated. The nonlinear state–space equation is linearized with the Jacobian matrix in the EKF algorithm. If the system is highly nonlinear, the conversion results may deviate greatly, which may lead to divergence of estimation results. The UKF uses an unscented transformation (UT) to approximate the probability density distribution of state variables of the system. Compared with the EKF, this transformation has certain advantages. Data transformation is not only more accurate but also simple in calculation and easy to implement, which is more suitable for state estimation of nonlinear systems.

Compared with commonly used electronic consumer products, electric vehicles have higher requirements for power battery performance. Power battery packs are often under discharge conditions with large currents and severe current fluctuations. Also, the components of battery capacity affected by temperature are relatively increased. These problems put forward stricter requirements for the SOC estimation algorithm.

a. The SOC estimation result should have high accuracy. To improve the energy utilization rate of power batteries and provide accurate battery life information, the maximum estimation error of SOC should generally be controlled by 10%. During the later period of the charge–discharge process, the estimated SOC value can converge to near the true value to ensure that the battery can be completely charge–discharge without overcharging or overdischarging.
b. SOC must be estimated in real time. The vehicle control strategy of electric vehicles is always adjusted according to the current SOC value of the power battery pack. The SOC estimation must have online real-time performance, which greatly increases the design requirements of the system data sampling unit and algorithm execution unit.
c. The current accumulation error is reduced. The long-term cumulative error makes the SOC estimation error increase and reaches the maximum in the later stage of charge–discharge. Therefore, the ampere–hour integration method cannot be used alone to estimate the SOC in the electric vehicle BMS.

Considering the above requirements, the improved methods based on the KF have great advantages in the SOC estimation. Compared with the neural network, the amount of calculation is smaller and the amount of preliminary data required is smaller. Compared with other algorithms, it has the ability of adaptive correction.

3.2 KALMAN FILTER AND MODIFICATIONS

Rudolf E. Kalman, a Hungarian mathematician, proposed and published the famous linear system for filtering and prediction in 1960. He gave a recursive form of the filtering algorithm, so later generations called it the KF algorithm. It was successfully applied to the navigation system for the Apollo Project by NASA in 1968–1972, which required estimation of the trajectories of manned spacecraft traveling to the moon and back. In the following decades, with the continuous development of computer technology, linear filtering technology has been widely used in navigation, measurement and control, radar system, and other fields. These wide-range applications have in turn proved the effectiveness of the KF [177,178]. The KF is a model-based state estimation algorithm. Its essence is a state estimator that optimally estimates the state of the system based on the previous state and current observations. The KF is different from the traditional filter. On the contrary, it is more like a state observer, which is based on a discrete state–space model.

3.2.1 Kalman Filter

The KF is an algorithm to optimize system state estimation, using linear system state equations and system input–output observation data. The optimal estimation can also be regarded as a filtering process because the observed data includes the influence of noise and interference in the system. Since the KF is a state observer, it is usually necessary to set the measured variable as the state variable of the system, to establish the state–space model [179,180]. The covariance of the process noise and the covariance of the observation noise are determined according to the actual situation. It should be pointed out that here, the process noise and the observation noise are generally considered to be Gaussian white noise with an expectation of their value equaling 0. If the expectation values of these two noises are not equal to 0, a series of conversions are to be required to make two kinds of noise expected to be 0 in the transformation model. The process noise is generally related to the accuracy of the model and the uncertainty of the process, while the observation noise is related to the noise of the sensor. Finally, the initial value of the state variable of the KF is determined. The initial value generally selects the expectation of the state variable.

After completing the above steps, the process model is established, the algorithm is written, and the KF algorithm's prototype is generated. However, to obtain a better estimation effect, a lot of experiments are needed to adjust the parameters of the KF, especially the covariance matrix of process noise and observation noise. The KF algorithm is mainly divided into two steps: prediction and verification. The algorithm steps are described, as shown in Figure 3.1.

Although the lithium-ion battery model is nonlinear, the KF algorithm is first introduced here with a linear model as an example. For a given linear system, the state variable $x \in R^n$ is defined, and the differential state equation for the system is established, as shown in Equation (3.5).

$$x_k = Ax_{k-1} + Bu_k + w_k \tag{3.5}$$

Using the sensor to measure the system, the measured value of the sensor can be defined as the observed variable of the system, so that the observation equation of the system can be listed, as shown in Equation (3.6).

$$z_k = Hx_k + v_k,\ z_k \in R^m \tag{3.6}$$

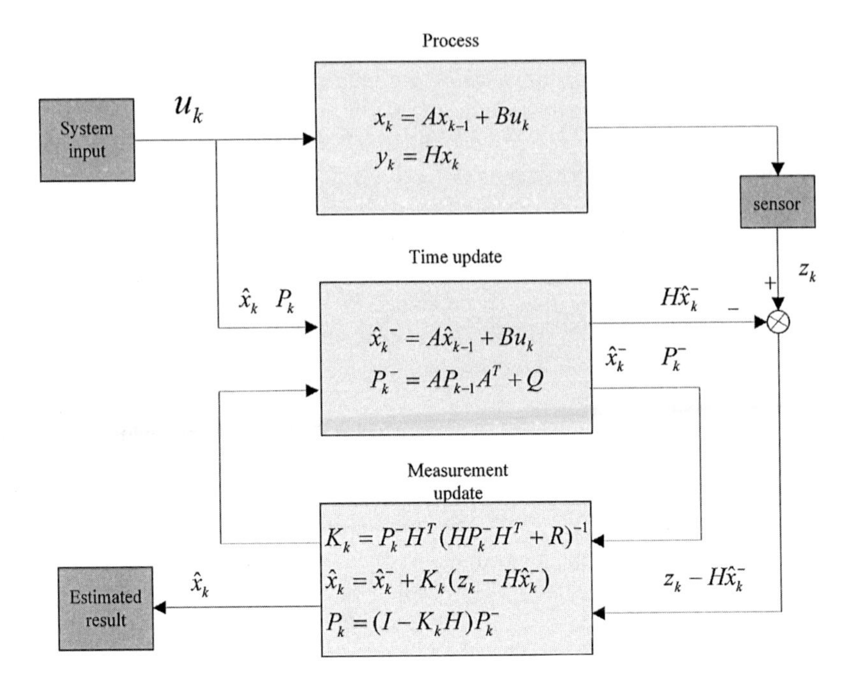

FIGURE 3.1 Basic principles of the KF algorithm.

In Equation (3.6), w_k is the process noise, and v_k is the observation noise. They are independent of each other and form normally distributed white noise. These two noises meet certain statistical characteristics, as shown in Equation (3.7).

$$\begin{cases} p(w) \sim N(0,Q) \\ p(v) \sim N(0,Q) \end{cases} \tag{3.7}$$

The optimal estimated value calculated by the KF is obtained by fusion between the predicted value obtained by the model recursion and the measured value obtained by the sensor. First, the state value at time point k is recursively calculated from the state value at time point $k-1$ according to the state equation, as shown in Equation (3.8).

$$\begin{cases} \hat{x}_k^- = A\hat{x}_{k-1} + Bu_k \\ P_k^- = AP_{k-1}A^T + Q \end{cases} \tag{3.8}$$

Equation (3.8) is called the time update, where $\hat{x}_{k-1}$ is the previous posteriori estimation, that is, the state estimated by the previous KF treatment. It is generally believed that the posterior estimation is the expectation of the system state at the time point of $k-1$, as shown in Equation (3.9).

$$E(x_{k-1}) = \hat{x}_{k-1} \tag{3.9}$$

In Equation (3.9), P_{k-1} is the covariance of the previous posterior estimation noise, and the relationship with the posterior estimation is shown in Equation (3.10).

$$E\left[\left(x_{k-1}-\hat{x}_{k-1}\right)\left(x_{k-1}-\hat{x}_{k-1}\right)^{T}\right]=P_{k-1} \tag{3.10}$$

In Equation (3.10), $\hat{x}_{k-1}$ is the posterior estimate at the time point of $k-1$, which is calculated after the last step of the filter measurement update. x_{k-1} is the state reference value at the time point of $k-1$. According to the time update equation shown above, the state of the system and the covariance of the noise have been updated. Then, the Kalman gain is calculated that is based on the prior covariance estimation of the sensor noise and corrected for the previous prediction. The second part is called measurement update, as shown in Equation (3.11).

$$\begin{cases} K_k = P_k^- H^T (HP_k^- H^T + R)^{-1} \\ \hat{x}_k = \hat{x}_k^- + K_k\left(z_k - H\hat{x}_k^-\right) \\ P_k = (I - K_k H)P_k^- \end{cases} \tag{3.11}$$

In Equation (3.11), K_k is the Kalman gain, and I is the identity matrix. The classical KF has many restrictions on its use, and it needs to meet the two conditions that the system is linear and that the noise is normally distributed. Although some experiments verify that the KF algorithm is still effective when the noise is not strictly normal. However, it can be observed from the KF equation that this algorithm is only suitable for linear systems rather than for nonlinear systems. To extend the scope of application of the KF and enable it to be applied to nonlinear systems, some improved KF algorithms have been also proposed.

3.2.2 Extended Kalman Filter

For linear dynamic systems, the classical KF is undoubtedly one of the best and most commonly used state estimation methods. However, the electric vehicle power battery pack itself is a nonlinear dynamic system. The battery SOC and the charge–discharge rate, battery operating voltage, and ambient temperature are also in a nonlinear relationship. Therefore, the EKF method is proposed to estimate the battery pack SOC online [181]. The basic idea of EKF is to use Taylor series expansion to linearize the nonlinear system, and then use the KF framework to filter the signal, so it is a suboptimal filter. Compared with the classical KF, the EKF algorithm linearizes the power battery nonlinear system by applying the first-order Taylor expansion to the system state–space model. Then, the cyclic iterative process of the standard KF algorithm is used to make the algorithm optimal estimation of the state variables [182–184]. The state–space model of the system is shown in Equation (3.12).

$$\begin{cases} x_{k+1} = f\left(x_k, u_k\right) + w_k \\ y_k = g\left(x_k, u_k\right) + v_k \end{cases} \tag{3.12}$$

In Equation (3.12), $f(x_k, u_k)$ and $g(x_k, u_k)$ correspond to the nonlinear state transition function and the nonlinear measurement function, respectively. In addition to the different selected state–space models, the EKF algorithm and the standard KF algorithm are essentially similar to each other. Both are composed of initialization, predictive estimation, and optimal estimation. The EKF uses a nonlinear model to estimate the state variables [185]. To use the KF algorithm for optimal estimation, the Taylor series expansion of $f(x_k, u_k)$ is carried out. The Taylor series expansion of the function $f(x)$ at a is shown in Equation (3.13).

$$f(x)=\frac{f(a)}{0!}+\frac{f'(a)}{1!}(x-a)+\frac{f''(a)}{2!}(x-a)^2+\cdots+\frac{f^{(n)}(a)}{n!}(x-a)^n+R_n(x) \tag{3.13}$$

In Equation (3.13), $R_n(x)$ is the remainder of the Taylor equation, which is the high–order infinitesimal of $(x-a)^n$, and $f^{(n)}(a)$ is the n derivative of the function $f(x)$. After the Taylor series expansion of the process function, only the first-order term of the expansion is retained, and the rest is omitted, the linearized approximate model of the battery model can be obtained. Assuming that $f(x,u)$ and $g(x,u)$ are first-order differentiable everywhere in the neighborhood of the expansion point, the linear approximate expressions of nonlinear equations $f(x,u)$ and $g(x,u)$ can be obtained, as shown in Equations (3.14) and (3.15).

$$f(x,u)\approx f(\hat{x},u)+\left.\frac{\partial f(x,u)}{\partial x}\right|_{x=\hat{x}} \tag{3.14}$$

$$g(x,u)\approx g(\bar{x},u)+\left.\frac{\partial g(x,u)}{\partial x}\right|_{x=\bar{x}} \tag{3.15}$$

Substituting into Equations (3.14) and (3.15), linear state equation and observation equation can be obtained, as shown in Equations (3.16) and (3.17).

$$x=f(\hat{x},u)+\left.\frac{\partial f(x,u)}{\partial x}\right|_{x=\hat{x}}[x=\hat{x}+w] \tag{3.16}$$

$$y=g(\bar{x},u)+\left.\frac{\partial g(x,u)}{\partial x}\right|_{x=\bar{x}}[x=\bar{x}+v] \tag{3.17}$$

Equivalently transforming Equations (3.16) and (3.17), the mathematical relationship can be obtained, as shown in Equation (3.18).

$$\begin{gathered}\Delta x=x-\hat{x},\Delta x'=x-f(\hat{x},u),\Delta y=y-g(\hat{x},u),\\ A(t)=\left.\frac{\partial f(x,u)}{\partial x}\right|_{x=\hat{x}},C(t)=\left.\frac{\partial g(x,u)}{\partial x}\right|_{x=\hat{x}}\end{gathered} \tag{3.18}$$

Then, the relationship can be expressed, as shown in Equation (3.19).

$$\begin{cases} \Delta x' = A(t)\Delta x + w \\ \Delta y = C(t)\Delta x + v \end{cases} \tag{3.19}$$

The discretized Equation (3.19) can be obtained, as shown in Equation (3.20).

$$\begin{cases} \Delta x_k = A_{k-1}\Delta x_{k-1} + w_{k-1} \\ \Delta y_k = C_k\Delta x_k + v_k \end{cases} \tag{3.20}$$

In Equation (3.20), the coefficients can be calculated, as shown in Equation (3.21).

$$\begin{cases} A_{k-1} = \left.\dfrac{\partial A(x,u)}{\partial x}\right|_{x=\hat{x}_{k-1}} T + I \\ C_k = \left.\dfrac{\partial C(x,u)}{\partial x}\right|_{x=\bar{x}} \end{cases} \tag{3.21}$$

In Equation (3.21), T is an infinitesimal quantity and I is the identity matrix. The calculation process of the EKF algorithm is described as follows.

1. Initialization of the system state–space equation

$$\begin{cases} \Delta x_k = A_{k-1}\Delta x_{k-1} + w_{k-1} \\ \Delta y_k = C_k\Delta x_k + v_k \end{cases} \tag{3.22}$$

2. Prior state estimation

$$\begin{cases} \bar{x}_k = \hat{x}_{k-1} + Tf\left(\hat{x}_{k-1}, u_{k-1}\right) \\ \Delta\bar{x}_k = A_{k-1}\Delta x_{k-1} \\ P_{k/k-1} = A_{k-1}P_{k-1}A_{k-1}^T + Q \\ K_k = P_{k/k-1}C_k^T\left[C_kP_{k/k-1}C_k^T + R\right]^{-1} \end{cases} \tag{3.23}$$

3. Posterior state estimation

$$\begin{cases} \Delta\hat{x}_k = \Delta\bar{x}_k + K_k\left[\Delta y_k - C_k\Delta\bar{x}_k\right] \\ P_k = P_{k/k-1} - K_kC_kP_{k/k-1} \\ \hat{x}_k = \Delta\hat{x}_k + \bar{x}_k \end{cases} \tag{3.24}$$

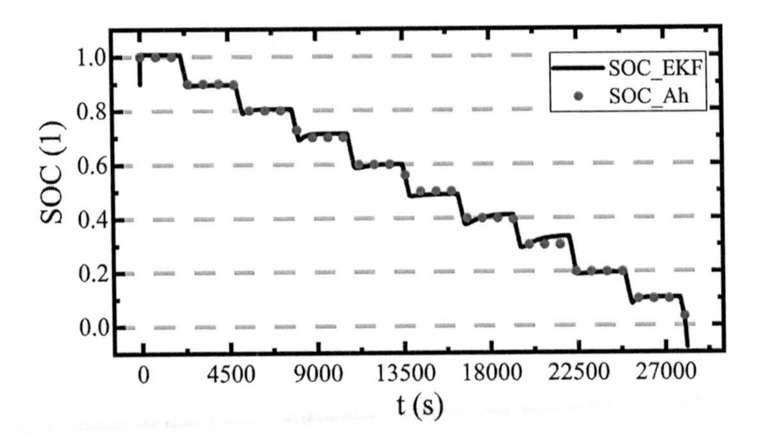

FIGURE 3.2 Comparison of SOC output curves under HPPC condition.

The EKF algorithm approximates the nonlinear system to a linear system through Taylor series expansion, so that the KF algorithm can be used to estimate the nonlinear system state variables. But for some systems with complex expressions, the approximation method is not only computationally expensive but also large errors occur in the linearization process. The more accurate unscented KF algorithm is proposed. In this experiment, U_{OC} is substituted into the algorithm with a fitted polynomial and the values of R_0, R_p and C_p are substituted into the algorithm in an average manner, the battery is discharged by the 1 C constant current in the experiment, and the EKF estimation of the lithium-ion battery SOC algorithm can be verified according to Figure 3.2.

In Figure 3.2, the red curve is the standard SOC estimation result, and the gray curve is the SOC estimation result of the EKF algorithm.

3.2.3 Improved Extended Kalman Filter

While the EKF increases the ability of the algorithm to deal with nonlinear processes, it also needs to be improved further for complex machine state monitoring problems in real-world applications [186]. The system noise in real-world battery packs, such as the constant divergence of the current and voltage sensors, is often non-Gaussian, particularly for large cubature battery packs. Furthermore, battery capacity loss, atmospheric temperature, and complex operating conditions are all closely linked to the characteristics, impacting the precision and robustness of the SOC calculation. As a result, the optimization treatment is used to boost the EKF algorithm to reduce the effects of the above variables [119].

Adaptive noise: signal noise can be divided into process noise and ambient noise with the KF. The ambient noise is caused by the error caused by the sensor and the environmental disturbance in the actual calculation process, and the process noise characterizes the accuracy of the model. Both noises have a significant effect on the efficiency of the filtering system, primarily affecting the convergence and accuracy of the system. The scheme is known to be converted to a constant value at an infinite time point for a converged KF, as shown in Equation (3.25).

$$\lim P_k = P_{\text{const}},\, k \to \infty \tag{3.25}$$

The Kalman gain, like the convergent covariance matrix, converges to a constant value according to the covariance matrix updating the EKF [187]. As a result, the KF degrades to a low-pass filter, with the Kalman gain calculated solely by machine noise, which affects precision. Process noise, on the other hand, affects the convergence level of the system. Where there is a divergence, the presence of noise helps the device eventually converge to the true value, with the speed of convergence depending on the amplitude of the noise. In real-world operating conditions, it is common to anticipate more noise in the initial stages of filtering to improve the effect of correcting initial value variance and filtering convergence speed. When the filtering is stabilized, the noise can be attenuated for improved accuracy. As a result, achieving adaptive machine noise is extremely important in practice. For discrete systems, the adaptive algorithm based on maximum posterior estimation can be defined, as shown in Equations (3.26) and (3.27).

$$\begin{cases} \hat{w}_k = (1-d_k)\hat{w}_{k-1} + d_k\left(\hat{x}_k - \left(A\hat{x}_{k-1} + Bu_k\right)\right) \\ \hat{v}_k = (1-d_k)\hat{v}_{k-1} + d_k\left(z_k - C_k x_k\right) \end{cases} \tag{3.26}$$

$$\begin{cases} \hat{Q}_k = (1-d_k)\hat{Q}_{k-1} + d_k\left(G_k\varepsilon\varepsilon^T G_k^T + P_k - A_{k-1}P_{k|k-1}A_{k-1}^T\right) \\ \hat{R}_k = (1-d_k)\hat{w}_{k-1} + d_k\left(\varepsilon\varepsilon^T - C_k P_{k|k-1} C_k^T\right) \end{cases} \tag{3.27}$$

In Equations (3.26) and (3.27), $d_k = (1-b)/\left(1-b^{k+1}\right)$; b is a constant between 0 and 1; $\varepsilon = z_k - C_k x_k$. The fading filter is a technique for improving the ability of a system to use observation calculation to filter out divergence phenomena. The covariance matrix may not be positive during the iterative process due to rounding error while computing with machines, causing the filtering scheme to oscillate or diverge. In addition, in the case the system model is severely mismatched, the system can only be corrected by observation. As a result, increasing the observation proportion, when updating the status, increases the reliability and robustness. For the arbitrary time point k in the time domain, traditional EKF algorithms conclude that the complete filtering time domain is N; thus, the mathematical relationship can be obtained, as shown in Equation (3.28).

$$P_{k|k-1}^N = A_{k-1}\left|P_{k-1}^N\, A_{k-1}^T + Q_{k-1}\right. \tag{3.28}$$

Taking s as the fading factor, and multiplying both sides by $s^{-(N-k)}$, the mathematical relationship can be obtained, as shown in Equation (3.29).

$$s^{-(N-k)}P_{k|k-1}^N = A_{k-1}\left(s^{-(N-k)}P_{k-1}^N\right)A_{k-1}^T + Q_{k-1} \tag{3.29}$$

The relationship between different parameters is described, as shown in Equation (3.30).

$$P_{k|k-1}^{N*} = s^{-(N-k)} P_{k|k-1}^{N} \tag{3.30}$$

$$P_{k-1}^{N*} = s^{-(N-(k-1))} P_{k-1}^{N} \tag{3.31}$$

Then, Equation (3.29) can be expressed, as shown in Equation (3.32).

$$P_{k|k-1}^{N*} = A_{k-1}\left(sP_{k-1}^{N*}\right)A_{k-1}^{T} + Q_{k-1} \tag{3.32}$$

To optimize the covariance matrix updating, Equation (3.32) may be substituted for Equation (3.28), and the covariance matrix has been expanded by s times as opposed to the initial equation, indicating that the calculation capacity has been improved. In Equation (3.32), the fading factor can be set to a value greater than 1. Also, it can adaptively be modified based on the degree of device model mismatch, but the error bars are visible.

Linear–nonlinear filter: as the EKF algorithm is increased by combining multiple solutions, the computation of algorithm cost rises, making quick estimation impossible. As a result, the linear–nonlinear $(L-N)$ filter is used to solve problems caused by the complexity of the algorithm. The second-order ECM has the following characteristics.

1. The state update process is linear, as shown in Equation (3.33).

$$x(k+1) = Ax(k) + w(k) \tag{3.33}$$

2. The observation process is nonlinear, as shown in Equation (3.34).

$$y(k) = f\left(x(k), w(k)\right) \tag{3.34}$$

In Equation (3.34), the observation method concentrates on the nonlinear proportion. As a result, the state update section is assumed to be processed independently from the observation portion. When working with linear systems that need little estimation, the traditional KF has higher precision. However, while the EKF can respond to nonlinear systems when dealing with linear systems, the accuracy is close to that of the classical KF, and a lot of computational power is wasted [188]. As a result, the $L-N$ filter is used to solve the problem for the special linear–nonlinear method described earlier. For the status updating process, the classical KF method is used, and the EKF algorithm is used for the observation process, which essentially decreases the amount of computation needed by the algorithm without impacting estimation accuracy, as shown in Table 3.1.

In the improved EKF, several methods are introduced to deal with the problem of filtering noise and divergence. The $L-N$ filter approach is used to minimize the sum of estimation. The improved EKF algorithm adopted can be described.

TABLE 3.1
Improved EKF Algorithm

	Initialization:
	$\hat{X}_0 = E[X_0]$
	$P_0 = E\left[\left(X - \hat{X}_0\right)\left(X - \hat{X}_0\right)^T\right]$
	$\hat{V}_0 = E(V_0)$
	$\hat{W}_0 = E(W_0)$
	$s = s_0$
	Prediction:
Status prediction	$x_{k\|k-1} = Ax_{k-1} + Bu_k + w_{k-1}$
Covariance matrix prediction	$P_{k\|k-1} = A_{k-1}\left(s \cdot P_k\right)A_{k-1}^T + Q_{k-1}$
	Updating:
Observation	$\hat{z}_k = g\left(x_k, u_k, v_k\right)$
Kalman gain matrix (K)	$K_k = P_{k\|k-1}C_k^T\left(C_k P_{k\|k-1}C_k^T + R_{k-1}\right)^{-1}$
Where:	$C = \left.\begin{bmatrix} \frac{\partial g_1}{\partial x_1} & \frac{\partial g_1}{\partial x_n} \\ \frac{\partial g_m}{\partial x_1} & \frac{\partial g_m}{\partial x_n} \end{bmatrix}\right\|_{x=\hat{x}_{k\|k-1}}$
Status update	$\hat{x}_k = \hat{x}_{k\|k-1} + G_k\left(z_k - \hat{z}_k\right)$
Covariance matrix update	$P_k = \left(I - G_k C_k\right)P_{k\|k-1}$
Noise adaptation	$\begin{cases} \hat{w}_k = (1-d_k)\hat{w}_{k-1} + d_k\left(\hat{x}_k - \left(A\hat{x}_{k-1} + Bu_k\right)\right) \\ \hat{v}_k = (1-d_k)\hat{v}_{k-1} + d_k\left(z_k - \hat{z}_k\right) \end{cases}$
	$\begin{cases} \hat{Q}_k = (1-d_k)\hat{Q}_{k-1} + d_k\left(G_k\varepsilon\varepsilon^T G_k^T + P_k - A_{k-1}P_{k\|k-1}A_{k-1}^T\right) \\ \hat{R}_k = (1-d_k)\hat{w}_{k-1} + d_k\left(\varepsilon\varepsilon^T - C_k P_{k\|k-1}C_k^T\right) \end{cases}$

3.2.4 Unscented Kalman Filter

The unscented Kalman filter (UKF) is an improved algorithm based on the combination of unscented transform and KF, which is combined with UT for linearization [189,190]. This algorithm mainly uses the idea of the KF; in the subsequent solving process, a sampling point is used to calculate the predicted and measured value. One-state point is converted into multiple-state points by using UT, it is passed to the following observation value according to the value of the certain weight, and then according to the error of the following observation value and the real value, perform forward feedback and continuous correction, and finally, the ideal state value is obtained. The basic principle of UT is to use sampling points to simulate nonlinear variables according to a certain probability distribution. By calculating the statistical value characteristics of nonlinear variables, multiple variable values conforming to statistical laws are obtained. It is the core and prerequisite part of the UKF algorithm [191,192].

As a nonlinear system estimation method, the UKF method abandons the processing idea of approximate linearization of nonlinear functions, and the UT is used to process the data and expand the data length. The original data is reconstructed with weights and the mean and variance of the data are kept unchanged. The extended data is used to estimate the parameters of the nonlinear system. Due to this special data processing method, the prediction accuracy of the UKF method is higher than that of the EKF method in the strong nonlinear system [193].

A limited number of sampling points are selected according to a certain method by the statistical characteristics of the variables. The probability distribution of the sampling points is made to be close to the probability distribution characteristics of the known variables, thereby transforming a variable value with a certain statistical characteristic into multiple combined variable values with the same statistical characteristic. According to the mean and variance of the estimator, the basic data of the estimator is used to generate a corresponding number of reference points using an asymmetric sampling strategy, and these reference points have the same mean and variance as the estimator. This process is called stigmatization of the estimated quantity [194]. The following briefly introduces the principle of the symmetric UT. The nonlinear function is $y = f(x)$, the dimension of the state variable x is n-dimensional, and the variance and the state mean are known to be P and $\bar{x}$ respectively; then it can be achieved through $2n+1$ sigma points X and the corresponding weights calculated by UT. The value ω is used to approximate the posterior probability distribution of the output variable y. The main process of UT is described as follows.

1. $2n+1$ sigma points are sampled according to the estimated points, as shown in Equation (3.35).

$$\begin{cases} X^0 = \bar{X}, i = 0 \\ X^i = \bar{X} + \left(\sqrt{(n+\lambda)P}\right)_i, i = 1 \sim n \\ X^i = \bar{X} - \left(\sqrt{(n+\lambda)P}\right)_i, i = n+1 \sim 2n \end{cases} \tag{3.35}$$

Among them, the ith column of the square root of the matrix can be represented, as shown in Equation (3.36).

$$\left(\sqrt{P}\right)^T\left(\sqrt{P}\right) = P, \tag{3.36}$$

2. Calculating the weights of $2n+1$ sigma points sampling, as shown in Equation (3.37).

$$\begin{cases} \omega_m^0 = \dfrac{\lambda}{n+\lambda} \\ \omega_c^0 = \dfrac{\lambda}{n+\lambda} + \left(1-\alpha^2+\beta\right) \\ \omega_m^i = \omega_c^i = \dfrac{\lambda}{2(n+\lambda)}, i = 1 \sim 2n \end{cases} \tag{3.37}$$

In Equation (3.37), ω_m^i is the mean weight of the ith sampling point, and ω_c^i is the corresponding variance weight. The parameter $1-\alpha^2+\beta$ in Equation (3.37) is a scaling factor used to reduce the total prediction error, where α affects the distribution of sigma point sampling. The parameter is used to merge the errors of higher-order terms.

3. Calculating the statistics of output variables
After the above-mentioned transformation of the sigma point set, the coefficient is weighted and the nonlinear transformation $y_i = f(x_i)$ is used, the corresponding output point set $\{y_i\}$ can be obtained and the equation is calculated, as shown in Equations (3.38) and (3.39).

$$\bar{y} = \sum_{i=1}^{2n+1} \omega_m^i y_i \tag{3.38}$$

$$P_y = \sum_{i=1}^{2n+1} \omega_c^i \left(y_i - \bar{y}\right)\left(y_i - \bar{y}\right)^T \tag{3.39}$$

The steps of the UKF algorithm are obtained as follows.

1. Setting the initial state, the initial value of the state variable $\bar{X}_0$ and the covariance matrix P_0 can be initialized, as shown in Equation (3.40).

$$\begin{cases} \bar{X}_0 = E(X_0) \\ P_0 = E\left[\left(X_0 - \bar{X}_0\right)\left(X_0 - \bar{X}_0\right)^T\right] \end{cases} \tag{3.40}$$

2. At time k, the UT sampled by the symmetrical distribution obtains $2n+1$ sigma sampling points $x_k^i, i = 1\ldots 2n+1$ from Equation (3.40) and calculates the weight of the corresponding mean value $\omega_m^i, i = 1\ldots 2n+1$ and the agreement $\omega_c^i, i = 1\ldots 2n+1$ by variance weights shown in Equation (3.37).

3. At time point k, $2n+1$ sigma points sampling x_k^i are substituted into the state parameters. The predicted value $X_{k+1|k}^i = f\left(X_k^i, U_{k+1}\right), i = 1\ldots 2n+1$ of the system state is calculated. The predicted mean value is obtained by weighted summation, as shown in Equation (3.41).

$$\overline{X}_{k+1|k}^i = \sum_{i=1}^{2n+1} \omega_m^i X_{k+1|k}^i \tag{3.41}$$

The state variance matrix can be calculated, as shown in Equation (3.42).

$$P_{k+1|k} = \sum_{i}^{2n+1} \omega_c^i \left(X_{k+1|k}^i - \overline{X}_{k+1|k}^i\right)\left(X_{k+1|k}^i - \overline{X}_{k+1|k}^i\right)^T + Q \tag{3.42}$$

4. At the time point k, substituting the above-mentioned set of points $x_{k+1|k}^i, i = 1\ldots 2n+1$ into the system observation, calculate the predicted value of the observation $Z_{k+1|k}^i = h\left(X_{k+1|k}^i, U_{k+1}\right), i = 1\ldots 2n+1$, and the weighted summation to obtain the mean value of the system observation can be calculated, as shown in Equation (3.43).

$$\overline{Z}_{k+1|k}^i = \sum_{i=1}^{2n+1} \omega_m^i Z_{k+1|k}^i, i = 1 \sim 2n+1 \tag{3.43}$$

The prediction variance matrix is shown in Equation (3.44).

$$P_z = \sum_{i=1}^{2n+1} \omega_c^i \left(Z_{k+1|k}^i - \overline{Z}_{k+1|k}^i\right)\left(Z_{k+1|k}^i - \overline{Z}_{k+1|k}^i\right)^T + R \tag{3.44}$$

5. The calculation of the system covariance matrix and Kalman gain is conducted, as shown in Equations (3.45) and (3.46).

$$P_x = \sum_{i=1}^{2n+1} \omega_c^i \left(X_{k+1|k}^i - \overline{X}_{k+1|k}^i\right)\left(Z_{k+1|k}^i - \overline{Z}_{k+1|k}^i\right)^T \tag{3.45}$$

$$K(k+1) = P_{xz} P_z^{-1} \tag{3.46}$$

6. The system state matrix and covariance matrix are updated according to Kalman gain, as shown in Equations (3.47) and (3.48).

$$X_{k+1} = \overline{X}_{k+1}^i + K(k+1)\left(Z_{k+1} - \overline{Z}_{k+1|k}^i\right) \tag{3.47}$$

$$P_{k+1|k+1} = P_{k+1|k} - K(k+1) P_z K(k+1)^T \tag{3.48}$$

7. At the time point $k = k+1$, the procedure goes to the second step for loop operation.

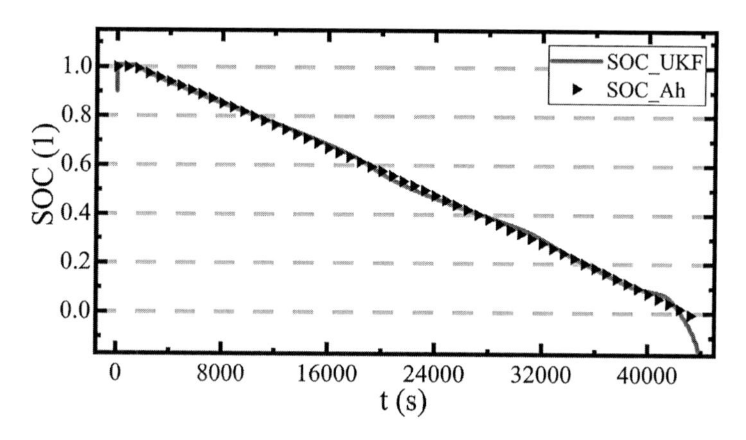

FIGURE 3.3 Comparison of SOC output curves under BBDST condition.

The UKF algorithm does not require Taylor series expansion of the nonlinear function but performs UT near the estimated point. The mean and covariance of the transformed sigma point set are equal to the original estimated point. It is subjected to a nonlinear state as well as observation and estimation, and the posterior probability distribution of the system is approximated. The estimation accuracy of the UKF algorithm in a nonlinear system is higher than that of the EKF algorithm, which applies to any state model. It is not necessary to calculate the Jacobian matrix at every sampling point, and the calculation complexity is lower. In this experiment, U_{OC} is substituted into the algorithm with a fitted polynomial and the values of R_0, R_p, and C_p are substituted into the algorithm in an average manner, the battery is discharged by a 1 C constant current in the experiment, and the EKF estimation of the lithium-ion battery SOC algorithm can be verified according to Figure 3.3.

In Figure 3.3, the red curve is the standard SOC estimation result, and the gray curve is the SOC estimation result of the UKF algorithm. It can be seen that the estimation result has higher accuracy and better robustness.

3.2.5 Adaptive Kalman Filter

The KF and its derived algorithms, such as UKF and EKF, have a premise that the system process and observation noises are known and remain unchanged from beginning to end. The internal structure of the battery and the environmental conditions may cause changes in noise, that is, the noise of the system is time varying, so the noise covariance of the system cannot be simply regarded as a constant. Sage–Husa adaptive filtering realizes the estimation and prediction of system noise in iterative operation through the addition of a noise estimator and avoiding the divergence of estimation results due to inaccurate noise [195,196]. The calculation process of the Sage–Husa adaptive filtering algorithm is described as follows. Initialization is conducted, as shown in Equation (3.49).

$$\begin{cases} \hat{x}_0 = E[x_0] \\ P_0 = E\left[(x_0 - \hat{x}_0)(x_0 - \hat{x}_0)^T\right] \\ \hat{Q}_0 = Q_0 \\ \hat{R}_0 = R_0 \end{cases} \tag{3.49}$$

In Equation (3.49), the iterated calculation is conducted for $k = 1, 2, \ldots$

1. Prediction stage

$$\begin{cases} \hat{x}_k = A_{k-1}\hat{x}_{k-1} + B_{k-1}u_{k-1} \\ P_k^- = A_{k-1}P_{k-1}A_{k-1}^T + Q_{k-1} \end{cases} \tag{3.50}$$

2. Update stage
   Estimating the statistical characteristics of measurement noise, as shown in Equation (3.51).

$$\begin{cases} e_k = y_k - C_k\hat{x}_k + D_k u_k \\ \hat{R}_k = (1 - d_k)\hat{R}_{k-1} + d_k e_k e_k^T - C_k P_k C_k^T \end{cases} \tag{3.51}$$

   Updating the state quantity and error variance matrix, as shown in Equation (3.52).

$$\begin{cases} K_k = P_k^- C_k^T \left(C_k P_k^- C_k^T + \hat{R}_k\right)^{-1} \\ \hat{x}_k = \hat{x}_k^- + K_k e_k \\ P_k = (I - K_k C_k) P_k^- \end{cases} \tag{3.52}$$

   Estimating the statistical characteristics of the process noise, as shown in Equation (3.53).

$$\hat{Q}_k = (1 - d_k)\hat{Q}_{k-1} + \left(K_k e_k e_k^T K_k^T + P_k - A_{k-1}P_{k-1}A_{k-1}^T\right) \tag{3.53}$$

   In Equation (3.53), $d_k = (1-b)/\left(1 - b^{k+1}\right)$, b is the forgetting factor generally taken as 0.9–1. The function of the method is to predict and correct the noise of the system online when iterative calculations are performed using experimental data, making the algorithm more adaptive and reliable. The method is relatively easy to be combined with other algorithms in practice, so it plays an important role in various fields.

3.2.6 Cubature Kalman Filter

The principle of the cubature Kalman filter (CKF) is similar to that of the UKF, and it is also a sampling filtering algorithm, which approximates the state value and state covariance of nonlinear systems by finding a set of cubature points. The CKF algorithm is first proposed by Canadian scholars Arasaratnam and Haykin in their master's thesis, which is based on the third-order spherical radial cubature criterion and uses a set of cubature points to approximate the state value and state covariance of highly nonlinear systems [197]. It is a powerful tool to solve the state estimation of nonlinear systems. Using the third-order spherical radial criterion to extend and calculate the standard Gaussian weighted integral is shown in Equation (3.54).

$$\int_{R^n} f(x)e^{-x^2}\, dx = \sum_{i=1}^{m} w_i f(x_i) \tag{3.54}$$

In Equation (3.54), m is the number of integration points, w_i is the weight of integration, and x_i is the point set of integration points. Generally speaking, the number of cubature points is twice the dimension n of the state vector. The nonlinear system is described mathematically, as shown in Equation (3.55).

$$\begin{cases} x_{k+1} = f(x_k, u_k) + w_k \\ y_k = g(x_k, u_k) + v_k \end{cases} \tag{3.55}$$

For the nonlinear system as shown in Equation (3.55), the system state is estimated by using the CKF algorithm, and the specific step expression is obtained, as shown in Equation (3.56).

$$\int_{R^n} f(x)e^{-x^2}\, dx = \sum_{i=1}^{m} w_i f(x_i) \tag{3.56}$$

1. Initialization state quantity, error covariance, process noise, and measurement noise.
2. Calculating the cubature point, as shown in Equation (3.57).

$$\begin{cases} p_k = S_k S_k^T \\ x_k = S_k x_i + \hat{x}_k,\ i = 1,2,\ldots\ldots,\ 2n \end{cases} \tag{3.57}$$

wherein, n is the dimension of state quantity, x_i is the cubature point set, and its specific values are shown in Equation (3.58).

$$x_i = \begin{cases} \sqrt{n}[e]_i,\ i = 1,2,\ldots\ldots n \\ -\sqrt{n}[e], i = n+1, n+2,\ldots\ldots, 2n \end{cases} \tag{3.58}$$

3. Propagation cubature point, as shown in Equation (3.59).

$$x_{k+1|k}^{i} = f\left(x_k^i, u_k\right) \tag{3.59}$$

4. Calculating the predicted value of state quantity and error covariance, as shown in Equation (3.60).

$$\begin{cases} \hat{x}_{k+1|k} = 1/(2n)\sum_{i=1}^{2n} x_{k+1|k}^{i} \\ P_{k+1|k} = 1/(2n)\sum_{i=1}^{2n} x_{k+1|k}^{i}\left(x_{k+1|k}^{i}\right)^T - \hat{x}_{k+1|k}\left(\hat{x}_{k+1|k}\right)^T + Q \end{cases} \tag{3.60}$$

5. Calculating the cubature point, as shown in Equation (3.61).

$$\begin{cases} P_{k+1|k} = S_{k+1|k}S_{k+1|k}^T \\ x_{k+1|k}^{i} = S_{k+1|k}x_i + \hat{x}_{k+1|k} \end{cases} \tag{3.61}$$

6. Propagation of the cubature point, as shown in Equation (3.62).

$$y_{k+1|k}^{i} = g\left(x_{k+1|k}^{i}, u_{k+1}\right) \tag{3.62}$$

7. Calculating the measured predicted value, as shown in Equation (3.63).

$$\hat{y}_{k+1} = 1/(2n)\sum_{i=1}^{2n} y_{k+1}^{i} \tag{3.63}$$

8. Calculating the covariance and cross-covariance of measurement error, as shown in Equation (3.64).

$$\begin{cases} P_{k+1}^{y} = 1/(2n)\sum_{i=1}^{2n} y_{k+1}^{i}\left(y_{k+1}^{i}\right)^T - \hat{y}_{k+1}\left(\hat{y}_{k+1}\right)^T + R \\ P_{k+1}^{xy} = 1/(2n)\sum_{i=1}^{2n} x_{k+1|k}^{i}\left(y_{k+1}^{i}\right)^T - \hat{x}_{k+1|k}\left(\hat{y}_{k+1}\right)^T + R \end{cases} \tag{3.64}$$

9. Calculating Kalman gain, the state quantity is updated as well as the corresponding error covariance, as shown in Equation (3.65).

$$\begin{cases} K_{k+1} = P_{k+1}^{xy}\left(P_{k+1}^{y}\right)^{-1} \\ \hat{x}_{k+1} = \hat{x}_{k+1|k} + K_{k+1}\left(y_{k+1} - \hat{y}_{k+1}\right) \\ P_{k+1} = P_{k+1} - K_{k+1}P_{k+1}^{y}K_{k+1}^{T} \end{cases} \tag{3.65}$$

Compared with the EKF algorithm and the UKF algorithm, the CKF algorithm has the advantages of good stability and accuracy in the state estimation process. Its advantages are described as follows.

1. As deterministic sampling filtering algorithms, the CKF algorithm generates point sets according to a certain sampling strategy when dealing with nonlinear equations, then directly propagates each sampling point in the point set nonlinearly, and finally calculates the mean and covariance of the posterior probability density distribution of system state by weighted summation. There is no need to linearize the nonlinear equation, which eliminates the linearization error and does not need to calculate the Jacoby matrix in the iterative process of the filtering algorithm, so it is easier to use in practice.
2. Based on the numerical integration theory, the CKF algorithm uses the third-order spherical radial cubature criterion to approximate Gaussian integration, which is a strict and complete theory. Therefore, from the perspective of the theoretical basis, the CKF algorithm is rigorous and stable.
3. As deterministic sampling filtering algorithms, both UKF and CKF need to generate point sets through a certain sampling method in each iteration. The point set size of the UKF is $2n+1$, while the cubature point set of the CKF algorithm is only $2n$, that is, one sampling point is calculated less than that of UKF in each iteration, so theoretically, the CKF algorithm has higher execution efficiency and better real time.
4. Stability analysis
   The stability factor is introduced as the standard to judge the stability of the algorithm. The closer the stability factor is to 1, the better the stability of the algorithm. The definition equation of stability factor λ is shown in Equation (3.66).

$$\lambda = \frac{\sum_i |w_i|}{\sum_i wi} \tag{3.66}$$

In Equation (3.66), w_i represents the value. It can be observed from the equation that the stability factor is the ratio of the sum of the absolute values of weights to the sum of weights. Because $w=1$ and the parameters of UKF need to be set to ensure $n+k=3$, the reliability factor λ of UKF is shown in Equation (3.67).

$$\lambda = \frac{\sum_i |w_i|}{\sum_i wi} = \begin{cases} 1, & n \leq 3 \\ \dfrac{2n}{3} - 1, n > 3 \end{cases} \tag{3.67}$$

In Equation (3.67), it can be observed that when the dimension of system state quantity $n > 3$, the reliability factor of UKF increases with the increase of dimension, the reliability becomes worse and worse, and the filtering accuracy greatly decreases. However, the weight of each cubature point of the CKF algorithm is the same, which is $1/2n$, so the reliability factor is always equal to 1, which is not restricted by the dimension of system state quantity, which has a high reliability and accuracy. The CKF is superior to EKF and UKF inapplicability, theoretical basis, time complexity, and reliability, and is suitable for SOC estimation of lithium-ion batteries [198,199]. However, the CKF algorithm also has some drawbacks. For the standard CKF algorithm, the error covariance matrix needs to be decomposed by Cholesky when calculating the cubature points. After multiple filterings, the truncation error of the computer easily leads to the loss of positive definition and symmetry of the error covariance matrix, which makes the filtering divergent. Because of this drawback, researchers put forward some improved algorithms based on CKF, such as square root cubature KF, strong tracking UKF, etc.

3.3 SOC ESTIMATION BASED ON THE SECOND-ORDER RC MODEL

3.3.1 Second-Order RC Modeling

The accuracy of the battery model has a great influence on whether the SOC can be accurately estimated. There are a wide variety of battery models, and currently commonly used models include the Rint model, Partnership for a New Generation of Vehicles (PNGV) model, and Thevenin model [200]. However, considering the polarization effect inside the battery, including electrochemical polarization and concentration polarization, a second-order RC equivalent model is adopted, as shown in Figure 3.4.

The RC circuit is formed by R_1 and C_1 and the figure represents the stage of rapid voltage change in the chemical reaction process within the batteries. The RC circuit consisting of R_2 and C_2 represents the phase in which the voltage changes slowly during a chemical reaction within the batteries. According to Kirchhoff's circuit law and Figure 3.4, the circuit dynamic model of the second-order differential network is shown in Equation (3.68).

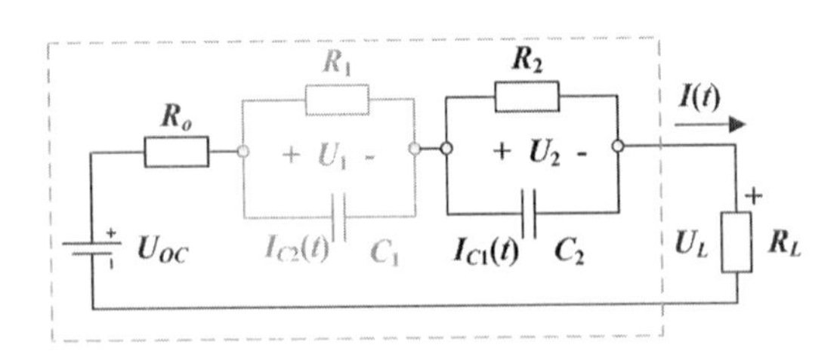

FIGURE 3.4 Second-order RC Equivalent model of lithium-ion battery.

$$\begin{cases} U_L = U_{OC}(\mathrm{SOC}) - i(t)R_0 - U_1 - U_2 \\ \dfrac{dU_1}{dt} = -\dfrac{U_1}{R_1C_1} + \dfrac{i}{C_1} \\ \dfrac{dU_2}{dt} = -\dfrac{U_2}{R_2C_2} + \dfrac{i}{C_2} \end{cases} \tag{3.68}$$

After discretization, its state-space equation can be shown in Equation (3.69).

$$\begin{cases} \begin{bmatrix} \mathrm{SOC}_{k+1} \\ U_{1,k+1} \\ U_{2,k+1} \end{bmatrix} = \begin{bmatrix} 1 & 0 & 0 \\ 0 & e^{\frac{\Delta t}{\tau_1}} & 0 \\ 0 & 0 & e^{\frac{\Delta t}{\tau_2}} \end{bmatrix} \begin{bmatrix} \mathrm{SOC}_k \\ U_{1,k} \\ U_{2,k} \end{bmatrix} + \begin{bmatrix} \dfrac{-\eta\Delta t}{Q} & 0 & 0 \\ 0 & R_1 e^{1-\frac{t}{\tau_1}} & 0 \\ 0 & 0 & R_2 e^{1-\frac{t}{\tau_2}} \end{bmatrix} I(t) \\ U_{L,k+1} = U_{OC}(\mathrm{SOC}, k+1) - U_1 - U_2 - IR_0 \end{cases} \tag{3.69}$$

Combining the hybrid pulse power characterization (HPPC) test and parameter identification in simulation software, the mathematical relationship can be obtained for the iterative calculation.

3.3.2 Parameter Identification

The HPPC experimental terminal voltage output waveform is obtained, as shown in Figure 3.5.

At the time points of discharge start and stop, the voltage of the lithium-ion battery changes suddenly, which is caused by R_0, so R_0 can be described in Equation (3.70).

$$R_0 = \frac{(U_1 - U_2) + (U_4 - U_3)}{2I} \tag{3.70}$$

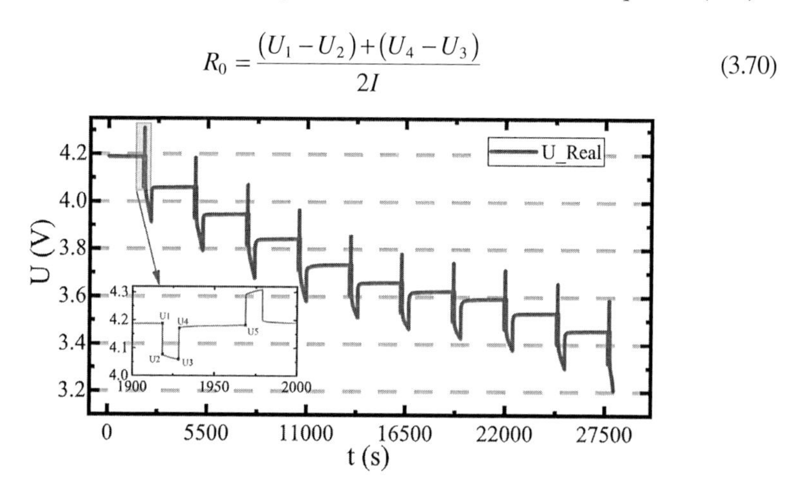

FIGURE 3.5 HPPC experimental voltage response curve.

The terminal voltage changes from U_2 to U_3 are the discharge process of the polarization capacitor to the polarization resistance, and it is the zero-input response of the second-order RC circuit. Therefore, the calculation procedure of U_L is described, as shown in Equation (3.71).

$$U_L = U_{OC} - IR_1 e^{-\frac{t}{\tau_1}} - IR_2 e^{-\frac{t}{\tau_2}} \tag{3.71}$$

The mathematical relationship of Equation (3.71) can be simplified, as shown in Equation (3.72).

$$U_L = f - ae^{-ct} - be^{-dt} \tag{3.72}$$

Through the parameter identification of Equation (3.72), the results are brought into the second-order RC model in Simulink to get the simulation results, which verify the correctness of the parameter identification. The results of the parameters are shown in Table 3.2.

The identified parameters are introduced into the Thevenin equivalent circuit model, as well as the same current as of the HPPC test to the model. By comparing the output voltage data of the model with the actual voltage data, the model is effective. The result is shown in Figure 3.6.

3.3.3 EKF-Based Calculation Procedure

The form of the state equation and observation equation of KF is shown in Equation (3.73).

$$\begin{cases} X_{k+1} = f(X_k, k) + w_k \\ Z_k = h(X_k, k) + v_k \end{cases} \tag{3.73}$$

TABLE 3.2
The Result of Parameter Identification

S	U_{OC}	R_0	R_2	C_2	R_1	C_1
1	4.1722	0.001696898	0.000966407	16776.27750	0.000076623	8557.997284
0.9	4.0336	0.001689400	0.001069370	16240.52975	0.000085195	8798.959277
0.8	3.9239	0.001693149	0.001278296	16914.40649	0.000105013	8015.703601
0.7	3.8228	0.001691768	0.001314432	16263.04342	0.000099957	9094.840909
0.6	3.7180	0.001696898	0.000967157	15370.27882	0.000081146	8044.018942
0.5	3.6445	0.001708144	0.001337775	25793.98922	0.000089643	13342.11066
0.4	3.6052	0.001725637	0.001040131	24519.70384	0.000070750	15112.04836
0.3	3.567	0.001754105	0.000805339	20105.45159	0.000062893	11505.09387
0.2	3.5029	0.001794363	0.000862444	15792.64231	0.000077997	7789.158458
0.1	3.4322	0.001907119	0.001406969	10231.00493	0.000178089	3217.855527

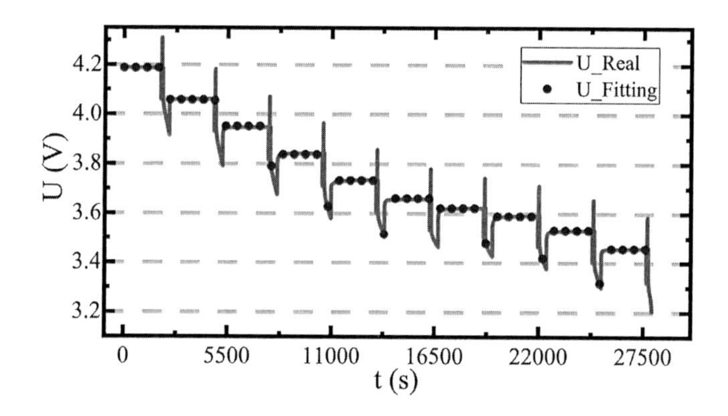

FIGURE 3.6 Output comparison.

To apply the KF, the nonlinear functions $f(*)$ and $h(*)$ are surrounded by first-order Taylor expansion, as shown in Equation (3.74).

$$\begin{cases} f(X_k,k) \approx f\left(\hat{X}_k,k\right) + \left.\dfrac{\partial f(X_k,k)}{\partial X_k}\right|_{X_k=\hat{X}_k} \left(X_k - \hat{X}_k\right) \\ h(X_k,k) \approx h\left(\hat{X}_k,k\right) + \left.\dfrac{\partial h(X_k,k)}{\partial X_k}\right|_{X_k=\hat{X}_k} \left(X_k - \hat{X}_k\right) \end{cases} \tag{3.74}$$

Defining A_k, B_k, C_k, and D_k, as shown in Equation (3.75).

$$\begin{cases} A_k = \left.\dfrac{\partial f(X_k,k)}{\partial X_k}\right|_{X_k=\hat{X}_k}, B_k = f\left(\hat{X}_k,k\right) - A_k\hat{X}_k \\ C_k = \left.\dfrac{\partial h(X_k,k)}{\partial X_k}\right|_{X_k=\hat{X}_k}, D_k = h\left(\hat{X}_k,k\right) - C_k\hat{X}_k \end{cases} \tag{3.75}$$

The updating process of the EKF is shown in Equation (3.76).

$$\begin{cases} \hat{X}_{k+1}^- = f\left(\hat{X}_k\right) \\ \hat{P}_{k+1}^- = A_k\hat{P}_kA_k^T + Q_{k+1} \\ K_{k+1} = \hat{P}_{k+1}^-C_{k+1}^T\left(C_{k+1}\hat{P}_{k+1}^-C_{k+1}^T + R_{k+1}\right)^{-1} \\ \hat{X}_{k+1} = X_{k+1}^- + K_{k+1}\left[Z_{k+1} - h\left(X_{k+1}^-\right)\right] \\ \hat{P}_{k+1} = \left[I - K_{k+1}C_{k+1}\right]P_{k+1}^- \end{cases} \tag{3.76}$$

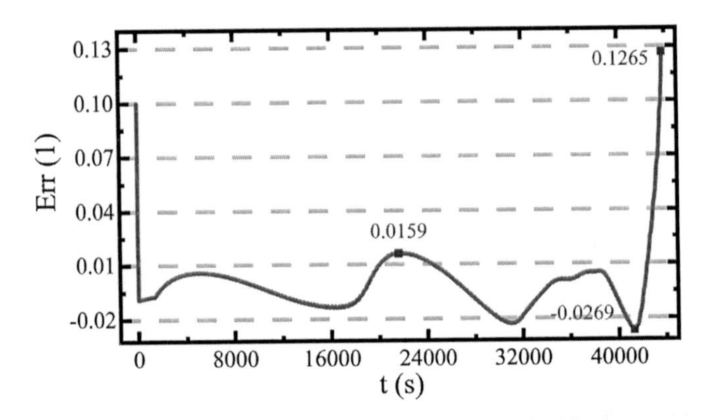

FIGURE 3.7 SOC estimation error curve.

The calculation results of the EKF algorithm are obtained and the error of SOC estimation between experiment and estimation is shown in Figure 3.7.

3.4 EKF-BASED SOC ESTIMATION

3.4.1 Parameter Identification

In the SOC estimation process, the accuracy and timeliness of the estimation results are affected by the quality of the equivalent model [201]. In the process of establishing the equivalent circuit model, it is necessary to comprehensively consider the accuracy and complexity of the model [202]. According to the analysis and research of the common equivalent circuit models, the Thevenin model is adopted. It has moderate structural complexity and few identification parameters among all models. Besides, it takes into account the abrupt and gradual voltage characteristics of the lithium-ion battery under the excitation of the load current during the charge-discharge process, which can better reflect the dynamic characteristics of the lithium-ion battery and static characteristics, as shown in Figure 3.8.

According to the Thevenin model shown in Figure 3.8, the discharge direction of the lithium-ion battery is the positive direction, and the voltage across the polarized capacitor is U_p. According to Kirchhoff's voltage law, the voltage relationship can be obtained, as shown in Equation (3.77).

$$U_{OC}(t) = U_L(t) + I(t)R_0 + U_p(t) \tag{3.77}$$

And then, according to Kirchhoff's current law, the current relationship can be obtained, as shown in Equation (3.78).

$$I(t) = \frac{U_p(t)}{R_p} + C_p \frac{dU_{OC}(t)}{dt} \tag{3.78}$$

The OCV U_{OC} is a function of SOC, and the mathematical relationship can be obtained, as shown in Equation (3.79).

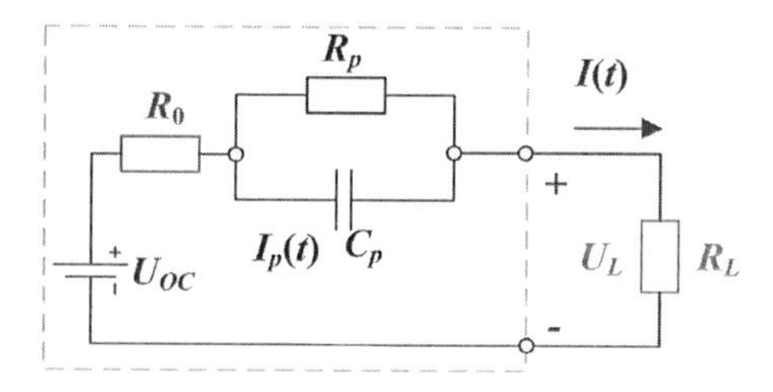

FIGURE 3.8 Thevenin model.

$$U_{OC}(t) = f(\mathrm{SOC}(t)) \tag{3.79}$$

Equations (3.78) and (3.79) characterize the circuit relationship of the model. In Equation (3.79), the SOC at the time point t is represented by $\mathrm{SOC}(t)$. Equation (3.79) can also be rewritten as Equation (3.80).

$$\dot{U}_p(t) = \frac{1}{C_p} I(t) - \frac{1}{C_p R_p} U_p(t) \tag{3.80}$$

Equation (3.80) is solved to obtain Equation (3.81).

$$U_p(t) = U_P(0) e^{\frac{1}{C_P R_P}} + R_p I(t) \left(1 - e^{\frac{1}{C_P R_P}} \right) \tag{3.81}$$

Combining Equations (3.79)–(3.81), the calculation of load terminal voltage U_L can be obtained, as shown in Equation (3.82).

$$U_L(t) = U_{OC}(t) - I(t)\left(R_o + R_p\right) + \left(I(t)R_p - U_P(0)\right) e^{\frac{1}{C_P R_P}} \tag{3.82}$$

The HPPC test is an experiment to test the performance of lithium-ion batteries through pulse charge-discharge. The HPPC experiment is to perform a pulse charge-discharge experiment on a lithium-ion battery and obtain the current voltage curve of the battery under the premise of fully considering the safety. According to the experimental data and curves, the parameters of the model are identified to obtain the corresponding relationship between them and the SOC value. Setting the experimental temperature at 25°C as an example, the experimental steps of HPPC are described as follows:

1. The lithium-ion battery is fully charged with constant current-constant voltage steps at a temperature of 25°C, which is shelved for 1 hour.
2. A constant current of 5 C (60 A) is discharged for 10 seconds, and the battery is shelved for 40 seconds after the discharge. And then, the battery is charged with a constant current of 3.75 C (45 A) for 10 seconds until the charging is stopped.

3. The battery is discharged at a constant current rate of 1 C (12 A) for 6 minutes, that is, the discharge is stopped after 10% of the capacity is discharged, which is shelved for 40 minutes to record the parameters until SOC equals to 0.9.
4. Repeating steps (2) and (3) in sequence, the capacity is reduced by 10% each time. And then, a group of charge-discharge pulse tests is carried out until the lithium-ion battery voltage is equal to 2.75 V, so the cycle is terminated and the experiment is stopped.

The current-voltage curve diagram of the HPPC experiment under the condition of 25°C constant temperature is shown in Figure 3.9.

The current curve in the HPPC experiment is shown in Figure 3.9a, and the voltage curve is shown in Figure 3.9b. The HPPC experiment is used to test the dynamic characteristics of the batteries. The experiment includes 10 pulses of charge-discharge treatment. After each pulse charge-discharge, the battery performs constant current discharge to reduce the capacity by 10% until the battery is discharged to the cutoff voltage, and the experiment is stopped. The parameter identification of the curve fitting method is an off-line parameter identification method. The model parameter results are obtained by curve-fitting experimental data. R_0, R_p, and C_p are the three parameters that need to be identified in the Thevenin model. Among these three parameters, R_0 is the easiest to identify. The pulse voltage and current curves are shown in Figure 3.10.

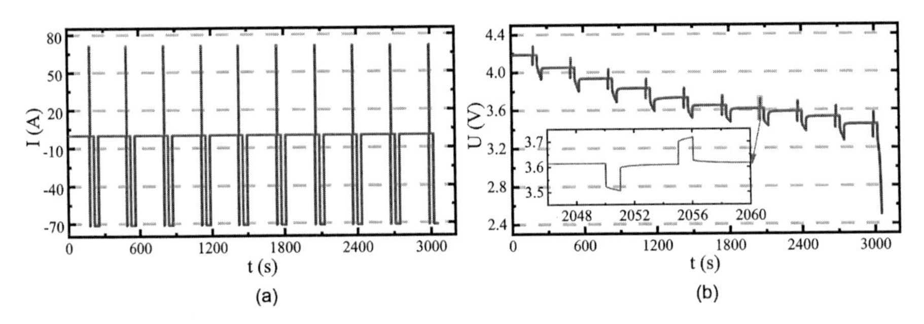

FIGURE 3.9 HPPC experiment current and voltage curve. (a) HPPC experiment current curve. (b) HPPC experiment voltage curve.

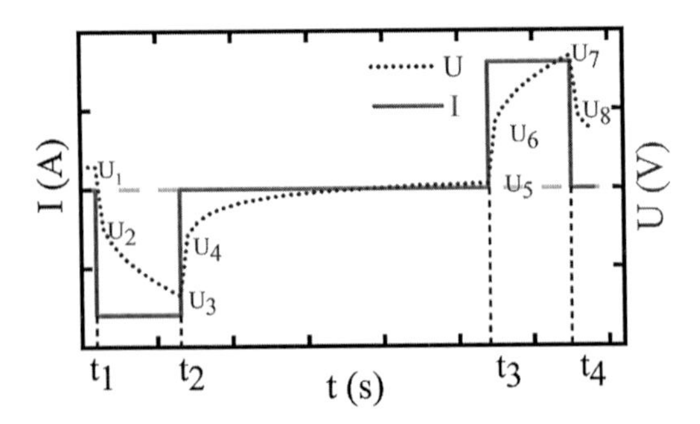

FIGURE 3.10 HPPC current–voltage curve when SOC=0.1.

As shown in Figure 3.10, the terminal voltage of the battery drops sharply at the time point the battery is discharged. This sudden voltage drop is due to the ohmic resistance that causes a voltage drop when the current flows through the lithium-ion battery at the time point of discharge [203,204]. Similarly, when the battery stops discharging, there is a sudden rise in the battery, and the voltage is also caused by the ohmic resistance. Therefore, the calculation method of ohmic resistance is shown in Equation (3.83).

$$R_0 = \frac{|\Delta U_{12}| + |\Delta U_{34}|}{2I} \tag{3.83}$$

In Equation (3.84), ΔU_{12} is the voltage drop at the time point of battery discharge, and ΔU_{34} is the sudden rise of the battery at the time point of stopping discharge. The value of the ohmic resistance at each SOC point can be calculated accordingly. The ohm internal resistance values from SOC=1–0.6 are shown in Table 3.3.

For the R_p and C_p, two parameters cannot be identified by simple equations, using the $t_2 \sim t_3$ section, where the RC loop of the Thevenin model is in a zero-state response. The expression of zero-state response can be obtained, as shown in Equation (3.84).

$$U_L(t) = U_{OC}(t) - I(t)R_0 - vI(t)R_p\left(1 - e^{-\frac{t}{\tau}}\right) \tag{3.84}$$

In Equation (3.84), U_L is the terminal voltage of the lithium-ion battery, U_{OC} is its OCV, and I is the current value at the time point. The parameter identification results of polarization resistance and polarization capacitance are shown in Table 3.4.

Through the above steps, the parameter identification of the curve-fitting method is completed for the Thevenin model, and the results are obtained for three parameters of R_0, R_p, and C_p. The method is an off-line identification method, which can effectively utilize experimental data [205]. The acquisition of model parameters is the basis for the SOC estimation of lithium-ion batteries.

TABLE 3.3
Resistance in Ohms

SOC(%)	$R_0(\Omega)$	SOC(%)	$R_0(\Omega)$
100	0.00848917	50	0.00744167
90	0.00789250	40	0.00748667
80	0.00766583	30	0.00757833
70	0.00748500	20	0.00764083
60	0.00737083	10	0.00867333

TABLE 3.4
Parameter Identification Results

SOC(%)	$R_p(\Omega)$	C_p(F)	SOC(%)	$R_p(\Omega)$	C_p(F)
100	0.002442	1528.634	50	0.002373	2438.982
90	0.002281	2004.831	40	0.002405	2289.561
80	0.002413	2018.122	30	0.002629	2456.258
70	0.002492	1965.556	20	0.003385	2150.565
60	0.002483	1996.233	10	0.005877	694.0455

3.4.2 Iterative Calculation Algorithm

The classic KF algorithm is often used to deal with linear random system problems, while the real systems are mostly nonlinear random systems [206]. The EKF algorithm is proposed to solve such problems. In the algorithm, Taylor expansion is performed on the nonlinear state equation and observation equation, ignoring the impact of higher-order terms on the system, and then performs calculations based on the classical KF algorithm. The state equation and observation equation of the nonlinear system are shown in Equation (3.85).

$$\begin{cases} x_{k+1} = f(x_k, u_k) + \Gamma_k w_k \\ z_k = g(x_k, u_k) + v_k \end{cases} \tag{3.85}$$

In Equation (3.85), the input variable is defined as $u_k \in R^r$, and uncorrelated noises are added to the two equations. Both $w_k \in R_p$ and $v_k \in R_q$ are Gaussian white noises, and statistical characteristics are described mathematically, as shown in Equation (3.86).

$$\begin{cases} E(w_k) = 0, \mathrm{Cov}(w_k, w_j) = Q_k \delta_{kj} \\ E(v_k) = 0, \mathrm{Cov}(v_k, v_j) = R_k \delta_{kj} \\ \mathrm{Cov}(w_k, v_j) = 0 \end{cases} \tag{3.86}$$

In Equation (3.86), the process n noise covariance is represented by Q_k. The observed noise covariance matrix is defined as R_k. The nonlinear state transition function is represented by $f(x_{k-1}, u_{k-1})$. The nonlinear measurement function is defined as $g(x_{k-1}, u_{k-1})$. At each time point, the functions $f(x_{k-1}, u_{k-1})$ and $g(x_{k-1}, u_{k-1})$ are expanded by the Taylor series, ignoring the influence of higher-order terms on the function, and two linearization processes are completed, that is, the nonlinear functions $f(x_{k-1}, u_{k-1})$ and $g(x_{k-1}, u_{k-1})$ expand Taylor series around the filter value and ignore the higher-order terms above the first order. The approximate linearization expressions of $f(x_{k-1}, u_{k-1})$ and $g(x_{k-1}, u_{k-1})$ are shown in Equation (3.87).

$$\begin{cases} f(x_k, u_k) \approx f(\hat{x}_k, u_k) + \left.\dfrac{\partial f(x_k, u_k)}{\partial x_k}\right|_{x_k = \hat{x}_k} (x_k - \hat{x}_k) \\ g(x_k, u_k) \approx g(\hat{x}_k, u_k) + \left.\dfrac{\partial g(x_k, u_k)}{\partial x_k}\right|_{x_k = \hat{x}_k} (x_k - \hat{x}_k) \end{cases} \tag{3.87}$$

$$\begin{cases} \hat{\phi}_k = \left.\dfrac{\partial f(x_k, u_k)}{\partial x_k}\right|_{x_k = \hat{x}_k} \\ \hat{H}_k = \left.\dfrac{\partial g(x_k, u_k)}{\partial x_k}\right|_{x_k = \hat{x}_k} \end{cases} \tag{3.88}$$

According to Equations (3.86)–(3.88), the nonlinear system is transformed into a linear system, and the transformed linear system is only related to state variables, as shown in Equation (3.89).

$$\begin{cases} x_{k+1} \approx \hat{\phi}_k x_k + \left[f(\hat{x}_k, u_k) - \hat{\phi}_k x_k \right] + \Gamma_k w_k \\ z_k \approx \hat{H}_k x_k + \left[g(\hat{x}_k, u_k) - \hat{H}_k x_k \right] + v_k \end{cases} \tag{3.89}$$

In Equation (3.89), note that $f(x_k, u_k) - \hat{\phi}_k x_k$ is not a function of x_k, and $g(\hat{x}_k, u_k) - \hat{H}_k x_k$ is not a function of x_k. According to the linear system obtained by Equation (3.89), take it into the classical KF algorithm.

Defining $\hat{\phi}_k = \left.\dfrac{\partial f(x_k, u_k)}{\partial x_k}\right|_{x_k = \hat{x}_k}$, $\hat{H}_k = \left.\dfrac{\partial g(x_k, u_k)}{\partial x_k}\right|_{x_k = \hat{x}_k}$, the following calculation procedure can be obtained as follows:

1. Initial conditions of the filter, as shown in Equation (3.90).

$$x_0 = E(x_0), P_0 = \text{var}(x_0) \tag{3.90}$$

2. State prior estimate, as shown in Equation (3.91).

$$\hat{x}_{k,k+1} = f(\hat{x}_{k-1}, u_{k-1}) \tag{3.91}$$

3. Prior estimate of error covariance, as shown in Equation (3.92).

$$P_{k,k+1} = \phi_{k,k+1} P_{k-1} \phi_{k,k+1}^T + \Gamma_{k,k-1} Q_{k-1} \Gamma_{k,k-1}^T \tag{3.92}$$

4. Gain matrix, as shown in Equation (3.93).

$$K_k = P_{k,k+1}H_k^T\left[H_k P_{k,k-1}H_k^T + R_k\right]^{-1} \tag{3.93}$$

5. State posterior estimate, as shown in Equation (3.94).

$$\hat{x}_k = \hat{x}_{k,k-1} + K_k\left[z_k - g\left(\hat{x}_{k-1}, u_k\right)\right] \tag{3.94}$$

6. The posterior estimate of the error covariance, as shown in Equation (3.95).

$$P_k = \left(I - K_k H_k\right)P_{k,k-1} \tag{3.95}$$

The main idea of the EKF algorithm is to convert the nonlinear system into a linear system by Taylor series expansion first and then calculate the linear system according to the process of the classical KF algorithm. It is an improvement of the classical KF, which makes it possible to solve the nonlinear system that the classical KF algorithm cannot solve. The premise of using EKF to estimate SOC is to establish a highly accurate battery model. The parameters of the Thevenin model have been identified and the establishment of the model has been completed. The observation of the EKF algorithm is established according to the state space of the Thevenin model, and the design flowchart of SOC estimation based on EKF is shown in Figure 3.11.

In Figure 3.11, the SOC value of the lithium-ion battery and the error covariance matrix P is initialized first, and then the terminal voltage U_L of the lithium-ion battery is detected by the BMS. It is judged whether the value of U_L reaches the cutoff condition, and the algorithm ends if it reaches the cutoff condition. If the cutoff condition is not reached, the parameter calculation is performed. The most important thing in the step is to calculate the three parameters R_0, R_p, and C_p. The third step is the prediction link, which predicts the three parameters of SOC, P, and U_L. The fourth step is to iterate the state transition matrix according to the OCV–SOC relationship. The fifth step is to calculate the gain matrix and modify the SOC to obtain the final predicted value. At this step, a cycle is completed. According to the above steps, the SOC value at each time point can be iterated continuously.

3.4.3 Experimental Analysis

To verify the accuracy and robustness of the SOC estimation result, the current and voltage data of the Beijing bus dynamic stress test (BBDST) operating conditions are used for testing. The current and voltage data of the working condition are input into the simulation system. And then, the parameter identification results are combined to estimate the SOC value. The verification result is obtained as shown in Figure 3.12.

In Figure 3.12, SOC_Ah is the SOC obtained by ampere–hour integration of the operating current, and the value is taken as the reference value of the experiment. The SOC_EKF in the figure is the SOC value estimated by the EKF algorithm. It

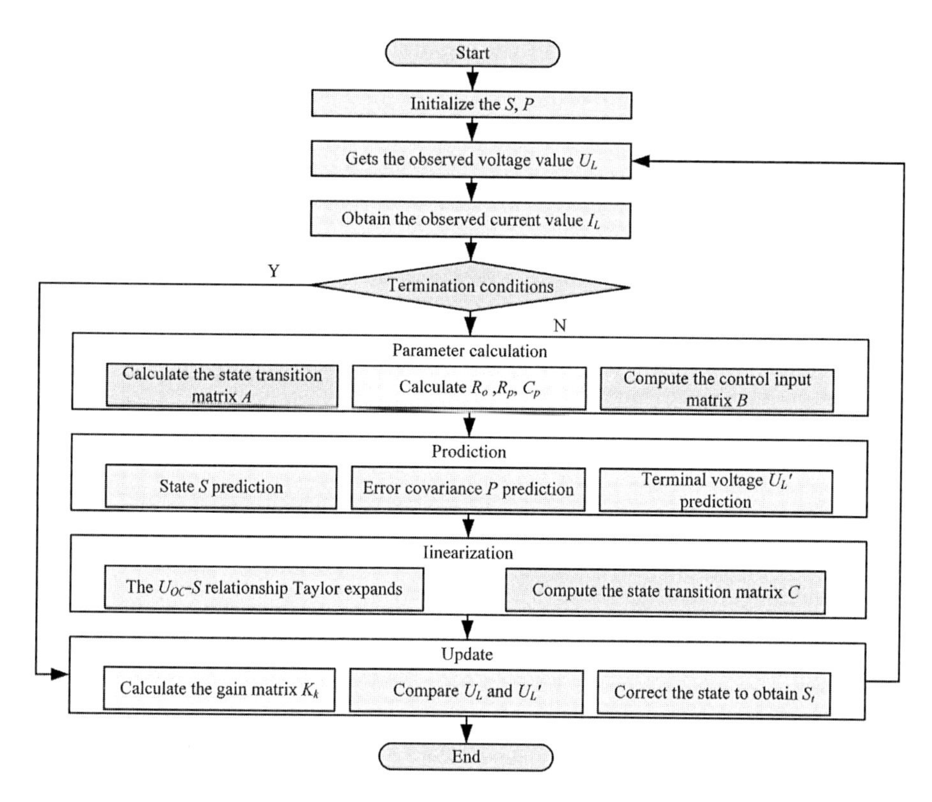

FIGURE 3.11 Flowchart design of SOC estimation based on EKF.

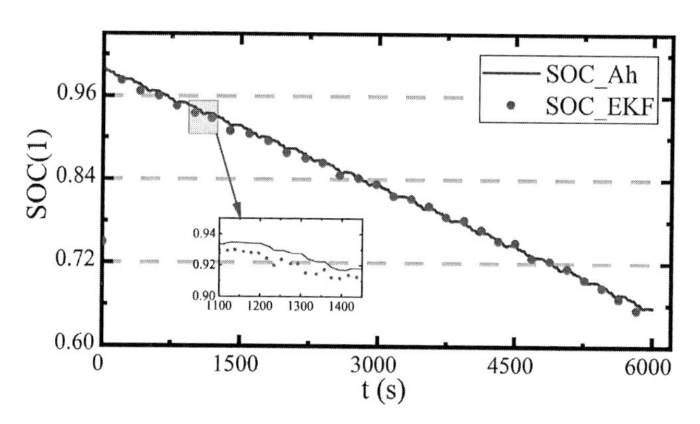

FIGURE 3.12 Real value and EKF-estimated value.

can be observed from the figure that the EKF algorithm can well follow the reference SOC estimation of the lithium-ion battery and has high accuracy. The error curve of using the EKF algorithm to estimate the SOC value is shown in Figure 3.13.

In Figure 3.13, it can be observed from the figure that in the early stage of the algorithm, there is a large error because the initial value of the lithium-ion battery SOC is not known, but as time accumulates, the algorithm continues to iterate to

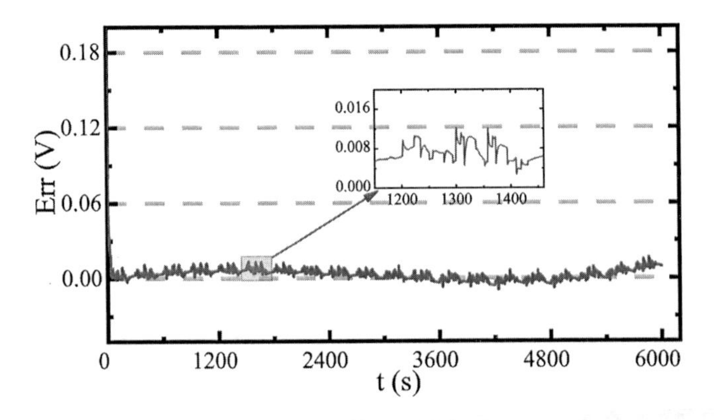

FIGURE 3.13 SOC estimation error.

the true value. The maximum error of the SOC estimation result based on the EKF algorithm is less than 2%.

3.4.4 Thevenin Model-Based Estimation

1. Equivalent modeling and parameter identification
   Thevenin model considers the self-discharge effect, so it has high accuracy [207]. It can overcome the error of the polarization effect. The steps are short and the principle is clear. It is very suitable for the transient analysis of power battery charge–discharge, so the Thevenin model is selected, as shown in Figure 3.14.

   The method of parameter identification is to carry out the HPPC experiment first and then analyze the results of the HPPC experiment. Starting from SOC = 1, the battery is discharged at a constant current every 6 minutes, and the battery is allowed to rest for 40 minutes. The voltage and current curves of the constant current discharge experiment are shown in Figure 3.15.

   The parameter identification is carried out by the method of curve fitting. The selected curve segment is the zero-state response curve. The results of ohm internal resistance, polarization resistance, and polarization capacitance are shown in Table 3.5.

   After getting the mathematical description of the model parameters, it is necessary to verify the accuracy of the model parameters. The identified parameters are introduced into the Thevenin equivalent circuit model, the same current as the HPPC experimental test is input, and the output voltage response data of the model is compared with the actual voltage data, verifying the model, and optimizing and improving the model according to the verification results. The results are shown in Figure 3.16.

   It can be observed from Figure 3.16 that the output voltage of the model is in good agreement with the actual value, which shows the rationality of the Thevenin equivalent circuit model and the feasibility and reliability

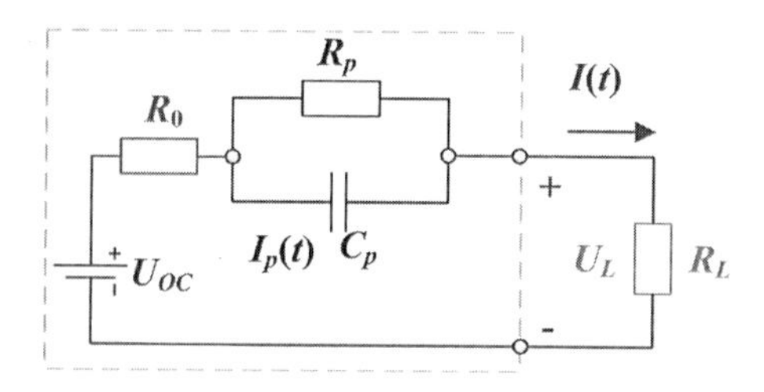

FIGURE 3.14 Thevenin equivalent circuit model.

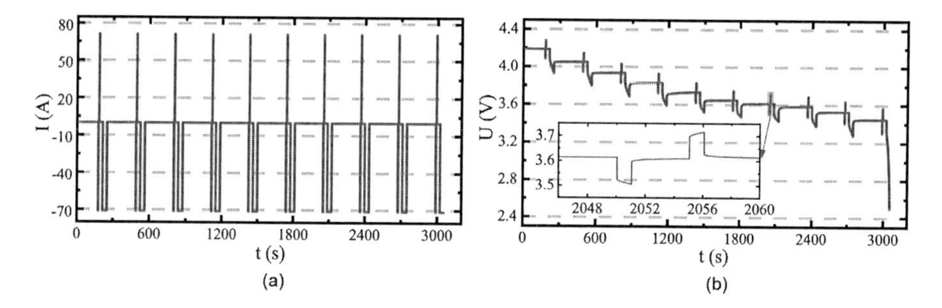

FIGURE 3.15 Current and voltage curves of HPPC condition. (a) Current curve of HPPC condition. (b) Voltage curve of HPPC condition.

TABLE 3.5
Parameter Identification Results

SOC	$R_0(\Omega)$	$R_p(\Omega)$	C_p(F)
1	0.058452	0.0015460	2636.1597
0.9	0.056504	0.0016020	2456.2946
0.8	0.054826	0.0015150	2518.3717
0.7	0.053346	0.0014600	2540.3209
0.6	0.051950	0.0014960	2413.1009
0.5	0.050962	0.0017230	2051.8802
0.4	0.050454	0.0015040	2325.7672
0.3	0.050034	0.0015054	2302.3392
0.2	0.049195	0.0016572	2053.4227
0.1	0.047883	0.0022139	1490.1568

of the parameter identification method. It can be observed from the error curve that there is no divergence phenomenon in the whole process of the model, and the larger error occurs in the dynamic pulse test stage. The reason is that the sudden change of battery input current leads to the concentration of chemical reactions in the battery, resulting from the rapid change of terminal voltage. At this time, the Thevenin model cannot perfectly show the response effect of the lithium-ion battery, but the error fluctuates in an acceptable and reasonable range. Even in the last stage of low SOC, when

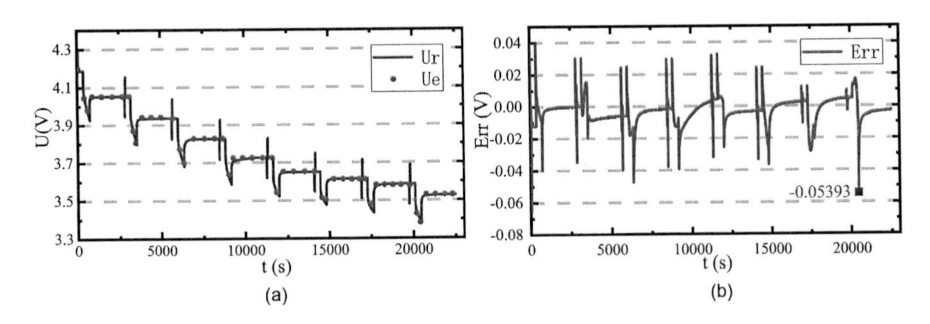

FIGURE 3.16 Comparison of terminal voltage output curves. (a) Output comparison. (b) Output error.

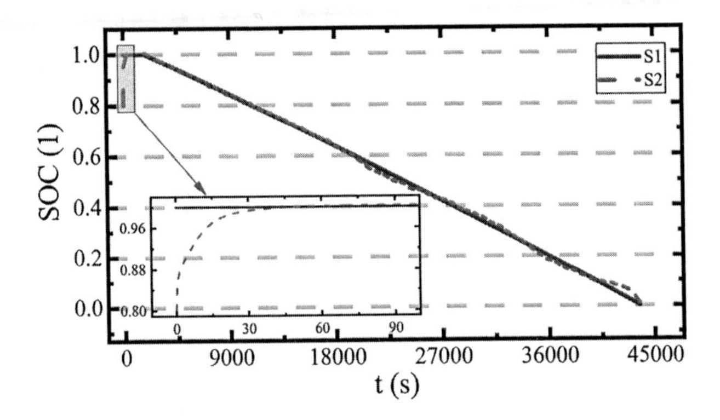

FIGURE 3.17 Comparison of SOC curves.

the model cannot well reflect the current state of the battery, the maximum error is not more than 0.03458 V.

2. Experimental analysis

   To verify the accuracy and robustness of the SOC estimation results, the current and voltage data of the BBDST operating condition are used for testing. The current and voltage data of the working condition is input into the simulation system. And then, the parameter identification results are combined to estimate the SOC value. The verification result is obtained, as shown in Figure 3.17.

   Figure 3.17 is used for EKF to estimate the SOC value, in which the black curve is the real SOC, and the red curve is the SOC curve of the EKF estimation. It can be observed from the figure that the EKF algorithm can well follow the true SOC of the lithium-ion battery and has high accuracy.

3.4.5 Improved PNGV-Based Estimation

1. Equivalent modeling and parameter identification

   To make the curve fitted by the model match the real voltage curve better, a group of RC circuits is added to the improved PNGV model, which can

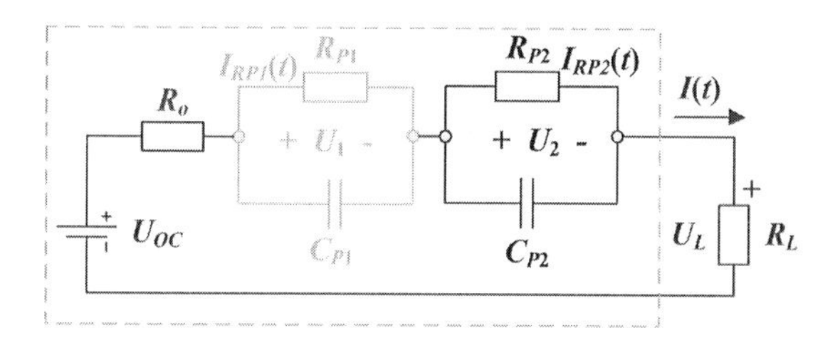

FIGURE 3.18 Lithium-ion battery-improved PNGV equivalent model.

achieve a better matching degree in the curve-fitting process [208,209]. The improved method of the PNGV model is shown in Figure 3.18.

The battery terminal voltage is shown in Equation (3.96).

$$U_L = U_{OC} - C_b\left(\int i(t)\,dt\right) - R_0 I_L - I_{p1}R_{p1} - I_{p2}R_{p2} \tag{3.96}$$

To more comprehensively express the external characteristics of the battery, we can continue to optimize based on the above typical battery model, such as the self-discharge factors, and hysteresis characteristics, expressed in the form of circuit, but this inevitably increases the complexity of the model and the difficulty of identification [210]. In the specific application of model selection and optimization, the typical model for parameter identification can be used firstly, and then make targeted improvements according to the difference between the model and the actual. The sudden change of terminal voltage at the beginning and stop of battery discharge is caused by ohmic resistance, so the ohmic resistance can be calculated by Ohm's law, as shown in Equation (3.97).

$$R_0 = \frac{(U_1 - U_2) + (U_4 - U_3)}{2I} \tag{3.97}$$

In Equation (3.96), C_b is the change of OCV caused by the accumulation of power supply current, and the calculation process is shown in Equation (3.98).

$$C_b = \frac{(t_2 - t_1)I}{U_1 - U_5} \tag{3.98}$$

During the battery discharge, the polarization capacitors C_{p1} and C_{p2} are used in the SOC estimation process, and the voltage of the RC parallel circuit increases exponentially. When the battery is put into the static state from the discharge state, the capacitors C_{p1} and C_{p2} discharge to their parallel resistors respectively, and the voltage decreases exponentially. The values of R and C in the model are related to the SOC of the battery. The data obtained from the HPPC experiment are fitted by using the simulation software, and then the values of R_{p1}, R_{p2}, C_{p1}, and C_{p2} in the improved PNGV model can be obtained by the method of undetermined parameters. The specific method is described as follows:

Firstly, the battery terminal voltage at different SOC values at the same stage as the shelving stage after the first discharge is extracted from the HPPC experimental data. The change of terminal voltage in this stage is the discharge process of polarization capacitance to polarization resistance and is the zero-input response of the double RC loop. In this stage, I_L is 0. Therefore, in this process, the output equation of battery terminal voltage is obtained as shown in Equation (3.99).

$$U_L = U_{OC} - IR_{P1}e^{-\frac{t}{\tau_1}} - IR_{P2}e^{-\frac{t}{\tau_2}} \tag{3.99}$$

To simplify the parameter identification, Equation (3.99) is replaced by the coefficient, as shown in Equation (3.100).

$$U_L = f - ae^{-ct} - be^{-dt} \tag{3.100}$$

In Equation (3.100), f, a, b, c, and d are taken as the undetermined parameters, and the experimental data are used, with the (3.100) single and double index curves fittings for the target equation.

3.5 UKF-BASED SOC ESTIMATION

3.5.1 Equivalent Circuit Modeling

To better reflect the dynamic and static characteristics of the battery, it is proposed to use multiple RC parallel circuits connected in series to imitate the polarization reduction process of lithium-ion batteries. Based on the Thevenin circuit model, multiple RC parallel circuits are connected in series to form a multistage Thevenin circuit model [211]. Relevant scholars identified the internal parameters of the Thevenin equivalent circuit model of different orders, combined with the experimental data to verify the accuracy of the battery model, and concluded that the second-order Thevenin circuit model can reflect the dynamic response changes between the two poles of the batteries. During the process, the dynamic performance characteristics of the battery model have strong clarity. The structure diagram of the second-order Thevenin equivalent circuit model of the batteries can be designed, as shown in Figure 3.19.

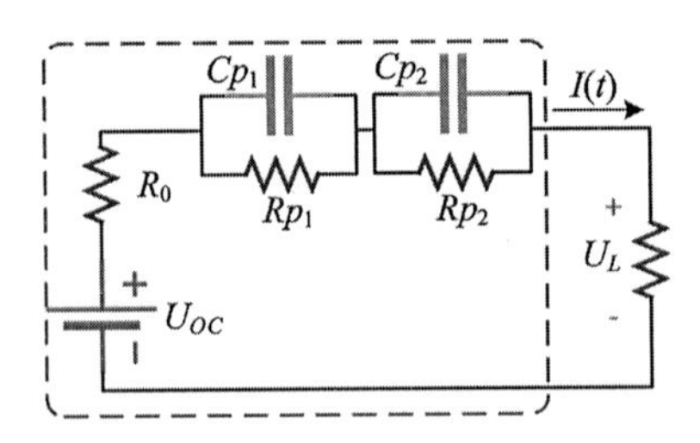

FIGURE 3.19 Second-order Thevenin equivalent circuit model.

In Figure 3.19, U_{OC} represents the OCV of the lithium-ion battery, U_L represents the output terminal voltage of the battery, and R_0 represents the ohmic resistance. The RC loop can accurately describe the dynamic characteristics of the battery and characterize the polarization effect of the working process of the lithium-ion batteries. The $R_{p1}C_{p1}$ loop refers to the electrochemical polarization effect of the positive and negative poles inside the battery, and the $R_{p2}C_{p2}$ loop reflects the concentration polarization of different materials in the battery under the electrolyte. According to Kirchhoff's voltage law, the voltage is written for the improved second-order Thevenin model, as shown in Equation (3.101).

$$\begin{cases} U_t = U_{OC} - I_t R_0 - U_{p1,t} - U_{p2,t} \\ I_t = \dfrac{U_{p1,t}}{R_1} + C_1 \dfrac{dU_{p1,t}}{dt} \\ I_t = \dfrac{U_{p2,t}}{R_1} + C_2 \dfrac{dU_{p2,t}}{dt} \end{cases} \tag{3.101}$$

In Equation (3.101), the discharge direction is taken as the reference direction (positive direction). The battery SOC can be obtained by the ampere-hour integration method, as shown in Equation (3.102).

$$\text{SOC}_t = \text{SOC}_{t_0} - \frac{\eta}{Q_N} \int_{t_0}^{t} I_t \, dt \tag{3.102}$$

In Equation (3.101), η is the coulomb efficiency coefficient. Q_N is the battery calibrated capacity. According to the definition of SOC and the structure of the second-order Thevenin model, the state–space equation of the battery is established, as shown in Equation (3.103).

$$\begin{cases} \begin{bmatrix} \text{SOC}_{k+1} \\ U_{p1,k+1} \\ U_{p2,k+1} \end{bmatrix} = \begin{bmatrix} 1 & 0 & 0 \\ 0 & e^{\frac{-T}{\tau_1}} & 0 \\ 0 & 0 & e^{\frac{-T}{\tau_2}} \end{bmatrix} \begin{bmatrix} \text{SOC}_k \\ U_{p1,k} \\ U_{p2,k} \end{bmatrix} + \begin{bmatrix} \dfrac{-\eta T_N}{Q_N} \\ R_{p1}\left(1 - e^{\frac{-T}{\tau_1}}\right) \\ R_{p2}\left(1 - e^{\frac{-T}{\tau_2}}\right) \end{bmatrix} I_k + \begin{bmatrix} w_{1,k} \\ w_{2,k} \\ w_{3,k} \end{bmatrix} \\ U_{L,k+1} = U_{OC}\left(\text{SOC}_{k+1}\right) - R_0 I_k - U_{p1,k} - U_{p2,k} + v_k \end{cases} \tag{3.103}$$

In Equation (3.103), $\tau_1 = R_{p1}C_{p1}$, $\tau_2 = R_{p2}C_{p2}$, according to which the state vector, observation vector, input vector, and system noise are renamed. Therefore, the mathematical relationship can be obtained, as shown in Equation (3.104).

$$\begin{cases} x_k = \left[\text{SOC}_k ; U_{p1,k} ; U_{p2,k} \right] \\ y_k = U_{L,k}, u_k = I_k \\ w_k = \left[w_{1,k} ; w_{2,k} ; w_{3,k} \right] \end{cases} \tag{3.104}$$

In Equation (3.104), the parameters of the second-order Thevenin model are identified by HPPC experimental data shown in Figure 3.10. It can be observed from the experimental curve that the abrupt changes in voltage at the beginning and end of the discharge are all caused by the ohmic resistance R_0. The value of R_0 can be identified by the point-taking calculation method, as shown in Equation (3.105).

$$R_0 = \frac{(U_A - U_B) + (U_D - U_C)}{2I} \tag{3.105}$$

In Equation (3.105), the terminal voltage of the lithium-ion battery drops slowly from U_B to U_C, which is due to the battery polarization effect. The discharge current charges the polarized capacitor, and the double RC series circuit is in a zero-state response. The circuit is analyzed in the time domain, and select the fitting curve BC section to obtain the terminal voltage U_L as a function of time point t, as shown in Equation (3.106).

$$U_L = U_{OC} - IR_0 - IR_{p1}\left(1 - e^{-\frac{t}{\tau_1}}\right) - IR_{p2}\left(1 - e^{-\frac{t}{\tau_2}}\right) \tag{3.106}$$

In Equation (3.106), the OCV U_{OC} is the suitable voltage at the positive and negative ends of the battery when the battery is left for a long time. Experiments show that the voltage of the battery after 40 minutes is suitable, which can be considered equal to the OCV of the batteries. Therefore, U_A can be used as the corresponding OCV at the SOC value, which corresponds to the OCV U_{OC} value at the SOC value at that temperature.

3.5.2 Unscented Transformation

The UT is a solution to nonlinear problems and an important part of the UKF algorithm. Its basic idea is to select a limited number of sampling points according to the statistical characteristics of state variables and a certain sampling method. These points need to meet the condition that their mean value and covariance are the same as the original state mean value and covariance, and then these points are substituted into the nonlinear function to obtain the corresponding nonlinear function value point set, and then obtain the mean and covariance of the transformed point set. Therefore, the key to the UT lies in the acquisition of sampling points and the determination of their corresponding weights. The commonly used sampling strategy is symmetric sampling. The dimension of the state variable x is n, P is the mean and covariance matrix respectively, and y is the observation variable. $2n+1$ sampling points are obtained in the nonlinear system, as shown in Equation (3.107).

$$\begin{cases} x^i = \hat{x}, i = 0 \\ x^i = \hat{x} + \left(\sqrt{(n+\lambda)P}\right)_i, i = 1,\ldots, \\ x^i = \hat{x} - \left(\sqrt{(n+\lambda)P}\right)_{i-n}, i = n+1,\ldots,2n \end{cases} \tag{3.107}$$

In Equation (3.107), the corresponding weight is shown in Equation (3.108).

$$\begin{cases} \omega_m^0 = \dfrac{\lambda}{n+\lambda} \\ \omega_c^0 = \dfrac{\lambda}{n+\lambda} + \left(1-\alpha^2+\beta\right) \\ \omega_m^i = \omega_c^i = \dfrac{\lambda}{2(n+\lambda)}, i = 1,\ldots,2n \end{cases} \tag{3.108}$$

In Equation (3.108), $\beta \geq 0$, $0.2 \leq \alpha \leq 1$, and λ all are scaling parameters. The above parameters generally satisfy the following relationship: $k = 3-n$, $\lambda = \alpha^2(n+k)-n$. The sampling points are transferred nonlinearly, and the function relationship obtained is shown in Equation (3.109).

$$y^i = f\left(x^i\right), i = 0 \sim 2n \tag{3.109}$$

In Equation (3.109), the mean value and covariance P_y are calculated after y transformation, as shown in Equation (3.110).

$$\begin{cases} \hat{y} = \sum_{i=0}^{2n} \omega_i f\left(x^i\right) \\ P_y = \sum_{i=0}^{2n} \omega_i \left[f\left(x^i\right) - \hat{y}\right]\left[f\left(x^i\right) - \hat{y}\right]^T \end{cases} \tag{3.110}$$

3.5.3 UKF Algorithm Analysis

The processing object of UKF is a type of nonlinear system. To facilitate the processing, the system is generally divided into two parts, namely the system state and the system observation part. To study the algorithm without losing generality, setting the nonlinear system to be processed is shown in Equation (3.111).

$$\begin{cases} x_{k+1} = f(x_k, u_k) + w_k \\ z_k = h(x_k, u_k) + v_k \end{cases} \tag{3.111}$$

In Equation (3.111), k is the time point, $x_k \in R^n$ is the state-space variable, and $z_k \in R^n$ is the observation space variable. Zero-mean Gaussian white noise w_k and v_k are state transition noise and observation noise, respectively, w_k and v_k are independent of each other, and the variances of w_k and v_k are Q_k and R_k, respectively. $f(*)$ is a nonlinear function representing the law of state transition and $h(*)$ is a nonlinear function representing the system state and observation. u_k is the system input at the time point k. The UT is the core part of the UKF algorithm, which constructs sampling points of the current state variable in a statistical sense and a limited number of sampling points as well as the corresponding weights. The mean and variance of these sampling points are equal to or only exist in the current state variable at the current time point. Acceptable limited tolerance, the collection of these sampling points is called the σ point set. For the nonlinear system given by Equation (3.111), the UT is performed on it.

The $2n+1$ sampling points of the state variable at the current time point are calculated by the method of distributed sampling, where n is the number of dimensions of the state-space x. The mean and variance of the point are set correspondingly equal to the state variable and state covariance at the current time point. The mean value of x for the sampling point is determined by Equation (3.112).

$$\begin{cases} x_0 = \bar{x} \\ x_i = \bar{x} + \left(\sqrt{(n+P)}\right)_i, i = 1, \ldots, n \\ x_i = \bar{x} - \left(\sqrt{(n+P)}\right)_i, i = n+1, \ldots, 2n \end{cases} \tag{3.112}$$

In Equation (3.112), x_i represents the ith sampling point, and $\left(\sqrt{(n+P)}\right)_i$ represents the i-th column of the matrix $\sqrt{(n+P)}$. Corresponding to each x_i, when calculating the mean value, its weight is given by Equation (3.113).

$$\begin{cases} \omega_0^m = \dfrac{\lambda}{n+\lambda} \\ \omega_0^c = \dfrac{\lambda}{n+\lambda} + 1 - \alpha^2 + \beta, i = 1, \ldots, 2n \\ \omega_i^m = \omega_i^c = \dfrac{1}{2(n+\lambda)} \\ \lambda = \alpha^2(n+\gamma) - n \end{cases} \tag{3.113}$$

In Equation (3.113), ω_i^m is the weight of the ith sampling point in calculating the mean value, and ω_i^c is the weight of the ith sampling point in calculating the covariance. n is the number of dimensions of the state variable x. The factor λ is the distribution state of the sampling points around its mean value. α controls the distance between the sampling points and the mean, which is generally a constant from $\alpha \in (10^{-6}, 1)$. The mathematical relationship is required to meet $\gamma + n \neq 0$. As $n \neq 0$, it can be generally chosen for calculation with the pre-distribution factor. The system given by Equation (3.113) has been assumed to obey the Gaussian distribution, so it is the optimal value. The parameter γ indicates the zoom ratio. The calculation process of UKF is to complete the specific and detailed implementation steps of core content from UT, before estimation-to-posterior estimation correction. The realization process can be subdivided into seven steps; the first step is the traceless transformation and the last step is the state update and covariance update.

1. No trace conversion
   The first step in the realization of the UKF algorithm is to perform a UT of the system state variables at the previous time point. For the system shown in Equation (3.111), using Equation (3.112) to calculate the sampling point of its state at the previous time point, Equation (3.113) is realized to calculate the weight corresponding to each sampling point in the untracked transformation process to complete the traceless conversion of the system.
2. One-step prediction of sigma point
   According to the sampling points of the state-space at the previous time obtained by the first step of the UT, the sampling point σ of the state-space at the next time is predicted, which is called the one-step prediction of the σ point. The one-step prediction of the point σ is given by Equation (3.113), which is given by Equation (3.114).

$$x_{i,k+1|k} = f\left(x_{i,k|k}, u_k\right) \tag{3.114}$$

3. Prediction of state variable and covariance
   The one-step prediction of σ point given by Equation (3.112) is multiplied by the corresponding weight given by Equation (3.113) to obtain the predicted value of x at the next time, as shown in Equation (3.114). Then, the covariance is calculated from the predicted value of x and the weight of the covariance. The calculation equation is given by Equations (3.115) and (3.116).

$$\hat{x}_{k+1|k} = \sum_{i=0}^{2n} \omega_i^m x_{i,k+1|k} \tag{3.115}$$

$$P_{k+1|k} = Q_k + \sum_{i=0}^{2n} \omega_i^c \left[\hat{x}_{k+1|k} - x_{i,k+1|k}\right] \times \left[\hat{x}_{k+1|k} - x_{i,k+1|k}\right]^T \tag{3.116}$$

4. One-step prediction of observations
   The KF algorithm not only needs to predict the state and covariance at $k+1$ from the state at the time point k but also needs to predict the state of the observation space at the time point $k+1$. Equations (3.115) and (3.116) have predicted the state and covariance at the time point $k+1$. In the UKF algorithm, to predict the observation space at the time point $k+1$, it is necessary to calculate a one-step prediction of the observation at the time point $k+1$ from the state at the time point k. The one-step prediction of the observation equation is shown in Equation (3.117).

$$z_{i,k+1|k} = h\left(x_{i,k+1,k}, u_k\right) \tag{3.117}$$

5. Prediction of system observation matrix and covariance matrix
   Observations obtained by nonlinear transmission based on sampling points given by Equation (3.117) are averaged to obtain the predicted amount of the observation matrix of the system and then respectively calculate the variance of the observed vector at $k+1$ given by the predicted value and the covariance between it and the state variable. Equation (3.118) gives the calculation method of the observation vector of the prediction system, and the variance of the observation vector is given by Equation (3.119). Finally, the covariance between the observation vector and the state vector is shown in Equation (3.120).

$$\overline{z}_{k+1|k} = \sum_{i=0}^{2n} \omega_i^m z_{i,k+1|k} \tag{3.118}$$

$$P_{z_k z_k} = R_k + \sum_{i=0}^{2n} \omega_i^c \left[z_{i,k+1|k} - \overline{z}_{k+1|k}\right]\left[z_{i,k+1|k} - \overline{z}_{k+1|k}\right]^T \tag{3.119}$$

$$P_{x_k z_k} = \sum_{i=0}^{2n} \omega_i^c \left[x_{i,k+1|k} - \overline{z}_{k+1|k}\right]\left[z_{i,k+1|k} - \overline{z}_{k+1|k}\right]^T \tag{3.120}$$

6. Kalman gain calculation
   Equations (3.119) and (3.120) give the calculation method for estimating the state and observation at $k+1$ from the information at time point k. The next step is to perform a posterior estimation based on the data at time point $k+1$, that is, to modify the prior estimation to obtain a more accurate estimation. Before the correction process starts, calculating the Kalman gain is a necessary step. The Kalman gain calculation equation is shown in Equation (3.121).

$$K_{k+1} = P_{x_k z_k} P_{z_k z_k}^{-1} \tag{3.121}$$

7. Calculating status update and covariance update
   The previous 6 steps are prepared for the status update of step 7. The covariance update is prepared for the estimation at the time point $k+2$. The status update is shown in Equation (3.122). The covariance update is shown in Equation (3.123).

$$\hat{x}_{k+1|k+1} = \hat{x}_{k+1|k} + K_{k+1}\left[z_{k+1} - \hat{z}_{k+1|k}\right] \tag{3.122}$$

$$P_{k+1|k+1} = P_{k+1|k} - K_{k+1}P_{z_k z_k}K_{k+1}^T \tag{3.123}$$

   In Equation (3.123), z_{k+1} is the observation value measured by the instrument at the time point $k+1$; $\hat{z}_{k+1|k}$ is the observation obtained from the state estimation at time point k; $\hat{x}_{k+1|k+1}$ is the posteriori state estimates at the time point $k+1$, The determination $\hat{x}_{k+1|k+1}$ is the optimal state estimation at the time point $k+1$.

Steps (1) to (7) give the realization process of the UKF algorithm to estimate the system state given by Equation (3.123). With the above method, the optimal estimation based on the criterion of minimum variance can be obtained, and the estimation accuracy is better than that of the EKF algorithm processed by linearization.

3.5.4 Unscented Kalman Filtering Steps

Aiming at the selected lithium-ion battery pack for electric vehicles, using the above-mentioned design method from Equations (3.122) to (3.123), an SOC estimation procedure based on the UKF algorithm is designed. The process diagram of the estimation method is shown in Figure 3.20.

Figure 3.20 shows the SOC estimation process based on the UKF algorithm. This process is a typical UKF from Equations (3.122) to (3.123) derived from the algorithmic process. At the beginning of the estimation, the initial value SOC_0 of the SOC is given by the dotted line in the figure. The principle of the UKF algorithm knows that the value is given arbitrarily. Whether it truly reflects the initial SOC value of the lithium-ion battery pack is not important [212]. Then, it enters the cyclic process of the UKF algorithm. First, the state value SOC_k at time point k is σ converted to obtain the sampling point $SOC_{i,k}$ that is at time point k, and then the sampling point is used to predict the SOC sampling point $SOC_{i,k+1|k}$ at the time point $k+1$.

According to $SOC_{i,k+1|k}$, the SOC value and its variance at the time point of $k+1$ can be predicted. Then, the estimated value and estimated variance are used to predict the observation vector and covariance matrix $SOC_{k+1|k}$ and P_{k+1} at the time point $k+1$. Then, the Kalman gain K is calculated. The state vector and its variance are updated according to the difference between the estimated value of the observation vector and the true value. Finally, the true value SOC_{k+1} of the state quantity at the time point $k+1$ is obtained, and the value is used as the initial SOC value of the next estimation cycle to promote the work of the next estimation cycle [213]. From Figure 3.20, it can be observed that the SOC estimation process based on UKF forms

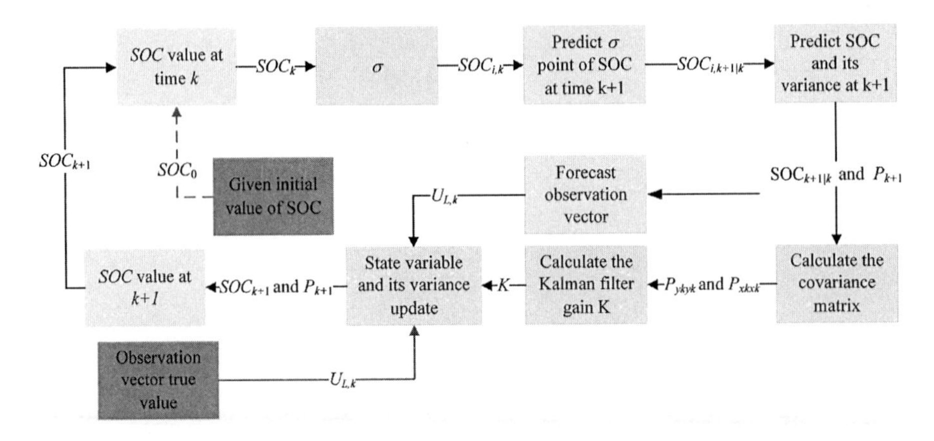

FIGURE 3.20 SOC estimation process diagram based on UKF.

a closed-loop system by itself. Once the initial value of the SOC is given and the estimation process is entered, the system can continuously iteratively estimate to achieve a continuous and real-time SOC estimation of the lithium-ion battery pack for electric vehicles.

3.5.5 Experimental Verification

1. Model validation

   Based on the previous research, through actual testing of various performance parameters of the lithium-ion battery, a multinomial fitting equation for each parameter of the second-order model is obtained. The terminal voltage and current obtained from the experiment and the identified parameters are used as the input of the second-order Thevenin equivalent circuit model, and the simulated voltage output by the model is compared with the simulated experimental voltage to verify the accuracy of the parameters of the equivalent model. The second-order Thevenin model simulation is shown in Figure 3.21.

   In Figure 3.21, the U_e is the actual output terminal voltage of the model output, and U_r is the terminal voltage of the selected lithium-ion batteries. It can be observed that the model output terminal voltage is in good agreement with the actual terminal voltage, which verifies the second-order Thevenin reasonableness and accuracy of the equivalent circuit model and also shows that the selected parameter identification method has certain feasibility and reliability. The high accuracy of the equivalent model can be extracted from the error curve. The verification deviation is toggled within the effective accuracy range, and the maximum error is 3.237% which verifies the feasible value of the model.

2. Algorithm verification

   To verify the applicability and stability of the algorithm, different working conditions are used to verify the estimation accuracy of the algorithm. The estimation results are shown in Figure 3.22.

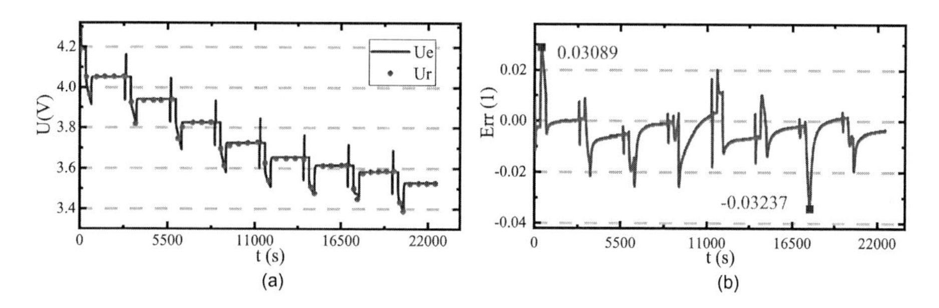

FIGURE 3.21 Model verification output curve. (a) Voltage comparison curve. (b) Estimation error.

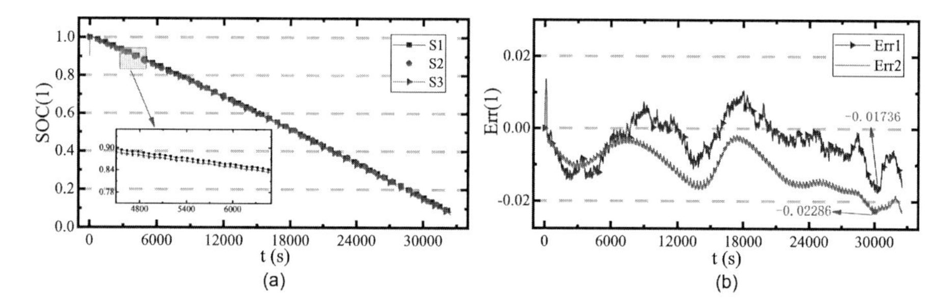

FIGURE 3.22 BBDST working condition simulation. (a) SOC diagram for BBDST conditions. (b) Error curve.

In Figure 3.22, the simulation effect is obtained under BBDST working conditions. In the figure, S1 is the reference value of SOC, S2 is the SOC estimation curve using the UKF algorithm, and S3 is the SOC estimation curve using the ampere-hour integration method. Err1 is the SOC estimation error under the UKF algorithm, and Err2 is the estimation error under the ampere-hour integration method. It can be observed from Figure 3.22a that under complex working conditions, the UKF algorithms can track the changes in the real value very well. During operation, the absolute value of the error is always less than 1%. From Figure 3.22b, it can be concluded that the UKF algorithms have strong self-correction capabilities, and the output waveform converges to the actual modulus curve within a finite sampling period, which verifies that the extension UKF algorithms are at the initial value. Under the wrong state, it can adjust adaptively, and the accuracy of SOC estimation does not decrease.

3.6 AEKF-BASED SOC ESTIMATION

3.6.1 Equivalent Circuit Modeling

1. Second-order RC equivalent circuit model

   At present, there are many types of lithium-ion battery equivalent circuit models. The second-order RC model is one of the commonly used ones,

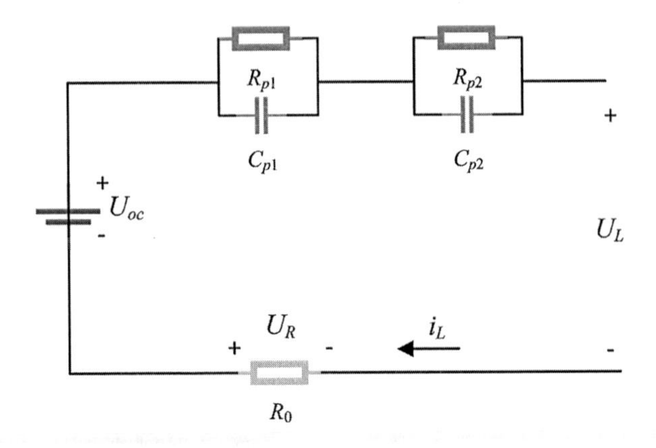

FIGURE 3.23 Second-order RC equivalent circuit model.

which can well reflect the dynamic and static characteristics of the battery, with moderate calculations and fewer identification parameters [214,215]. Therefore, the second-order RC model is used as the battery equivalent circuit model, as shown in Figure 3.23.

In Figure 3.23, U_{OC} represents the OCV of the circuit; U_L represents the terminal voltage. R_0 is the ohmic resistance and U_R is the ohmic voltage, which characterizes the battery voltage drop effect at the time point the battery is discharged and the end of the discharge. i_L is the battery charge-discharge current, and the discharge direction is set to positive. The RC loop is composed of R_{p1} and C_{p2} characterizes the internal chemical reaction of the battery, the stage where the voltage changes rapidly during the process. According to Kirchhoff's circuit law, the equivalent model circuit can be listed, as shown in Equation (3.124).

$$U_L = U_{OC} - U_R - U_{p1} - U_{p2} \tag{3.124}$$

The RC loop is composed of R_{p2} and C_{p2} represents the stage where the voltage changes slowly during the chemical reaction within the batteries, as shown in Equation (3.125).

$$\begin{cases} \dfrac{dU_{p1}}{dt} = -\dfrac{U_{p1}}{R_{p1}C_{p1}} + \dfrac{i}{C_{p1}} \\ \dfrac{dU_{p2}}{dt} = -\dfrac{U_{p2}}{R_{p2}C_{p2}} + \dfrac{i}{C_{p2}} \end{cases} \tag{3.125}$$

The equation for the SOC definition is obtained, as shown in Equation (3.126).

$$\text{SOC}_t = \text{SOC}_{t0} - \frac{\int_{t0}^{t} I(t)\eta\, dt}{Q_0} \tag{3.126}$$

Combined with the SOC definition, select state–space variables, input variables, and output variables, as shown in Equation (3.127).

$$\begin{cases} x_k = \left[\mathrm{SOC}_k, U_{p1,k}, U_{p2,k}\right]^T \\ U_k = \left[I_k\right], y_k = \left[U_{L,k}\right] \end{cases} \tag{3.127}$$

The discrete state-space expression can be obtained, as shown in Equations (3.128) and (3.129).

$$\begin{cases} \begin{bmatrix} \mathrm{SOC}_{k+1} \\ U_{p1,k+1} \\ U_{p2,k+1} \end{bmatrix} = \begin{bmatrix} 1 & 0 & 0 \\ 0 & e^{-\frac{\Delta t}{\tau_1}} & 0 \\ 0 & 0 & e^{-\frac{\Delta t}{\tau_2}} \end{bmatrix} \begin{bmatrix} \mathrm{SOC}_k \\ U_{p1,k} \\ U_{p2,k} \end{bmatrix} \\ \quad + \begin{bmatrix} -\frac{\Delta t}{Q_N} \\ R_1\left(1 - e^{-\frac{T}{\tau_1}}\right) \\ R_2\left(1 - e^{-\frac{T}{\tau_2}}\right) \end{bmatrix} I_k + \begin{bmatrix} w_{1,k} \\ w_{2,k} \\ w_{3,k} \end{bmatrix} \\ \tau_1 = R_1 C_1,\ \tau_2 = R_2 C_2 \end{cases} \tag{3.128}$$

$$U_{L,k} = U_{\mathrm{OC},k} - R_{0,k} I_k + \begin{bmatrix} 0 \\ -1 \\ -1 \end{bmatrix}^T \begin{bmatrix} \mathrm{SOC}_k \\ U_{p1,k} \\ U_{p2,k} \end{bmatrix} + v_k \tag{3.129}$$

In Equations (3.128) and (3.129), Δt is the sampling time interval, w is the state error, and v is the measurement error.

2. HPPC experiment and parameter identification

The test platform is described as follows: (1) The experimental lithium-ion battery is a ternary lithium iron phosphate battery with a nominal capacity of 72 Ah and an actual calibrated capacity of 69.23 Ah. (2) The battery test system (NEWAREBTS-4000) is used to test the charge-discharge of the tested lithium-ion batteries. The equipment can record the voltage, current, capacity, temperature, and other data during the experiment and the sampling interval is 0.1 seconds. (3) A constant temperature (25°C) environment incubator is provided for the test batteries.

HPPC is performed on the battery as an identification experiment. The battery model parameters are identified by analyzing the working characteristics of the battery during the experiment. The parameters to be identified are U_{OC}, R_0, R_{p1}, C_{p1}, R_{p2}, and C_{p2}. The procedure for the HPPC experiment is described below. (1) Charge to maximum capacity with constant voltage and constant current and set it aside completely. (2) Discharge with 1 C (69.23 A) current pulse for 10 seconds; after discharge, the battery is shelved for 40 seconds. (3) Use 1 C current pulse to charge for 10 seconds and rest for 40 seconds after the end of the charge. (4) Discharge at a constant current of 1 C for 6 minutes, so that the SOC value drops by 0.1, and rest for 30 minutes after the discharge ends. (5) Cycle the test steps (2)–(4) until the SOC value drops to 0.1. The HPPC experiment current curve and voltage response curve are shown in Figure 3.24.

Single HPPC voltage and current change curve can be obtained, as shown in Figure 3.25.

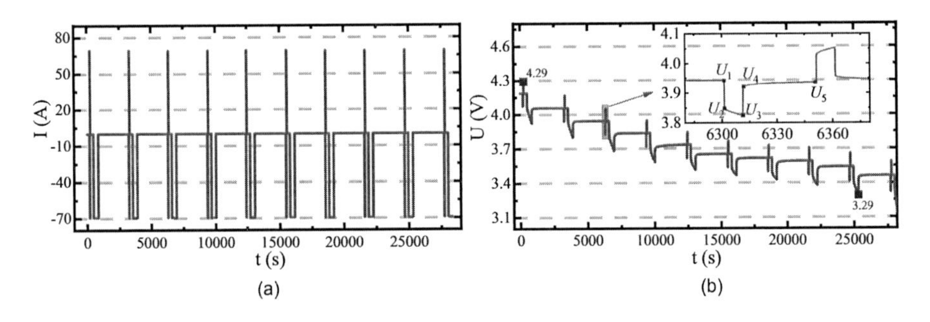

FIGURE 3.24 HPPC voltage and current curve. (a) Current curve. (b) Voltage curve.

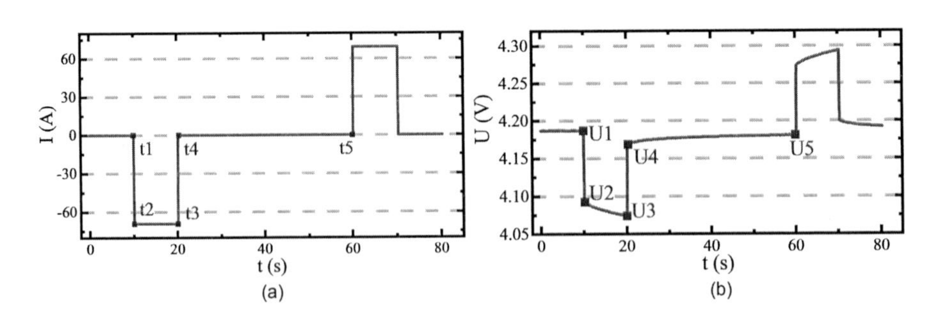

FIGURE 3.25 Single HPPC voltage and current change curve. (a) Single HPPC current change curve. (b) Single HPPC voltage change curve.

The graph of the HPPC experimental voltage variation curve is shown, and the variation characteristics can be extracted to obtain the model parameters. Analyzing Figure 3.25, the following 4 stage characteristics can be obtained. (1) Discharge starts at t_1. The sudden drop of the voltage at the lithium-ion battery terminal from U_1 to U_2 is mainly due to the voltage change caused by the ohmic resistance of the lithium-ion batteries. (2) During the period from t_2 to t_3, the voltage of the lithium-ion battery terminal slowly drops from U_2 to U_3, which is due to the existence of the battery polarization effect. The process of charging the polarization capacitor by the discharge current is the zero-state response of the double RC series circuit. (3) During the period from t_3 to t_4, the voltage of the lithium-ion battery terminal rises suddenly from U_3 to U_4, which is also due to the voltage change caused by the ohmic resistance of the lithium-ion batteries. (4) During the period from t_4 to t_5, the voltage of the lithium-ion battery terminal slowly rises from U_4 to U_5, which is the process of discharging the polarizing capacitor to the polarizing resistance, which is the zero-input response of the dual RC circuit.

The terminal voltage of a lithium-ion battery suddenly changes at the time point of discharging and at the time point of stopping, both of which are caused by the ohmic resistance. Therefore, the value of the ohmic resistance can be calculated by Equation (3.130).

$$R_0 = \frac{(U_2 - U_1) + (U_4 - U_3)}{2I} \tag{3.130}$$

During the period from t_2 to t_3, the voltage of the lithium-ion battery terminal slowly drops from U_2 to U_3. The result is due to the existence of the battery polarization effect. The process of charging the polarized capacitor by the discharge current is the zero-state response of the double RC series circuit. The time-domain analysis is carried out to the circuit by choosing the data from t_2 to t_3. And then, a function relation of U_L and time point t can be obtained, as shown in Equation (3.131).

$$U_L(t) = U_{OC} - IR_0 - i_L R_{p1}\left(1 - e^{-\frac{t}{\tau_1}}\right) - i_L R_{p2}\left(1 - e^{-\frac{t}{\tau_2}}\right) \tag{3.131}$$

In Equation (3.131), curve fitting is conducted for the HPPC experimental data by using the fitting toolbox to fit Equation (3.131) as the target equation, and then the U_{OC}, R_{p1}, C_{p1}, R_{p2}, and C_{p2} parameters can be identified. The parameter identification results are shown in Table 3.6.

The SOC value is a variable x, and $f(x)$ is a polynomial function fitted to the parameter and the SOC value. Taking into account the error and correlation between the fitted curve and the discrete data, the fitting equation is obtained, as shown in Equation (3.132).

$$f(x) = a_0 + a_1 x + a_2 x^2 + a_3 x^3 + a_4 x^4 + a_5 x^5 + a_6 x^6 + a_7 x^7 \tag{3.132}$$

TABLE 3.6
Equivalent Model Parameter Identification Result

SOC(1)	U_{OC}(V)	R_0(mΩ)	R_{p1}(mΩ)	C_{p1}(F)	R_{p2}(mΩ)	C_{p2}(F)
1	4.1838	1.3095	0.0410	25058	0.5043	32260
0.9	4.0581	1.3716	0.0541	19712	0.6114	30896
0.8	3.9434	1.3759	0.0432	20625	0.5859	25449
0.7	3.8380	1.3846	0.0586	19176	0.7824	28540
0.6	3.7354	1.3911	0.0615	19140	0.8830	29978
0.5	3.6527	1.4524	0.0299	27093	0.3942	34396
0.4	3.6162	1.4740	0.0402	29987	0.5798	43135
0.3	3.5904	1.5058	0.0488	29257	0.7104	43919
0.2	3.5408	1.5411	0.0389	15665	0.4792	29026
0.1	3.4641	1.6169	0.0992	3476	0.5655	18549

TABLE 3.7
Parameter Fitting Polynomial Coefficients

Para	R_0	R_{p1}	R_{p2}	C_{p1}	C_{p2}	U_{OC}
a_0	0.001006	0.001006	0.009941	1.567e+04	1.702e+05	3.626
a_1	0.001583	−0.01929	−0.2116	−3.468e+05	−3.495e+06	−5.006
a_2	−0.02599	0.1496	1.75	2.744e+06	2.888e+07	51.47
a_3	0.1317	−0.5873	−7.111	−4.517e+06	−1.116e+08	−219.1
a_4	−0.3275	1.272	15.69	−7.005e+06	2.318e+08	481
a_5	0.4266	−1.538	−19.15	2.757e+07	−2.68e+08	−566.6
a_6	−0.2782	0.9704	12.13	−2.784e+07	1.627e+08	341.9
a_7	0.07154	−0.2488	−3.113	9.402e+06	−4.041e+07	−83.16

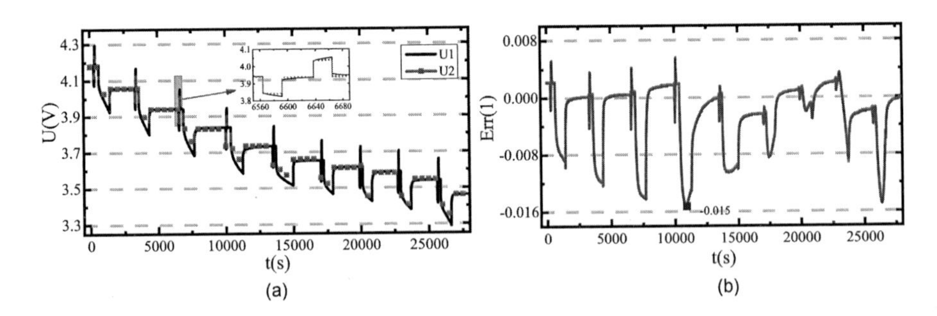

FIGURE 3.26 Model verification. (a) Voltage comparison curves. (b) Error curve.

The identification results are shown in Table 3.7.

The parameter identification results are input into the battery simulation model, and the HPPC experimental voltage is compared with the model simulation voltage to verify the accuracy of the model. Subsequently, the output voltage comparison and error curves are obtained, as shown in Figure 3.26.

In Figure 3.26a, U_1 is the HPPC experimental voltage curve and U_2 is the model voltage simulation curve. Figure 3.26b is the error curve. It can be observed that the overall tracking of the model is relatively accurate, the maximum error is 1.5%, and the overall error remains between 1%, indicating that the selected second-order RC equivalent circuit model can more accurately reflect the operating characteristics of the lithium-ion batteries.

3.6.2 AEKF Algorithm Analysis

Combining the second-order equivalent circuit model and its state-space in Figure 3.27, the adaptive EKF algorithm (AEKF) based on the Sage-Husa algorithm is proposed, so that the AEKF algorithm can be obtained. The flowchart for estimating SOC is shown in Figure 3.27.

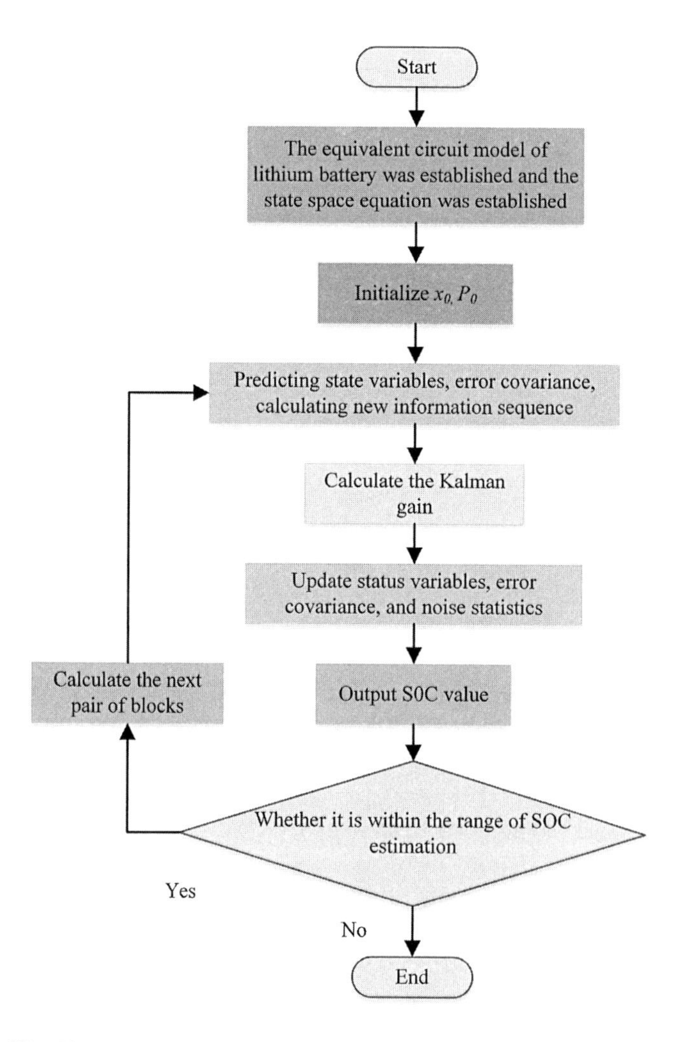

FIGURE 3.27 Flowchart of SOC estimation based on AEKF algorithm.

The state and observation of the system are shown in Equation (3.133).

$$\begin{cases} x_{k+1} = Ax_k + Bu_k + \Gamma w_k \\ y_k = Cx_k + v_k \end{cases} \tag{3.133}$$

In Equation (3.133), k is the discrete time, x_k is the state of the system at time point k, and y_k is the observation value of the corresponding state. A is the state transition matrix, B is the system control matrix, and C is the observation matrix. Γ is the noise driving matrix. w_k is the system noise and v_k is the measurement noise, which respectively obeys the following distribution, as shown in Equation (3.134).

$$\begin{cases} w_k \sim N(q_k, Q_k) \\ v_k \sim N(r_k, R_k) \end{cases} \tag{3.134}$$

The specific steps of combining the Sage-Husa adaptive algorithm based on the KF are described as follows:

1. Initialization, as shown in Equation (3.135).

$$\begin{cases} \hat{x}_0 = E[x_0] \\ P_0 = E\left[(x_0 - \hat{x}_0)(x_0 - \hat{x}_0)^T\right] \end{cases} \tag{3.135}$$

2. Prediction stage, as shown in Equation (3.136).

$$\begin{cases} \hat{x}_{k+1/k} = Ax_k + Bu_k + \Gamma w_k \\ P_{k+1/k} = AP_k A^T + \Gamma \hat{Q}_k \Gamma^T \end{cases} \tag{3.136}$$

3. Calculating the Kalman gain, as shown in Equation (3.137).

$$K_k = P_{k+1/k} C^T \left(CP_{k+1/k} C^T + \hat{R}_k\right)^{-1} \tag{3.137}$$

4. Update stage, as shown in Equation (3.138).

$$\begin{cases} \tilde{y}_{k+1} = y_{k+1} - \hat{y}_{k+1} \\ \hat{x}_{k+1} = \hat{x}_{k+1/k} + K\tilde{y}_{k+1} \\ P_{k+1} = (E - K_k C) P_{k+1/k} \end{cases} \tag{3.138}$$

5. Noise update stage, as shown in Equation (3.139).

$$\begin{cases} \hat{q}_{k+1} = (1-d_k)\hat{q}_k + d_k G(\hat{x}_{k+1} - A\hat{x}_k - Bu_k) \\ \hat{Q}_{k+1} = (1-d_k)\hat{Q}_k + d_k G(K_k \tilde{y}_k \tilde{y}_k^T K_k^T + P_k - AP_{k+1/k}A^T)G^T \\ \hat{r}_{k+1} = (1-d_k)\hat{r}_k + d_k (y_{k+1} - C\hat{x}_{k+1/k}) \\ \hat{R}_{k+1} = (1-d_k)\hat{R}_k + d_k (\tilde{y}_{k+1}\tilde{y}_{k+1}^T - CP_{k+1/k}C^T) \end{cases} \tag{3.139}$$

In Equation (3.139), the iterative equation of the forgetting factor d_k is described, as shown in Equation (3.140).

$$d_{k-1} = (1-b)/(1-b^k) \tag{3.140}$$

In Equation (3.140), b is the forgetting factor, $0.95 < b < 1$. From the above process, the online real-time estimation of $\hat{q}_k$, $\hat{r}_k$, $\hat{Q}_k$, and $\hat{R}_k$ can achieve real-time adjustment of the system changing noise by the algorithm under complex working conditions, so that the real-time noise can be more adapted to the state of the system, thereby realizing adaptive filtering and improving the SOC estimation accuracy.

3.6.3 Experimental Analysis

1. Constant current working condition verification
   A 1 C constant current discharge experiment is performed on a full-capacity lithium-ion battery to verify the accuracy of the AEKF algorithm under simple conditions. The ampere-hour integration method, EKF, and AEKF algorithms are used to compare the tracking effect, the SOC initial value of the algorithm is set as 0.8, and the convergence effect of the algorithm under the unknown initial state value and the tracking effect in the discharge process are verified. The comparison chart of the algorithm operation is shown in Figure 3.28.

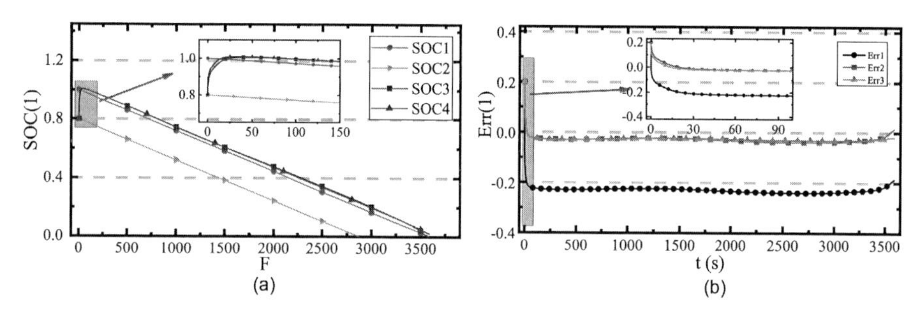

FIGURE 3.28 Comparison chart of SOC estimation. (a) SOC Estimated graph. (b) Error comparison chart.

In Figure 3.28a, SOC1 refers to the reference SOC value, SOC2 is the SOC value calculated using the ampere-hour integration method, SOC3 is the SOC value estimated using the EKF algorithm, and SOC4 is the SOC value estimated using the AEKF algorithm. In the case of an inaccurate initial value of SOC = 0.8, the ampere-hour integration method cannot track the real state of the system; EKF can track the real state of the system faster. The time required for the EKF algorithm to track the theoretical value is about 30 seconds, but the estimation accuracy is high enough. The tracking effect of the AEKF algorithm is almost the same as that of the EKF algorithm, and the later estimation accuracy of AEKF is higher than that of EKF. Figure 3.28b is the error curve, Errl is the ampere-hour integration error curve, Err2 is the EKF error curve, and Err3 is the error curve of the AEKF algorithm. The overall error of Errl exceeds 20% with a small error, the tracking speed of Err2 and Err3 is fast, and the overall tracking effect of Err3 is better, indicating that the AEKF algorithm has a better SOC estimation effect than the EKF algorithm.

2. DST working condition verification

   To verify the estimation effect of the AEKF algorithm under complex working conditions, the DST algorithm is used to simulate the actual operating conditions, and the specific working steps are set as follows. (a) The experimental lithium-ion battery is charged with constant current and constant voltage to full capacity. (b) The battery is shelved for 30 minutes. (c) The constant current discharge is conducted at 0.5 C rate for 4 minutes with 30 seconds of shelved treatment after stopping the discharging treatment. (d) The constant current charge is investigated at a 0.5 C rate for 4 minutes, stop charging, and put it on hold for 30 seconds. (e) Steps of (c) and (d) are repeated until the battery discharge is over.

   After obtaining the battery discharge data, the ampere-hour integration method, EKF, Sage-Husa AEKF, and improved Sage-Husa AEKF algorithms are used to compare the tracking effect. The initial value of the algorithm is set as SOC = 0.8 to verify the convergence effect and tracking effect during the discharge process. The running effect diagram of the algorithm is shown in Figure 3.29.

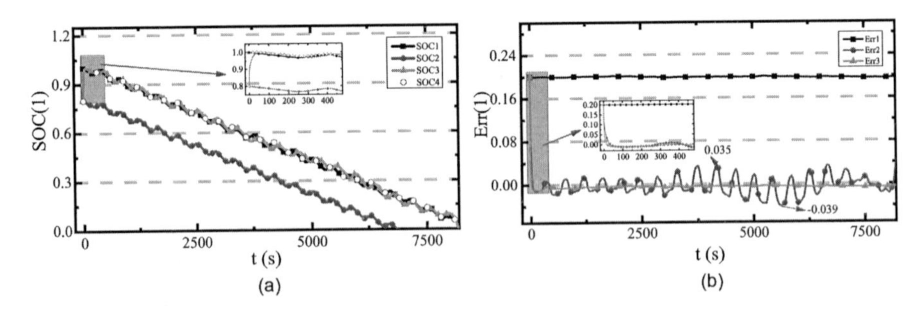

FIGURE 3.29 Comparison curves of SOC estimation under DST condition. (a) SOC-estimated graph. (b) Error comparison curves.

In Figure 3.29a, SOC1 refers to the SOC value, SOC2 is the SOC value calculated using the ampere–hour integration method, SOC3 is the SOC value calculated using the EKF algorithm, and SOC4 is the AEKF algorithm. Under the condition of an inaccurate initial value of SOC=0.8, the ampere–hour integration method cannot track the real state of the system. EKF can track the real state of the system faster. The time required for the EKF algorithm to track the theoretical value is about 30 seconds, but the estimation accuracy is not high enough. The convergence effect of the AEKF algorithm in the early stage is almost the same as that of the EKF algorithm, and the estimation accuracy of AEKF in the later stage is higher. Figure 3.29b is the error curve, Errl is the ampere-hour integral error curve, Err2 is the EKF error curve, and Err3 is the error curve of the AEKF algorithm. The overall error of Errl exceeds 20%, the Err2 and Err3 tracking systems are fast, and the convergence speed is less than 50 seconds. Among them, the error of Err2 is small, the error of the entire tracking process is less than 2%, and the overall error is small; while the maximum error of Err3 is 3.5%, the overall tracking effect is fluctuating and the accuracy is low. It shows that AEKF has high accuracy and strong applicability in estimating SOC under complicated working conditions.

3.7 SOC ESTIMATION BASED ON OTHER ALGORITHMS

3.7.1 Particle Filtering

Particle filtering (PF) is also called the approximate Bayesian filtering algorithm. It is a probability density function based on Monte Carlo simulation. The algorithm implementation process is mainly based on the Bayesian filtering operation principle to weigh discrete random samples to complete the integral operation with the sample average. PF algorithm detailed derivation of the theory is described as follows: the iterative calculation process is obtained and the following steps are used for description [216,217].

1. Particle filtering calculation
Initialization: the prior probability is used to generate N initial SOC particles, parameter weight, as shown in Equation (3.141).

$$p(x_0),\{\mathrm{SOC}_0^i\}_{i=1}^N,\{q_0^i\}_{i=1}^N=\frac{1}{N} \tag{3.141}$$

The algorithm cycle process is described as follows:

Update: according to the system update, get the prior probability sample at the next time point and update the parameter weight, as shown in Equation (3.142).

$$w_k^i=w_{k-1}^i p\left(U_{L(k)}\middle|\mathrm{SOC}_k^i\right)=w_{k-1}^i p\left(U_{L(k)}-h\left(\mathrm{SOC}_k^i\right)\right),\{\mathrm{SOC}_k^i\}_{i=1}^N \tag{3.142}$$

In Equation (3.142), $i = 1,2,\ldots,N$.

Normalization of weights is conducted as well as the normalized weights, as shown in Equation (3.143).

$$w_k^i = \frac{w_k^i}{\sum_{i=1}^{N} w_k^i} \tag{3.143}$$

The least mean square estimation is conducted, as shown in Equation (3.144).

$$\widehat{\mathrm{SOC}}_k \approx \sum_{i=1}^{N} w_k^i \mathrm{SOC}_k^i \tag{3.144}$$

Resampling: calculating the effective number of particles and determining the conditions, a new parameter set $\{i = 1,2,\ldots,N\}$ can be obtained, as shown in Equation (3.145).

$$N_{\text{eff}} = \frac{1}{\sum_{i=1}^{N} \left(w_k^i\right)^2}, N_{\text{eff}} \le N_s, \mathrm{SOC}_{0;k}^{i^*} \tag{3.145}$$

Prediction: the state parameter SOC_{k+1}^i is used to predict unknown parameters. Judging the end condition of the program, if it is not over, time point $k = k+1$, go to step (1). When the particle filter is used to estimate the battery SOC, the state-space model of the lithium-ion battery is established, and the estimation of process noise and observation noise is completed without affecting the system. During the estimation simulation process, the state–space model of the lithium-ion battery is obtained from the battery process and observation model. When setting the model boundary bars, the observed variable is equal to the lithium-ion battery load voltage, and the state variable is the SOC estimation of the lithium-ion battery, as shown in Equation (3.146).

$$\begin{cases} x_{k+1} = f\left(x_k, i_k, w_k\right) = x_k - \dfrac{n i_k \Delta t}{\eta_i \eta_T \eta_n Q_n} + w_k \\ y_{k+1} = f\left(y_k, i_k, v_k\right) = k_0 - R i_k - \dfrac{k_1}{x_k} - k_2 x_k + k_3 \ln\left(x_k\right) + k_4 \ln\left(1 - x_k\right) + v_k \end{cases} \tag{3.146}$$

In Equation (3.146), w_k is the process noise of the system, v_k is the observation noise of the system, set $w_k - N(0,Q)$, $v_k - N(0,R)$, Δt is the sampling period of the system.

Initialization: $k = 0$. From the initial probability distribution $p(x_0)$, N SOC initial particles are randomly generated to form a new parameter set $\{x_0, i\}$.

Updating status: To find the right particles, all of the new particles generated in Step 1 are screened by the established equation of state, and then a new parameter can be set as $\{x_0, i\}^+$.

Parameter weight calculation and normalization: After the new particle set is formed, the predicted value of the observation value can be obtained through the observation $\{y_0, i\}^+$, and then the error is calculated. The size of the error is the difference between the observed value added to the system and the predicted value obtained by each equation, and the parameter weight w_i is obtained through the difference. The parameter weight is shown in Equation (3.147).

$$\bar{x}_k = \frac{1}{\sqrt{2\pi R}} e^{\frac{(y_k - y_{i,k})^2}{2R}} \tag{3.147}$$

The normalization is conducted, as shown in Equation (3.148).

$$w_i^* = \frac{1}{\sum_{1}^{N} w_i} \tag{3.148}$$

Resampling is mainly for the fact that the threshold between the effective samples and the set number of samples in the random samples obtained in the above steps should be small. In this case, the sampling method is to remove particles with lower weights and retain particles with higher weights, and recalculate the weights of the filtered particles. The new particle swarm estimation is conducted to obtain $\bar{x}_k$, as shown in Equation (3.149).

$$\bar{x}_k = \sum_{i=1}^{N} w_k(i) x_k(i) \tag{3.149}$$

Judging whether the program is over. If it is not over, steps (2)–(4) are repeated by $k = k + 1$ to realize the recursive estimation of the state quantity x_k.

2. Particle degradation

As for particle degradation, when the PF algorithm is completed, the extracted particles have a great influence on the accuracy of the calculation. The main influences are divided into two steps. First, the imbalance is caused by different parameter weights; some particles have low weights and some have high weights. Unbalanced weights reduce the calculation accuracy of the algorithm. Second, if the number of particles with small weights is large, a large amount of calculation time can be used for the calculation of particles with small weights during the calculation process, which greatly wastes calculation time and reduces the computational efficiency.

In the calculation process, the imbalance of the parameter weights wastes the calculation time of the algorithm. Also, in more serious cases, there may be too many particles with smaller weights, undercomputing large particles, and omissions. This greatly reduces the computational accuracy of the algorithm. In other words, for particles with smaller weights, the effectiveness is low, which not only increases the amount of calculation but also takes longer to compute, causing serious particle degradation. In practical applications, effective particles can be counted. If the value is low and cannot meet the calculation requirements, it can be controlled by some means to reduce the phenomenon of particle degradation.

3. Resampling

   Resampling is aimed at particle degradation caused by imbalance. Resampling of particles is carried out by polynomial methods. Resampling filters the particle swarms with unbalanced weights and eliminates particles with low weights and does not enter the next algorithm calculation. Through the above steps, a new particle swarm is formed. The particle swarm is mainly composed of high-weight particles, which can greatly reduce the particle degradation effect.

   Although resampling can reduce the particle degradation effect and improve the calculation accuracy, it also increases the amount of calculation of the algorithm, and the calculation efficiency of the algorithm also decreases. Also, too many resampling times eliminate a large number of particles and greatly reduce the particle library. The particle library is depleted. Therefore, to ensure that the effective number of particles cannot be depleted due to a large reduction due to resampling, the number of particles can be guaranteed by setting a threshold in the application. The resampling method is performed by polynomials, and the polynomial resampling has the advantages of simplicity and low complexity. Taking $U[0,1]$ in the uniform distribution of $[0,1]$ and satisfying $x_k(i)$, a new parameter set can be formed with the obtained particles, as shown in Equation (3.150).

$$\sum_{j=1}^{j-1} w_k(j) \leq u \leq \sum_{j=1}^{j} w_k(j) \tag{3.150}$$

   The particle polynomial is resampled by the above equation, and the weights of the particles are rechecked and screened to form a new particle library. In the parameter set obtained by the above screening, the weights are evenly distributed. The specific steps of resampling are designed for the basic PF algorithm.

4. Prediction

   First, new particles are extracted and used as a new parameter set. The part of the particles is obtained from the important density function of the system: $x_k(i) - p(x_k \mid \chi_{k-1}(i)), i = 1,2,\ldots,N$.

   The weight of each particle $w_k^*(i)$ is calculated, as shown in Equation (3.151).

$$q\left(\chi_k \middle| x_{0:k}(i), y\right) = p\left(x_k \middle| \chi_{k-1}(i)\right) \tag{3.151}$$

The normalization processing is performed, as shown in Equation (3.152).

$$w_k(i) = \frac{w_k^*(i)}{\sum_{i=0}^{N} w_k^*(i)} \tag{3.152}$$

The state estimation is conducted, as shown in Equation (3.153).

$$\bar{x}_k = \sum_{i=1}^{N} w_k(i)\chi_x(i) \tag{3.153}$$

The polynomial resampling is performed accordingly. To avoid the depletion of the particle library, it is necessary to set a threshold value for the particle library to ensure the number of particles. The resampled number is counted and compared with the threshold value. If the number of particles is smaller than the threshold value, a new selection is performed. In this way, the reselected particles form a new collection of particles. Resampling uses a PF algorithm, and additional random variance is added after sampling. Due to this reason, it is necessary to estimate the system such as the posterior of the system. Aiming at the depletion phenomenon of the parameter set because of pre-sampling, the parameter α is added to the resampling design of the seat to suppress the possible depletion phenomenon of the parameter set. The parameter needs to satisfy Equation (3.154).

$$w_t(i) = \left(w_{t-1}^i\right)^\alpha \frac{p\left(z_t \mid x_t(i)\right)p\left(x_t(i) \mid \chi_{t-1}(i)\right)}{q\left(x_t(i) \mid x_{t-1}(i), z_{1:t}\right)} \tag{3.154}$$

In Equation (3.154), the parameter α has a very important meaning in the genetic algorithm (GA). Its function is to control it according to the weight of the importance in the PF algorithm, which can reduce the depletion phenomenon in the sample. The algorithm used to estimate the battery SOC is the PF algorithm. For the two types of noise in the system, one is process noise and the other is observation noise, the algorithm does not put forward any requirements. First, create the battery state model and then use the software to write a program to estimate and predict the SOC estimation of the lithium-ion batteries. The specific algorithm international flowchart is shown in Figure 3.30.

According to the PF algorithm flowchart, the code implementation of the topic using PF for SOC estimation is shown below.

1. Initialization
   In the process of system initialization, memory space needs to be allocated to variables. At that time, the memory space provided inside the program is randomly allocated. In the algorithm design, assign initial values to the

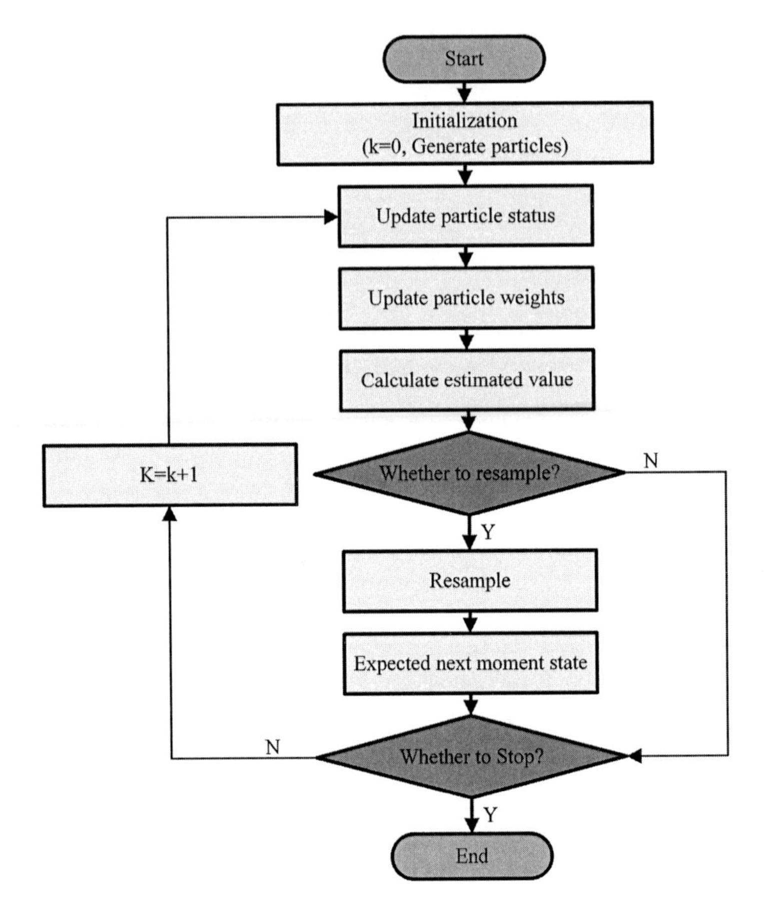

FIGURE 3.30 Particle filter flowchart.

variables of the system firstly and store arrays, such as particle filter estimation state, collection, and weights, to avoid serious deviations in the estimation result because the value of the initial variable itself in the memory has not been cleared.

2. Cycle start

The programming language is similar to the C language style, and the expression of loop statements can also be used for and while. When using PF to estimate SOC, use it for the program statement. The purpose of the cyclic test is to correspond to an SOC value at each time point and continuously approximate the measured SOC value through the PF resampling process.

```
%System initial value
M=100;
X0=[1;0];
%Filter initialization
Xpf=zeros (2, N);
```

```
Xpf (:, 1)=X0;
% Parameter set initialization
Xm=zeros (2, M, N);
X1=[1;0];
% Weight initialization
W=zeros (N, M);
```

3. Importance sampling → calculation weight → normalized weight
   For each pin in the parameter set, the importance weight needs to be calculated. The weight of each pin is constantly updated and iterative during the running of the algorithm program, which is related to the weight at the previous time point. In the core part of the PF algorithm—the resampling process, it is obtained according to the weight, which can make the particles appear more in the selected segment and improve the efficiency of the use of the particles.

```
for i=1:M
Zm (1, i, k)=feval ('hfun', Xm (:, i, k));
W (k, i)=exp (- (Z (1, k)-Zm (1, i, k))^2/2/R)/ (2*R*pi)^0.5+1e-99;
end
W (k, :)=W (k, :)./sum (W (k, :));
```

4. Resampling

```
for i=1:M
u=rand;
qtempsum=0;
for j=1: M
qtempsum=qtempsum+W (k, j);
if qtempsum >= u
Xm (:, i, k)=Xm (:, j, k);
```

   Selecting a sampling strategy and resampling the parameter set according to the size of the weight, the weight and parameter sets have a one-to-one correspondence. In the strategy, particles with larger weights are more likely to be copied, while particles with relatively low weights are gradually eliminated during the continuous resampling process. The process can effectively solve the problems in the Monte Carlo method with a lack of particles.
5. Output
   Finally, the experimental results and filtering results are displayed in the form of graphs, and the estimation error is obtained.
6. End of loop
   Through the previous learning, the parameter values required by the experiment can be obtained, and then these parameters can be simulated by the process recurrence and the observation noise to truly simulate the current change and then perform SOC prediction and estimation. Through the above simulation image analysis, it is concluded that the PF algorithm has a good filtering effect on tracking the SOC estimation of the lithium-ion batteries. Through calculation, the error can be stabilized within 6%, as shown in Figure 3.31.

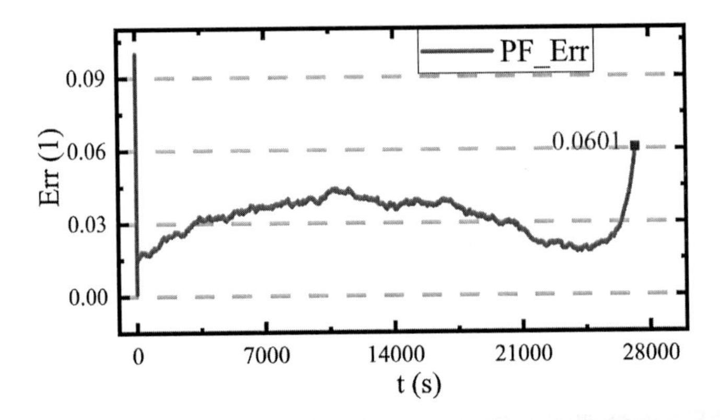

FIGURE 3.31 PF tracking error.

The core part of PF particles is resampled. Through continuous selection and training of particles, the tracking error of PF is suitable within 5%, and the algorithm has reliable adaptability. This section analyzes in detail the basic principles and specific recurrence process of the PF algorithm. To verify the effectiveness of the PF algorithm proposed in the chapter to estimate the SOC value of lithium-ion batteries, a simulation estimation experiment was carried out on the battery SOC under constant current conditions. The simulation results show that the accuracy of the lithium-ion battery SOC estimated based on the PF algorithm can reach 95%, with high accuracy, fast convergence, and good stability.

3.7.2 Extended Particle Filtering

The EKF algorithm is a local linearization method realized by the first-order Taylor expansion, which makes the system a Gaussian distribution model according to the nonlinear system state and observation equation. At each time point, the functions $f(x_k, u_k)$ and $h(x_k, u_k)$ are expanded by the Taylor series and the influence of higher-order terms on the function is ignored, and two linearization processes are completed to obtain Equations (3.155) and (3.156).

$$\begin{cases} f(x_k, u_k) \approx f(\hat{x}_k, u_k) + \left.\dfrac{\partial f(x_k, u_k)}{\partial x_k}\right|_{x_k = \hat{x}_k} (x_k - \hat{x}_k) \\ h(x_k, u_k) \approx h(\hat{x}_k, u_k) + \left.\dfrac{\partial h(x_k, u_k)}{\partial x_k}\right|_{x_k = \hat{x}_k} (x_k - \hat{x}_k) \end{cases} \tag{3.155}$$

$$\begin{cases} \hat{F}_k = \left.\dfrac{\partial f(x_k, u_k)}{\partial x_k}\right|_{x_k = \hat{x}_k} \\ \hat{H}_k = \left.\dfrac{\partial h(x_k, u_k)}{\partial x_k}\right|_{x_k = \hat{x}_k} \end{cases} \tag{3.156}$$

According to Equations (3.155) and (3.156), the nonlinear system is transformed into a linear system, and the transformed linear system is only related to state variables, as shown in Equation (3.157).

$$\begin{cases} x_{k+1} \approx \hat{F}_k x_k + \left[f\left(\hat{x}_k, u_k\right) - \hat{F}_k x_k \right] + w_k \\ z_k \approx \hat{H}_k x_k + \left[h\left(\hat{x}_k, u_k\right) - \hat{H}_k x_k \right] + v_k \end{cases} \tag{3.157}$$

In Equation (3.157), it is noted that functions $f\left(\hat{x}_k, u_k\right) - \hat{F}_k x_k$ and $h\left(\hat{x}_k, u_k\right) - \hat{H}_k x_k$ are not the x_k functions. The linear system is brought into the classical KF algorithm. In the sampling stage, sampling is achieved through the mean and variance of each particle obtained by the EKF algorithm. The resampling method is carried out by polynomial, which has simplicity and low complexity advantages, taking a particle $u \sim U[0,1]$ from the $[0,1]$ uniform distribution and making the obtained particles into a new particle set, as shown in Equation (3.158).

$$\sum_{j=1}^{j-1} w_k(j) \le u \le \sum_{j=1}^{j} w_k(j) \tag{3.158}$$

The particle polynomial is resampled by Equation (3.158), and the particle weights are rechecked and screened to form a new particle library. The weights in these particle sets screened by Equation (3.158) are evenly distributed. The process at this stage is described as follows:

1. Initialization
   Firstly, new particles are extracted and used as a new particle set. This part of particles makes it possible to obtain the important density function of the system, as shown in Equation (3.159).

$$x_k(i) \sim p\left\{x_k \middle| x_{k-1}(i)\right\} \; (i = 1, 2, \ldots, N) \tag{3.159}$$

   The initial conditions of the filter equation state X_0 and the covariance P_0, as shown in Equations (3.160) and (3.161).

$$X_0 = E(X_0) \tag{3.160}$$

$$P_0 = E\left[\left(X_0 - \bar{X}_0\right)\left(X_0 - \bar{X}_0\right)^T\right] \tag{3.161}$$

2. Update
   Updating the particle set and using the EKF algorithm to update the mean and variance of each independent particle in the sample. The calculation process is described as follows:
   i. State prior estimation, as shown in Equation (3.162).

$$\overline{X}^{i}_{k,k+1} = f\left(X^{i}_{k-1}, u^{i}_{k-1}\right) \tag{3.162}$$

ii. A prior estimate of the error covariance, as shown in Equation (3.163).

$$P^{i}_{k,k+1} = F^{i}_{k,k+1} P^{i}_{k-1} \left(F^{i}_{k,k+1}\right)^{T} + Q_{k-1} \tag{3.163}$$

iii. Kalman gain matrix, as shown in Equation (3.164).

$$K_k = P^{i}_{k,k+1} \left(H^{i}_{k}\right)^{T} \left[H^{i}_{k} P^{i}_{k,k-1} \left(H^{i}_{k}\right)^{T} + R_k \right]^{-1} \tag{3.164}$$

iv. State posterior estimation, as shown in Equation (3.165).

$$\overline{X}^{i}_{k} = \overline{X}^{i}_{k,k-1} + K_k \left[Z_k - h\left(\overline{X}^{i}_{k-1}, u^{i}_{k}\right) \right] \tag{3.165}$$

v. The posterior estimate of the error covariance, as shown in Equation (3.166).

$$P^{i}_{k} = P^{i}_{k,k-1} - K_k H^{i}_{k} P^{i}_{k,k-1} \tag{3.166}$$

To get the sample mean $\overline{X}^{i}_{k}$ and covariance P^{i}_{k}. Where F^{i}_{k} is the Jacobian matrix of the model state transition equation, H^{i}_{k} is the observation matrix, and K_k is the Kalman gain. Use the particle set updated by the algorithm to update the state of the ith particle, as shown in Equations (3.167)–(3.169).

$$\hat{X}^{i}_{k} \sim q\left(\overline{X}^{i}_{k} \mid X^{i}_{0:k-1}, Z_{1:k}\right) = N\left(\overline{X}^{i}_{k}, P^{i}_{k}\right) \tag{3.167}$$

$$\hat{X}^{i}_{0:k} \triangleq \left(X^{i}_{0:k-1}, \overline{X}^{i}_{k}\right) \tag{3.168}$$

$$\hat{P}^{i}_{0:k} \triangleq \left(P^{i}_{0:k-1}, \overline{X}^{i}_{k}\right) \tag{3.169}$$

Therefore, according to Equations (3.167)–(3.169), each particle's weight w^{i}_{k} is calculated, as shown in Equations (3.170) and (3.171).

$$q\left(\chi_k \mid X^{i}_{0:k}, y\right) = P\left(x_k \mid \chi^{i}_{k-1}\right) \tag{3.170}$$

$$w^{i}_{k} = w^{i}_{k-1} \frac{p\left(Z_k \mid \overline{X}^{i}_{k}\right) p\left(\overline{X}^{i}_{k} \mid X^{i}_{k-1}\right)}{q\left(\hat{X}^{i}_{k} \mid X^{i}_{0:k-1}, Z_{1-k}\right)} \tag{3.171}$$

3. Normalized weight, as shown in Equation (3.172).

$$w_k^i = w_k^i \Big/ \sum_{i=0}^{N} w_k^i \tag{3.172}$$

4. State estimation, as shown in Equation (3.173).

$$\overline{x}_k = \sum_{i=1}^{N} w_k^i \chi_k^i \tag{3.173}$$

3.7.3 Unscented Particle Filter

The UKF algorithm is utilized to improve the PF algorithm with the UT algorithm [218,219]. Compared with the first-order Taylor expansion of the EKF algorithm, this method can theoretically calculate the posterior variance accuracy to the third order and has higher accuracy. It is also an effective means to calculate mean and covariance. Based on the PF algorithm estimation framework, the core calculation of the UPF algorithm is described as follows:

a. Initialization

   Extract particles from the prior distribution $P(X_0)$ as the new particle set initial state. These particles are obtained from the important density function, as shown in Equations (3.174)–(3.177).

$$X_0^i = E\left(X_0^i\right) \tag{3.174}$$

$$P_0^i = E\left[\left(X_0^i - \overline{X}_0^i\right)\left(X_0^i - \overline{X}_0^i\right)^T\right] \tag{3.175}$$

$$X_0^{i,a} = E\left(\overline{X}_0^{i,a}\right) = \begin{bmatrix} \left(\overline{X}_0^i\right)^T & 0 & 0 \end{bmatrix}^T \tag{3.176}$$

$$P_0^{i,a} = E\left[\left(X_0^{i,a} - \overline{X}_0^{i,a}\right)\left(X_0^{i,a} - \overline{X}_0^{i,a}\right)^T\right] = \begin{bmatrix} P_0^i & 0 & 0 \\ 0 & Q & 0 \\ 0 & 0 & R \end{bmatrix} \tag{3.177}$$

b. Importance sampling stage
   i. Unscented transform.

The $2n+1$ sampling points of the state variable at the current time point are calculated by distributed sampling method, where n is the state-space X dimension. For the nonlinear system, the UT process is performed on it. The mean and variance of the sigma point are set correspondingly equal to the state variable at the current time point [220]. The mean value of X is supposed to be $\bar{X}$ and then the sampling point is determined, as shown in Equation (3.178).

$$\begin{cases} X_0 = \bar{X}_0 \\ X_i = \bar{X}_i + \sqrt{(n_a + \lambda)P_i}, i = 1,\ldots,n \\ X_i = \bar{X}_i - \sqrt{(n_a + \lambda)P_i}, i = n+1,\ldots,2n \end{cases} \tag{3.178}$$

In Equation (3.178), X_i represents the ith sampling point, and $\sqrt{(n_a + \lambda)P_i}$ represents the ith column of the matrix $\sqrt{(n_a + \lambda)P}$. Corresponding to each X_i, its weight is given when calculating the mean coefficient, as shown in Equation (3.179).

$$\begin{cases} \omega_0^m = \dfrac{\lambda}{n+\lambda} \\ \omega_0^c = \dfrac{\lambda}{n+\lambda} + \left(1 - \alpha^2 + \beta\right) \\ \omega_i^m = \omega_i^c = \dfrac{\lambda}{2(n+\lambda)}, i = 1,\ldots,2n \\ \lambda = \alpha^2(n+\gamma) - n \end{cases} \tag{3.179}$$

In Equation (3.179), ω_i^m is the weight of the ith sampling point in calculating the mean value, and ω_i^c is the weight of the ith sampling point in calculating the covariance. The n is the dimension degree of state variable x. The α factor is the distribution state of the sampling points around the mean. The α controls the distance between the sampling points and the mean. The α generally takes a constant of $\alpha \in (10^{-6}, 1)$. γ is required to satisfy $\gamma + n \neq 0$. Because $n \neq 0$, so $\gamma = 0$ can generally be selected. β is a prior distribution factor. Since it has been assumed that the system obeys Gaussian distribution, $\beta = 2$ is optimal. The λ parameter represents the zoom ratio.

ii. Sigma point set update.
Sigma point set update is the process of using the state-space sampling points obtained by the UT at the previous time point to predict the

sampling points at the current time point. It is also the sigma further prediction and $\overline{X}_{k|k-1}^{i,a}$ is given, as shown in Equation (3.180).

$$\overline{X}_{k|k-1}^{i,a} = f\left(\overline{X}_{k-1}^{i,x}, \overline{X}_{k-1}^{i,v}\right) \tag{3.180}$$

iii. State variable X and covariance P prediction.
Multiply the sigma point predicted values by the corresponding weight in (3.180) to obtain the x predicted value at the next time point. Then, the covariance is calculated from the x predicted value and the covariance weight, and the calculation equation is given, as shown in Equations (3.181) and (3.182).

$$\overline{X}_{k|k-1}^{i} = \sum_{j=0}^{2n_a} \omega_i^m \overline{X}_{j,k-1}^{i,x} \tag{3.181}$$

$$P_{k|k-1}^{i} = Q_k + \sum_{j=0}^{2n_a} \omega_j^c \left[\overline{X}_{j,k-1}^{i,x} - \overline{X}_{k|k-1}^{i}\right]\left[\overline{X}_{j,k-1}^{i,x} - \overline{X}_{k|k-1}^{i}\right]^T \tag{3.182}$$

iv. Observation of further prediction.
In the UKF algorithm, to predict the observation space at time point $k+1$, the observation predicted value at time point $k+1$ needs to be calculated from the state at time point k, as shown in Equation (3.183).

$$Z_{k|k-1}^{i} = h\left(X_{k|k-1}^{i,x}, X_{k-1}^{i,n}\right) \tag{3.183}$$

v. System observation matrix and covariance matrix prediction.
Observations obtained by nonlinear transfer based on sampling points given by (3.139) are averaged to obtain the predicted amount of the system observation matrix. Then calculate the variance of the observed measurement at $k+1$ given by the predicted value and the covariance between it and the state variable. The calculation method is provided for the prediction system observations, as shown in Equation (3.184).

$$\overline{Z}_{k|k-1}^{i} = \sum_{j=0}^{2n_a} \omega_j^m Z_{k|k-1}^{i} \tag{3.184}$$

The variance of the observations is given by Equation (3.185).

$$P_{\overline{Z}_k} = P_k + \sum_{j=0}^{2n_a} \omega_j^c \left[Z_{j,k|k-1}^{i} - Z_{k|k-1}^{i}\right]\left[Z_{j,k|k-1}^{i} - Z_{k|k-1}^{i}\right]^T \tag{3.185}$$

Finally, the covariance between the observed quantity and the state quantity is determined, as shown in Equation (3.186).

$$P_{X_k,Z_k} = \sum_{j=0}^{2n_a} \omega_j^c \left[X_{j,k|k-1}^i - X_{k|k-1}^i\right]\left[Z_{j,k|k-1}^i - Z_{k|k-1}^i\right]^T \tag{3.186}$$

vi. Status and covariance update.

The calculation method is obtained to estimate the state and observation at $k+1$ from the information at time point k. Then, a posterior estimation is performed by using the data at $k+1$, that is, to modify the prior estimation to get a more accurate estimation value. Before the correction process starts, calculating the Kalman gain is a necessary step, as shown in Equation (3.187).

$$K = P_{X_k,Z_k} / P_{\overline{Z}_k} \tag{3.187}$$

The previous five steps are prepared for the status update in step vi. The covariance update is to prepare for the estimation at time point $k+2$. The status update is shown in Equation (3.188). The covariance update is shown in Equation (3.189).

$$\overline{X}_k^i = \overline{X}_{k|k-1}^i + K\left(Z_k - \overline{Z}_{k|k-1}^i\right) \tag{3.188}$$

$$\hat{P}_k^i = P_{k|k-1}^i - KP_{\overline{Z}_k}K^T \tag{3.189}$$

In Equation (3.189), Z_k is the observation measured by the instrument at $k+1$. $\overline{Z}_{k|k-1}^i$ is the observation obtained from the state estimation at the time point k. $\overline{X}_k^i$ is the posterior state estimate at $k+1$, and $\overline{X}_k^i$ is the optimal state estimation at the current time. Using the UKF algorithm to estimate the SOC of the lithium battery can obtain the optimal estimation based on the minimum variance criterion, and the estimation accuracy is better than that of the EKF algorithm. The subsequent steps are consistent with the EPF algorithm flow.

3.7.4 Backpropagation Neural Network

The backpropagation (BP) neural network is a multilayer network that generalizes the W-H learning rules and performs weight training on nonlinear differentiable functions [221,222]. The BP neural network includes an input layer, a hidden layer, and an output layer. The hidden layer can be one or multiple layers. BP network is mainly used for function approximation, pattern recognition, classification, and data compression. The simplest BP neural network structure is a three-layer structure, as shown in Figure 3.32.

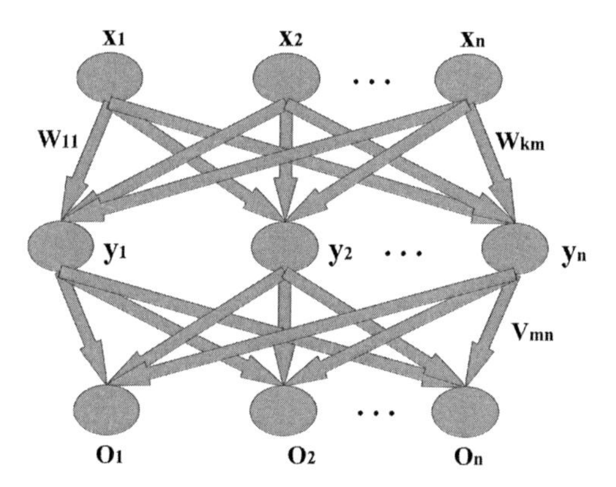

FIGURE 3.32 Three-layer BP neural network structure.

In Figure 3.32, X represents the input layer, Y represents the hidden layer, O represents the output layer, W is the weight from the input layer to the hidden layer, and V is the weight from the hidden layer to the output layer. The BP neural network consists of two parts—that is the forward transmission of information and the backward propagation of errors.

1. The output of the ith neuron in the hidden layer is obtained, as shown in Equation (3.190).

$$h_i = f1\left(\sum_{j=1}^{r} w1_{ij} p_j + b1_i\right) \tag{3.190}$$

In Equation (3.190), h represents the value of the hidden layer. $f1$ represents the activation function of the hidden layer. The activation function of the BP network must be differentiable everywhere, generally using the S-type logarithmic or tangent activation function and linear function. $w1$ is the weight between the input and output layers, and p is the input of $b1$ is the deviation of the hidden layer neuron. r is the number of input neurons. The output of the kth neuron in the output layer is obtained, as shown in Equation (3.191).

$$o_k = f2\left(\sum_{i=1}^{m} w2_{kj} h_j + b2_k\right) \tag{3.191}$$

2. The output of the kth neuron in the output layer
o represents the value of the output layer, $f2$ is the activation function of the output layer, $w2$ is the weight between the hidden layer and the output layer, $b2$ is the deviation of the output layer neurons, and m is the number of neurons in the hidden layer.

3. The error function is defined, as shown in Equation (3.192).

$$E(W,B) = \frac{1}{2}\sum_{k=1}^{n}(t_k - o_k)^2 \tag{3.192}$$

In Equation (3.192), E is the error function, t is the expected value, and n is the number of neurons in the output layer.

The gradient descent method is used to find the weight change and the BP of the error.

1. The weight change of the output layer
The weights can be obtained from the ith input to the kth output, as shown in Equation (3.193).

$$\Delta w2_{ki} = -\eta\frac{\partial E}{\partial w2_{ki}} = -\eta\frac{\partial E}{\partial ok}\cdot\frac{\partial ok}{\partial w2_{ki}} = \eta\cdot\delta_{ki}\cdot h_i \tag{3.193}$$

In Equation (3.193), the parameters are calculated, as shown in Equation (3.194).

$$\begin{cases}\delta_{ki} = (t_k - o_k)\cdot f2' = e_k\cdot f2' \\ e_k = t_k - o_k\end{cases} \tag{3.194}$$

The same can be obtained, as shown in Equation (3.195).

$$\Delta b2_{ki} = -\eta\frac{\partial E}{\partial b2_{ki}} = \eta(t_k - o_k)\cdot f2' = \eta\cdot\delta_{ki} \tag{3.195}$$

2. Hidden layer weight change can be obtained, as shown in Equation (3.196).

$$\Delta w1_{ij} = -\eta\frac{\partial E}{\partial w1_{ij}} = -\eta\frac{\partial E}{\partial o_k}\cdot\frac{\partial o_k}{\partial h_i}\cdot\frac{\partial h_i}{\partial w1_{ij}} = \eta\cdot\delta_{ij}\cdot p_j \tag{3.196}$$

In Equation (3.196), the mathematical relationship can be obtained, as shown in (3.197).

$$\begin{cases}\delta_{ij} = e_i\cdot f1' \\ e_i = \sum_{k=1}^{n}\delta_{ki}w2_{ki} \\ \delta_{ki} = e_k\cdot f2' \\ e_k = t_k - o_k\end{cases} \tag{3.197}$$

The same can be obtained, as shown in Equation (3.198).

$$\Delta b1_i = \eta \delta_{ij} \tag{3.198}$$

The method of BP neural network training is to use the error between the actual output of the network and the expected output to carry out BP, thereby training the weights and reducing the sum of squares of the error as much as possible. Each time the weight and deviation changes are proportional to the influence of the network error. The parameters are transmitted to each layer in the way of BP. Before training the neural network, the content of the neural network parameters needs to be determined, including the number of the network layers. Generally, the three-layer neural network can approximate any nonlinear function. The number of neurons can be adjusted. The number of neurons in the input layer is the number of input vectors, and the same is true for the output layer. If there is only one hidden layer, it can be adjusted according to Equation (3.199).

$$m = \sqrt{n+k} + i \tag{3.199}$$

In Equation (3.199), n is the number of input neurons, k is the number of output neurons, and m is the number of hidden layer neurons.

3. The activation function of each layer. Generally, the activation function of the hidden layer selects the S-type activation function. The S-type function can convert any input function signal into the interval of [−1, 1], so the S-type activation function can handle and approximate the nonlinear function. The output layer generally uses a linear activation function, which can make the network output any value. However, if the output layer requires a 0–1 output type, the activation function needs to be changed to a threshold type.
4. Expected error.
5. The maximum number of training sessions.
6. Learning rate. The learning rate is generally between [0.01, 0.7].
7. Training function.

The limitations and shortcomings of ordinary BP neural networks: long training time, inability to train, falling into local optimal, overfitting, and no generalization ability. In response to these shortcomings, some improved BP neural networks are proposed.

1. Additional time point
   The basic principle is shown in Equation (3.200).

$$\text{mc} = \begin{cases} 1.04, & \text{SSE}(k) > \text{SSE}(k-1) \\ 0.95, & \text{SSE}(k) < \text{SSE}(k-1) \\ \text{mc}, & \text{SSE}(k) < \text{SSE}(k-1) \end{cases} \tag{3.200}$$

The principle of the whole function can be directly called the trading function.

2. Adaptive learning rate.
   The basic principle is shown in Equation (3.201).

$$\eta(k+1) = \begin{cases} 1.05\eta(k), & \text{SSE}(k+1) < \text{SSE}(k) \\ 0.7\eta(k), & \text{SSE}(k+1) > \text{SSE}(k) \\ 1.04\eta(k), & \text{other} \end{cases} \tag{3.201}$$

   In Equation (3.201), the principle of the whole function can directly call the trading function.
3. Elastic BP algorithm.
   Based on the above, the conjugate gradient method, Broyden–Fletcher–Goldfarb–Shanno quasi-Newton method, secant quasi-Newton method, and Levenberg-Marquardt method were derived. These function training methods can greatly reduce the training time and the number of iterations and can avoid local optima to a certain extent. BP nerve is used to estimate the SOC value, and the input vector can contain easily obtained external measurement values, such as current, voltage, temperature, etc. The output vector is the SOC value. In the measured dataset, one part is used as the training set and the other part is used as the test set.

   Also, BP neural network can be combined with EKF to estimate SOC together and use BP neural network to make up for some of the shortcomings of EKF [223,224]. EKF uses the first-order Taylor expansion, and the nonlinear system is transformed into a linear system. The higher-order terms are ignored, which causes certain errors. Especially for highly nonlinear batteries, the resulting error is greater, and the error of using the adaptive extended Kalman alone to estimate the SOC is greater. Therefore, it is proposed to use BP neural network to optimize and compensate for the error generated by EKF. The neural network toolbox is used to build a four-input one-output three-layer BP neural network. The hidden layer selects ten neurons, the expected value is set to 10-3, the maximum number of iterations is 4000, and the L-M method is used for training. The input vector of the experiment is the estimated value $S(k \mid k-1)$ and polarization voltage $U(k \mid k-1)$ of the initial battery SOC at the time point k by adaptively extending KF. The Kalman gain K_{SOC} of SOC and the polarization capacitance is the gain K_c. The output vector is the difference between the improved EKF and the theoretical SOC value. The overall flowchart of SOC estimation is shown in Figure 3.33.

   The specific experimental steps of HPPC are described as follows. First, the battery is charged at constant current (69.27 A) and constant voltage (4.20 V), and then the battery is laid aside for 40 minutes after charging. Constant discharge at 69.27 A should be conducted for 10 seconds, and the battery shelved for 40 seconds after the discharging treatment is stopped.

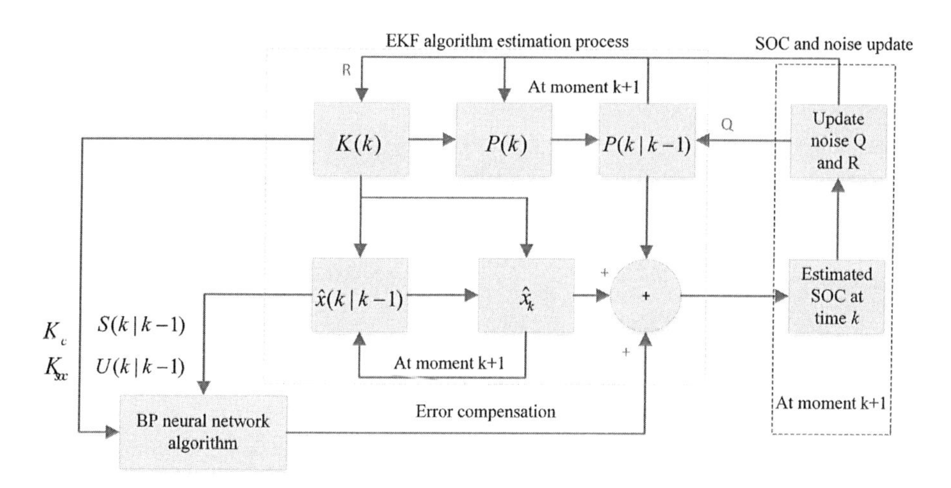

FIGURE 3.33 Overall flowchart of SOC estimation.

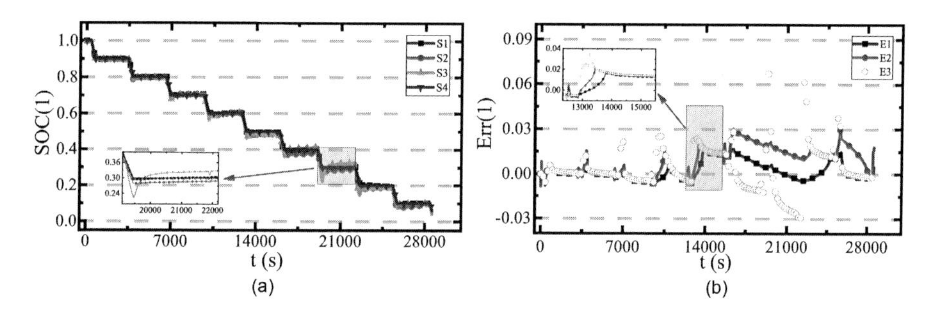

FIGURE 3.34 HPPC test results. (a) SOC estimation of each algorithm. (b) The error of each algorithm.

The battery is charged with a constant current for 10 seconds with a current of 69.27 A. And then, the battery is shelved for 5 minutes after stopping charging. And then, the constant discharge is carried out with the current of 69.27 A for 6 minutes and then rested for 40 minutes. Three steps are repeated until the end of the battery discharge. According to the battery model carried in the front, the HPPC experiment is carried out on the battery, and the simulation experiment is built by programming in simulation software. The simulation results are shown in Figure 3.34.

In Figure 3.34a, S1 is the estimated SOC value of the improved EKF based on BP, S2 is the estimated SOC value of the improved EKF, S3 is the estimated SOC value of the EKF, and S4 is the theoretical value of the SOC value. In Figure 3.34b, E1 is the error curve of the improved EKF based on BP, and E2 is the error curve of the improved EKF. E3 is the error curve of EKF. Data can be obtained by sorting out the data in Figure 3.34. The estimation results of different algorithms are shown in Table 3.8.

TABLE 3.8
HPPC Experimental Data Analysis

Algorithms	Maximum Error	Average Error	Root Mean Square Error
EKF	7.712%	1.036%	1.5361%
Improved EKF	3.074%	1.01%	1.2877%
Improved EKF-BP	1.642%	0.4499%	0.6353%

According to Table 3.8, the maximum error of the improved EKF based on BP is 1.642%, which is 6.07% less than that of EKF. The average error is 0.4499%, which is 0.5891% less than that of EKF. The root-mean-square error is 0.6353%, which is 0.9008% less than EKF. From the error diagram in Figure 3.34b and Table 3.8, it can be observed that the maximum error, average error, and root-mean-square error of the improved EKF based on BP are smaller than that of the EKF algorithm, and the error is more stable, and the error range is within 2%.

The specific experimental steps of DST are described as follows. First, the battery is charged at a constant current (69.27 A) and constant pressure (4.20 V). After charging, the battery is put on hold for 40 minutes to stabilize the chemical reaction inside the batteries. Then, the battery is discharged at 34.35 A for 4 minutes, and then the battery is left resting for 40 seconds after the discharge is stopped. The battery is charged with a constant current of 34.35 A for 2 minutes, then rest the battery for 40 seconds after stopping charging. Constant discharge is carried out at 69.27 A current for 4 minutes and then stayed for 40 seconds. Three steps are repeated until the battery is discharged. Under DST conditions, the SOC estimation results of different algorithms are shown in Figure 3.35.

In Figure 3.35a, S1 is the estimated SOC value of the improved EKF based on BP; S2 is the estimated SOC value of the improved EKF; S3 is the estimated SOC value of the EKF; and S4 is the theoretical SOC value. In Figure 3.35b, E1 is the error curve of the improved EKF based on BP; E2 is the error curve of the improved EKF; and E3 is the error curve of EKF. According to the results, the indicators of different algorithms are shown in Table 3.9.

As shown in Table 3.9, the maximum error of the improved EKF based on BP is 1.433%, and it is 4.877% less than that of EKF. The average error is 0.7073%, which is 1.8917% less than that of EKF. The root-mean-square error is 0.8387%, which is 2.1661% less than that of EKF. According to the error diagram in Figure 3.35b and Table 3.9, the error of the improved EKF algorithm based on BP is smaller than that of the EKF algorithm in all aspects, and the maximum error is within the range of 2%. It is more accurate than the EKF algorithm, which can better track the theoretical value of SOC and is more suitable for the estimation of SOC value.

The BP neural network is used to perform error compensation on the improved EKF. The compensation equation is $SOC_r = SOC_{EKF} + Err_{BP}$,

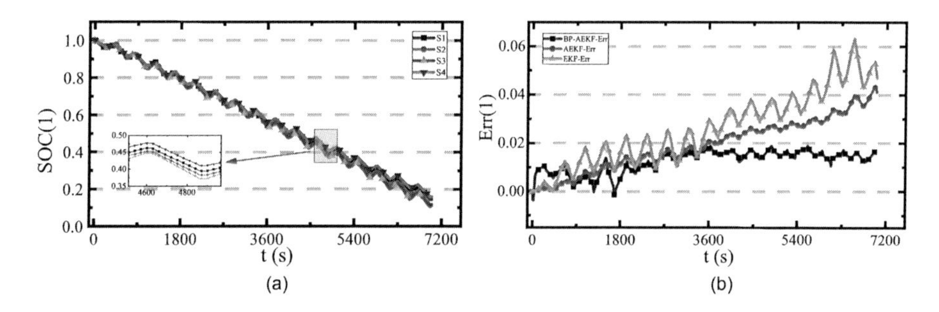

FIGURE 3.35 DST test results. (a) SOC estimation of each algorithm. (b) The error of each algorithm.

TABLE 3.9
DST Experimental Data Analysis

Algorithms	Maximum Error	Average Error	Root Mean Square Error
EKF	6.31%	2.599%	3.0048%
Improved EKF	2.53%	1.715%	1.8654%
Improved BP-EKF	1.433%	0.7073%	0.8387%

which can further improve the accuracy of SOC and reduce the requirements for model accuracy. In addition, BP neural network can also be combined with a GA. The optimization of the GA is used to find the optimal initial weight of the BP neural network. The basic principle is to treat each weight and threshold as an individual and use the test error norm of the BP neural network as the output of the objective function. Consequently, the best individual through selection, crossover, and mutation can be obtained. After the GA, other optimization methods can also be used to optimize BP neural network, such as the ant colony algorithm, particle swarm algorithm, etc.

4. Additional time point method
To overcome the situation that the network falls into local minima, the additional time point is used to slide through these minima. When using the additional time point method, the correction of its weight takes into account not only the effect of the error on the gradient but also the effect of the error trend. The weight and threshold adjustment equations with additional time point factors are shown in Equation (3.202).

$$\Delta w_{ij}(k+1) = (1 - \text{mc})\eta\delta_i\rho_i + \text{mc}\Delta w_{ij}(k) \tag{3.202}$$

In Equation (3.128), k is the training time of the BP neural network, mc is the regulating time point factor, and mc is generally valued at about 0.95. Additional time point method is based on the BP method; in each weight change, add a value proportional to the previous weight change, and according to the BP method, generate a new weight change. At that time, the new weight changes are generated by the steepest descent method. The new weight change is the last. In this way, the adjustment of the weight is changed toward the average direction at the bottom of the error surface. When the weight approaches the flat region at the bottom of the error surface, the appearance of 0 and AB = 0 can make the training jump out of the local minimum value of the error surface and effectively solve the situation that the training process stops because the weight error changes are very small, as shown in Equation (3.203).

$$\Delta w_{ij}(k+1) = \text{mc}\Delta w_{ij}(k) \tag{3.203}$$

5. Combined with a genetic algorithm
   The error of the BP algorithm is reduced, and it is carried out in the direction of the anti-gradient. Therefore, it is easy to fall into the predicament of local minimum points. Once the number of training samples is large, the input–output relationship is complex, and the convergence speed of the network becomes slow. It is shown that the initial value of the network structure is very high. The unreasonable initial value causes the convergence swing of the BP algorithm or even non-convergence.

   The initial value of the BP neural network structure is optimized by GA in random point concentration, and then the BP neural network is used for self-learning. GA is different from other optimization algorithms in that it introduces the mechanism of "natural selection" into the optimization process and has few restrictions on the optimization problem. It requires neither continuity nor differentiability for the constraint and objective function. Therefore, its search process always covers the whole search space, and the global optimal solution can be obtained.

   In this research, according to the characteristics of the GA solution space throughout the whole search space, the optimal solution within the optimal solution fixed range is inherited from the random solution set to form the initial value of the A-N structure, and then the A-N structure is quickly searched by the negative gradient direction to reach the global minimum. In this way, both the global minimum point and the speed of convergence can be guaranteed. GA optimizes the weight of the neural network repeatedly until the mean value no longer increases in a meaningful way. At this time, the parameter combination obtained by decoding is sufficiently close to the optimal parameter combination. On this basis, the BP algorithm is used to fine-adjust them, which has good generality.

3.7.5 Dual-Extended Kalman Filter

As an open-loop estimation method, the ampere–hour integration method can better track the change of battery SOC in a short time, but without the correction feedback function, as time goes by, the SOC estimation error becomes larger and larger.

The EKF algorithm can reduce the system error and measurement error to a certain extent [225–227]. The SOC can be estimated more accurately through the battery voltage correction, but its accuracy is greatly affected by the accuracy of the battery model. To solve the above-mentioned problems, a layer of KF filter is reconstructed to form the DEKF algorithm based on these two algorithms, and Kalman fusion is performed on the estimation results of the two algorithms to obtain a more suitable and accurate estimation value.

The idea of the DEKF algorithm is to filter out system output noise and model internal noise by constructing a double-layer KF. First, the EKF algorithm is used to estimate the SOC, and then EKF_SOC is used as the input of the s-level KF algorithm to correct the Ah_SOC calculated by the ampere–hour integral method. In the process of double-layer KF, the error of the ampere–hour integration method is easily affected by current measurement accuracy, battery capacity, and other factors. The parameter is taken as the internal noise of the model, and the error caused by the EKF algorithm is taken as the output noise of the system. The schematic diagram of the algorithm is shown in Figure 3.36.

When performing a double-layer KF for fusion, the established linear system state and observation are shown in Equation (3.204).

$$\begin{cases} X(k+1) = AX(k) + BI(k) + W(k) \\ Y(k+1) = x(k+1) + V(k+1) \end{cases} \tag{3.204}$$

The second-layer filter takes the SOC estimated by the first-layer filter as the input state quantity, which uses the ampere–hour integration method as the output. The battery state and output are defined, and the initial covariance matrix is established. After obtaining the covariance matrix and predicted value of the state variable at the current time point, the Kalman gain is obtained, and then the state variable and covariance are updated based on the DEKF estimation to obtain the new SOC value. The complete process of the DEKF algorithm is shown in Figure 3.37.

In Figure 3.37, A = 1, B = 1, $x(k)$ represents the SOC state value at time point k, $Y(k)$ represents the system observation value at time point k, $I(k)$ represents the system input current at the current time, $W(k)$ represents the internal noise of the model, and $V(k)$ represents the output noise of the system. EKF_SOC(k) represents the estimated SOC value obtained by the EKF algorithm.

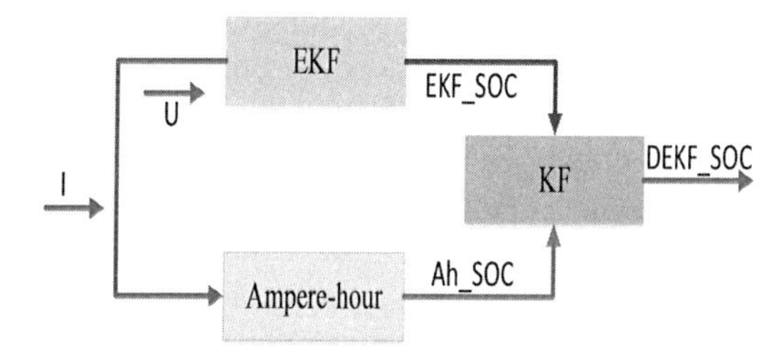

FIGURE 3.36 DEKF algorithm schematic diagram.

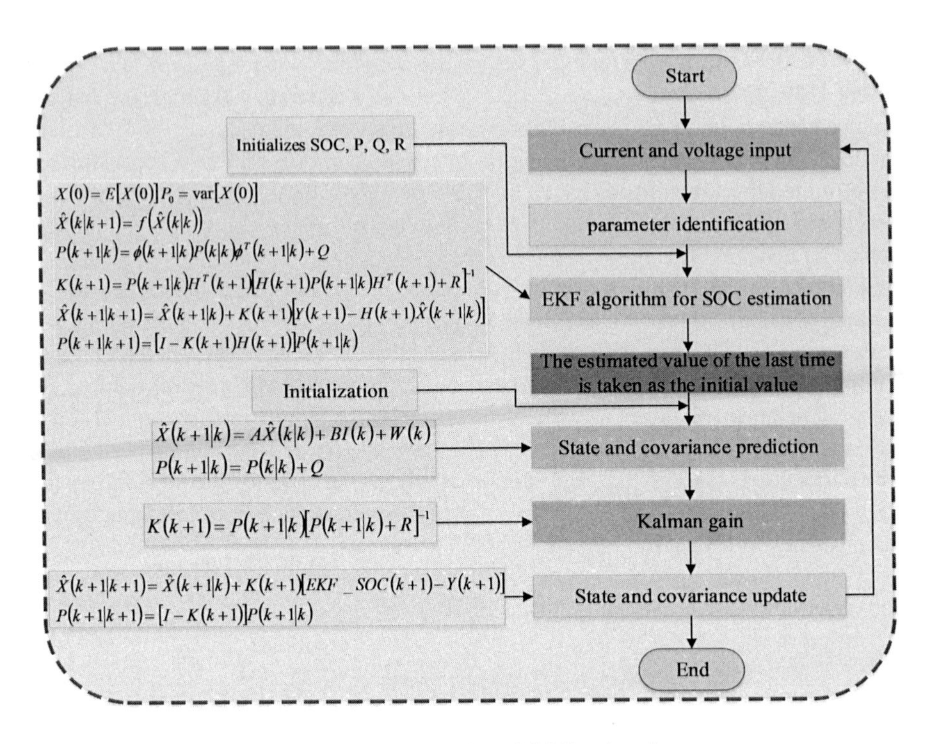

FIGURE 3.37 The whole process of DEKF-based SOC estimation.

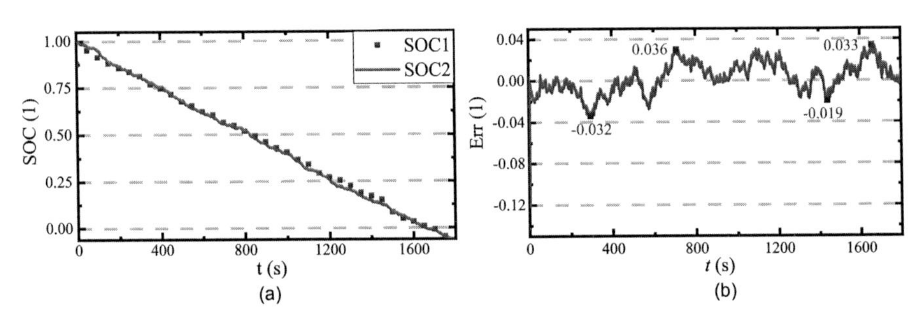

FIGURE 3.38 SOC estimation result based on DEKF. (a) Comparison of SOC estimation results. (b) SOC estimation error.

To verify the feasibility of the DEKF algorithm, the lithium-ion battery SOC estimation model is established. SOC estimation based on DEKF is carried out under different working conditions such as constant current discharge and HPPC, and the ampere–hour integration method under HPPC working conditions. The estimation accuracy of EKF and DEKF are compared to verify the superiority of the proposed algorithm. In the experiment, R_0, R_1, R_2, C_1, and C_2 are substituted into the algorithm in the form of average values, and the test battery model is a ternary 50 Ah lithium-ion battery. The 2 C constant current discharge experiment is used to verify the effectiveness of the EKF algorithm in battery SOC estimation. The simulation effect is shown in Figure 3.38a, and the error is shown in Figure 3.38b.

In Figure 3.38a, the SOC1 is the actual SOC curve, and SOC2 is the DEKF-estimated SOC curve. The experimental results in Figure 3.38b show that the maximum estimation error of SOC during 2C constant current discharge is 0.036, which is less than 3.6%, and it is within the allowable error range. It can be observed from the figure that as the number of filtering increases, the predicted value curve gradually approaches the true value curve and fluctuates near the true value curve. The error value also gradually decreases from the initial larger error and finally fluctuates around 0, showing a trend of convergence.

After deriving and analyzing the principles and steps of the algorithm, the DEKF algorithm is used to estimate the battery SOC, and the results are compared and analyzed with the ampere–hour integral method and the EKF algorithm. The parameter design of the first-level EKF algorithm remains unchanged. In the s-level KF algorithm, the internal noise R to the model is 0.001, and the system output noise Q is obtained by the posterior estimated covariance of the EKF algorithm. The simulation results under HPPC working conditions are shown in Figure 3.39.

According to Figure 3.39a, SOC1 is the SOC curve estimated by DEKF, and SOC2 is the actual SOC curve. The experimental results in Figure 3.39b show that under HPPC conditions, the maximum estimation error of SOC is 0.0039, which is less than 0.39%. The ampere integral method is a simple and widely used SOC estimation method. By integrating the current flowing through the battery within a certain working period, the charge–discharge capacity of the battery can be obtained. The simulation results of HPPC operating conditions based on Ah points are shown in Figure 3.40.

In Figure 3.40a, the SOC1 is the actual SOC curve, and SOC2 is the ampere–hour integral estimated SOC curve. The experimental results in Figure 3.40b show that under HPPC conditions, the maximum estimation error of SOC is 0.0382, which is less than 3.82%, and it is within the allowable error range. It can be observed from the above figure that the estimation error of the ampere–hour integration method continues to increase during the entire discharge process. This is closely related to factors such as current detection accuracy, fluctuations, battery capacity, and temperature. Under HPPC working conditions, the SOC estimation results based on the EKF algorithm are shown in Figure 3.41.

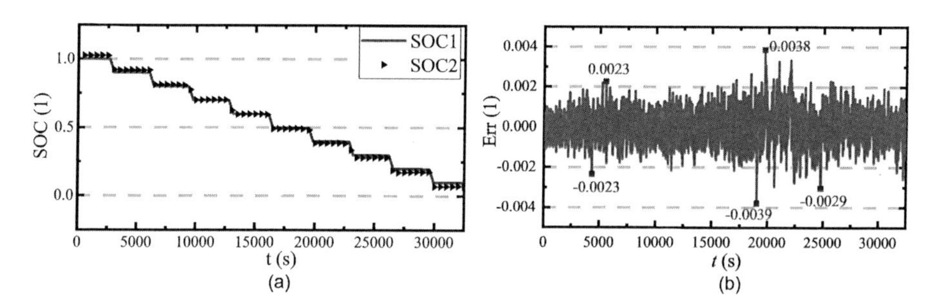

FIGURE 3.39 SOC estimation results based on DEKF under HPPC conditions. (a) Comparison of SOC estimation results. (b) SOC estimation error.

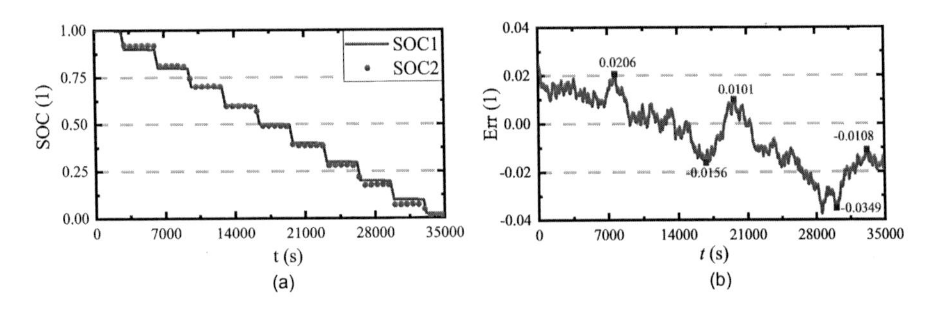

FIGURE 3.40 SOC estimation under HPPC working conditions. (a) Comparison of SOC estimation results. (b) SOC estimation error.

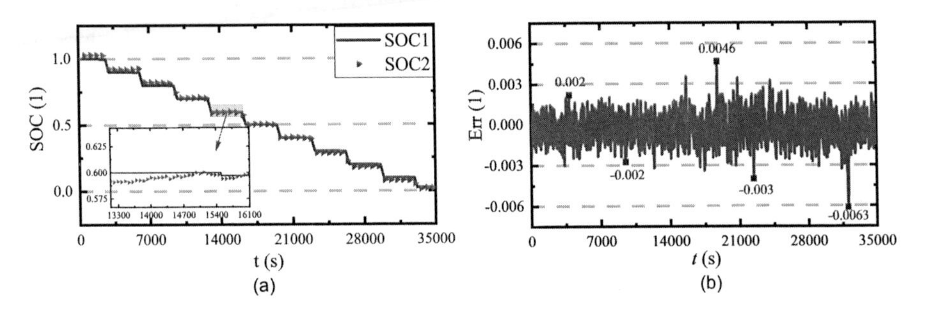

FIGURE 3.41 SOC estimation results based on EKF under HPPC conditions. (a) Comparison of SOC estimation results. (b) SOC estimation error.

TABLE 3.10
Comparison of Experimental Results of Different Algorithms

Estimation Method	Ampere–Hour Integration	EKF Algorithm	DEKF Algorithm
Maximum error	3.82%	0.63%	0.39%

In Figure 3.41a, SOC1 is the SOC value estimated by the EKF algorithm, and SOC2 is the actual SOC value. Figure 3.41b shows that the maximum estimation error of SOC is 0.0063 under the HPPC condition, which is less than 0.63%. It is worth noting that when the battery is at the end of discharge, the SOC estimation error fluctuates greatly, which is mainly caused by the nonlinearity of the battery itself. The comparison of the experimental results of the ampere–hour integration method, EKF, and DEKF algorithm to estimate SOC under HPPC conditions is shown in Table 3.10.

Through the above comparison, although the ampere–hour integration method is simple and easy to implement, the initial value cannot be obtained, the accuracy requirements are strict, and the lack of feedback correction links leads to a large accumulation of errors. The EKF algorithm has good robustness, but the DEKF estimation accuracy is high; it is not easily disturbed by the accuracy of the battery model with strong stability and convergence.

3.7.6 Strong Tracking-Adaptive Extended Kalman Filter

The core idea of the strong tracking-adaptive extended Kalman filtering (ST-AEKF) algorithm is that for nonlinear Gaussian systems, based on the EKF, it combines a strong tracking filter and an adaptive filter. To make the system have the ability to track sudden changes, a fading factor is introduced into the error covariance matrix of EKF to strengthen the proportion of current observation data. For achieving the strong tracking characteristics of the filter, a couple of conditions need to be met during recursion, as shown in Equation (3.205).

$$\begin{cases} E\left(X_k - \hat{X}_k\right)\left(X_k - \hat{X}_k\right)^T = \min \\ E\left(\gamma_k \gamma_{k+j}^T\right) = 0, \ \ k = 1,2,3,\ldots; \ j = 1,2,3,\ldots \\ \gamma = Z_{k+1} - h\left(X_{k+1}^-\right) \end{cases} \tag{3.205}$$

If the noise characteristic is always assumed to be Gaussian white noise, the actual estimated error value of the system state is different from the theoretically calculated error value. Therefore, the introduction of the adaptive filter can continuously estimate and correct the statistical characteristics of noise through measurement data to reduce estimation errors. The error covariance matrix after optimization is shown in Equation (3.206).

$$\hat{P}_{k+1}^- = \lambda_{k+1} A_k \hat{P}_k A_k^T + \Gamma Q_{k+1} \Gamma^T \tag{3.206}$$

In Equation (3.206), λ_{k+1} represents the fading factor and Γ is the noise drive matrix. When using the orthogonal principle of Equation (3.124) to calculate the fading factor, the gradient method is required. This method uses nonlinear programming to solve the optimal attenuation factor, and the calculation amount is too large to realize online calculation. Therefore, this research adopts the calculation method of suboptimal fading factor, as shown in Equation (3.207).

$$\lambda_k = f(x) = \begin{cases} 1, & e_k \le 1 \\ e_k = \dfrac{tr\left(N_k\right)}{tr\left(M_k\right)}, & e_k > 1 \end{cases} \tag{3.207}$$

In Equation (3.207), N_k and M_k can be expressed in Equation (3.208).

$$\begin{cases} N_k = V_k - \beta R_k - C_k \Gamma Q_{k-1} \Gamma^T C_k^T \\ V_k = f(x) = \begin{cases} \gamma_1 \gamma_1^T, \ k = 1 \\ \dfrac{\rho V_{k-1} + \gamma_k \gamma_k^T}{1+\rho}, \ k > 1 \end{cases} \\ M_k = C_k A_{k-1} P_{k-1} A_{k-1}^T C_k^T \end{cases} \tag{3.208}$$

In Equation (3.208), V_k is the residual covariance matrix, ρ is the forgetting factor with $0 < \rho \leq 1$, and β is the weakening factor with $\beta \geq 1$. In this research, the adaptive filter is a statistically large posterior suboptimal unbiased estimator of noise based on measured values. The recursive process of input process noise Q and observation noise R is shown in Equation (3.209).

$$\begin{cases} \hat{Q}_{k+1} = \dfrac{1}{k+1} G \displaystyle\sum_{i=0}^{k} \left(K_{k+1} \tilde{\gamma}_{k+1} \tilde{\gamma}_{k+1}^{T} K_{k+1}^{T} + P_{k+1} - A P_{k+1} A^{T} \right) G^{T} \\ G = \left(\Gamma^{T} \Gamma \right) \Gamma^{T} \\ \hat{R}_{k+1} = \dfrac{1}{K+1} \displaystyle\sum_{i=0}^{k} \tilde{\gamma}_{k+1} \tilde{\gamma}_{k+1}^{T} - C P_{k+1} A^{T} C^{T} \end{cases} \quad (3.209)$$

The adaptive filter can estimate Q and R online in real time and achieve the goal of continuous correction of the SOC estimation value, thereby realizing the adaptive correction function to achieve the effect of improving the accuracy of the SOC estimation. Therefore, the recursive process of the second improved KF combined with a strong tracking filter and an adaptive filter is shown in Equation (3.210).

$$\begin{cases} \hat{X}_{k+1}^{-} = f\left(\hat{X}_k \right) \\ \hat{P}_{k+1}^{-} = \lambda_{k+1} A_k \hat{P}_k A_k^{T} + \Gamma Q_{k+1} \Gamma^{T} \\ K_{k+1} = \hat{P}_{k+1}^{-} C_{k+1}^{T} \left(C_{k+1} \hat{P}_{k+1}^{-} C_{k+1}^{T} + R_{k+1} \right) \\ \hat{X}_{k+1} = X_{k+1}^{-} + X_{k+1} \left[Z_{k+1} - h\left(X_{k+1}^{-} \right) \right] \\ \hat{P}_{k+1} = \left[I - K_{k+1} C_{k+1} \right] P_{k+1}^{-} \end{cases} \quad (3.210)$$

In Equation (3.210), the fading factor λ_{k+1} is introduced into the EKF algorithm to enhance the tracking ability of sudden changes. Meanwhile, the adaptive filter which can statisticize the characteristics of time-varying noise is used to adjust the noise parameters of the system for continuously estimating and modifying the statistical characteristics of noise through the measurement data to reduce the error of estimation.

3.7.7 Cubature Kalman Filter

The core idea of the cubature Kalman filtering (CKF) algorithm is that for nonlinear Gaussian systems, the posterior mean and covariance of the state are approximated by the third-order spherical radial cubature criterion to ensure that the third-order polynomial is theoretically approximate to the posterior mean and variance of the nonlinear Gaussian state [228–231]. The state and observation values of the system are shown in Equation (3.211).

$$\begin{cases} x_k = f(x_{k-1}, u_{k-1}) + w_k \\ y_k = h(x_{k-1}, u_{k-1}) + v_k \end{cases} \tag{3.211}$$

In Equation (3.211), w_k is the process noise and v_k is the measurement noise. Then, the SOC estimation based on the CKF algorithm can be divided into three steps:

Step 1: initialization, as shown in Equation (3.212).

$$\begin{cases} \hat{X}_k = E(X_0) \\ P_k = E\left(X_0 - \hat{X}_k\right)\left(X_0 - \hat{X}_k\right)^T \end{cases} \tag{3.212}$$

Step 2: time update

Calculating the cubature point, as shown in Equation (3.213).

$$\begin{cases} P_{k-1|k-1} = S_{k-1|k-1} S_{k-1|k-1}^T \\ X_{i,k-1|k-1}^i = S_{k-1|k-1}\xi_i + \hat{x}_{k-1|k-1},\ i = 1,2,\ldots 2n \end{cases} \tag{3.213}$$

In Equation (3.213), the calculation equation of ξ is shown in Equation (3.214).

$$\xi_i = \begin{cases} \sqrt{n}I,\ i = 1,2,\ldots n \\ -\sqrt{n}I,\ i = n+1, n+2,\ldots 2n \end{cases} \tag{3.214}$$

Spread cubature point, as shown in Equation (3.215).

$$X_{k+1|k}^* = f\left(X_{i,k+1|k}\right) \tag{3.215}$$

The predicted value of the state variable and the predicted value of the estimated error covariance can be calculated, as shown in Equation (3.216).

$$\begin{cases} \hat{x}_{k+1|k} = \dfrac{1}{2n}\displaystyle\sum_{i=1}^{2n} X_{i,k+1|k}^* \\ P_{k+1} = \dfrac{1}{2n}\displaystyle\sum_{i=1}^{n} X_{i,k+1|k}^* \left(X_{i,k+1|k}^*\right)^T - \hat{x}_{k+1|k}\left(\hat{x}_{k+1|k}\right)^T + Q \end{cases} \tag{3.216}$$

Step 3: measurement update

Calculating the cubature point, as shown in Equation (3.217).

$$\begin{cases} P_{k|k-1} = S_{k|k-1} S_{k|k-1}^{T} \\ X_{i,k|k-1}^{i} = S_{k|k-1}\xi_i + \hat{x}_{k|k-1}, i = 1,2,\ldots 2n \end{cases} \tag{3.217}$$

Spreading the cubature point, as shown in Equation (3.218).

$$Z_{k|k+1}^{i} = h\left(X_{i,k+1|k}\right) \tag{3.218}$$

Calculating and measuring the predicted value, as shown in Equation (3.219).

$$\hat{z}_{k|k+1} = \frac{1}{2n}\sum_{i=1}^{2n} Z_{i,k+1} \tag{3.219}$$

Calculating the innovation covariance matrix and cross-covariance matrix, as shown in Equation (3.220).

$$\begin{cases} P_{k|k+1}^{z} = \frac{1}{2n}\sum_{i=1}^{2n} Z_{k|k+1}^{i}\left(Z_{k|k+1}^{i}\right)^{T} - \hat{z}_{k|k+1}\left(z_{k|k+1}^{i}\right)^{T} + R \\ P_{k|k+1}^{xz} = \frac{1}{2n}\sum_{i=1}^{n} x_{k|k+1}^{i}\left(x_{k|k+1}^{i}\right)^{T} - \hat{x}_{k|k+1}\left(\hat{z}_{k|k+1}\right)^{T} \end{cases} \tag{3.220}$$

Calculating the Kalman gain, state quantity, as well as the error covariance correction, as shown in Equation (3.221).

$$\begin{cases} K_k = P_{k|k-1}^{xz}\left(P_{k|k-1}^{z}\right)^{-1} \\ \hat{x}_{k|k} = \hat{x}_{k|k-1} + K_k\left(z_k - \hat{z}_{k|k-1}\right) \\ P_{k|k} = P_{k|k-1} - K_k P_{k|k-1}^{z} K_k^{T} \end{cases} \tag{3.221}$$

Taking a ternary lithium-ion battery with a rated capacity of 70 Ah as the research object, using a second-order RC equivalent model. The type parameters are identified by the offline method and verified under different working conditions. The SOC estimation result under HPPC working conditions is shown in Figure 3.42.

In Figure 3.42a, S1 is the reference SOC value; S2 and S3 represent the SOC values based on the EKF and CKF algorithms, respectively; and Err1 and Err2 represent the SOC estimation error values of the EKF algorithm and the CKF algorithm, respectively. In Figure 3.42b, the maximum of Err1 and Err2 are 0.0847 and 0.0650, respectively. As can be observed from the experimental results, the error of the estimation result of the CKF algorithm is smaller and the tracking performance is better.

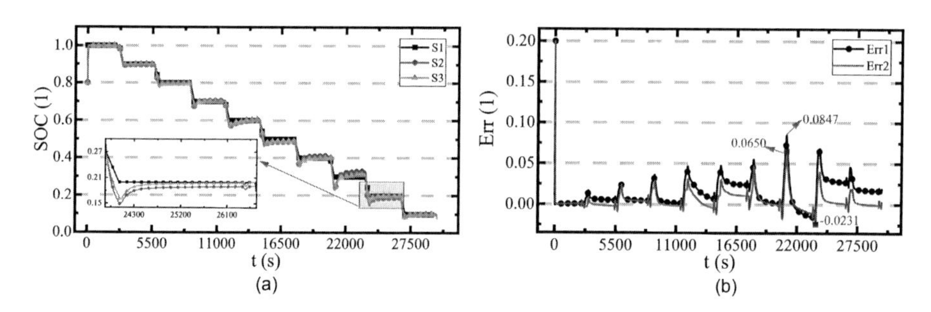

FIGURE 3.42 SOC estimation results in HPPC working conditions. (a) Comparison of estimation results. (b) SOC estimation errors.

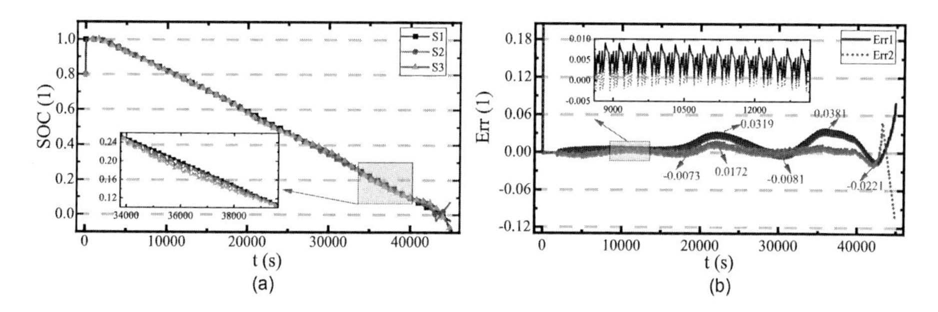

FIGURE 3.43 SOC estimation results of BBDST working conditions. (a) Comparison of SOC estimation. (b) SOC estimation errors.

To further verify the superiority of CKF algorithm SOC estimation, the BBDST working condition is used for verification, and the result is shown in Figure 3.43.

In Figure 3.43, S1 is the reference SOC value, S2 and S3 represent the SOC values based on CKF and EKF respectively, and Err1 and Err2 represent the SOC estimation error values of the CKF algorithm and the EKF algorithm, respectively. In Figure 3.43b, when the SOC estimation error becomes larger, the discharge ends, which is the result of the violent chemical reaction inside the batteries. Except for the end of discharge, the CKF algorithm SOC estimation has a good overall effect.

3.7.8 Square Root Cubature Kalman Filter

When the CKF algorithm is used for lithium-ion battery SOC estimation, the state quantity error covariance matrix needs to meet two conditions, which are symmetry and positive definiteness. However, in each iteration process, due to the square root operation of the matrix, matrix inversion calculation, matrix squaring operation to amplify the rounding error, and the subtraction operation of the two matrices when updating the error covariance matrix, both the symmetry and the positive definiteness of the covariance cannot be guaranteed, which can eventually lead to divergence or instability of the system.

To solve the above problems, a square-root filter is introduced into the CKF algorithm to form the square-root CKF algorithm (SR-CKF). The SR-CKF algorithm directly calculates the square-root factor of the predicted value of the state error covariance and the estimated value of the state error covariance, which can avoid the step of finding the square root of the matrix. Compared with the CKF algorithm, the SR-CKF algorithm has two advantages: the first is that the recursive update is directly performed in the form of the square root of the covariance matrix in the filtering process, which reduces the computational complexity and obtains higher efficiency, and the second is that it can guarantee the nonnegative qualitative nature of the covariance matrix, effectively avoid the divergence of the filter, and improve the convergence speed and numerical stability of the filter [232]. Setting the state-space expression of the nonlinear discrete system, as shown in Equation (3.222).

$$\begin{cases} x_k = f(x_{k-1}, u_{k-1}) + w_k \\ y_k = h(x_{k-1}, u_{k-1}) + v_k \end{cases} \tag{3.222}$$

In Equation (3.222), x_k and y_k are the state vector and measurement vector of the system, $f(x)$ and $h(x)$ are known functions, w_k and v_k are process noise and observation noise, respectively, and both of the two are Gaussian white noise, that is, the mean value is 0 and is not correlated with each other. The SOC estimation process based on SR-CKF is described as follows:

Step 1. Initialization of state estimation

Defining the initial state variable $\hat{X}_{0|0}$ and the initial value $P_{0|0}$ of the error covariance matrix and performing Cholesky decomposition on the error covariance matrix, the result is shown in Equation (3.223).

$$\begin{cases} \hat{X}_{0|0} = E(X_0) \\ S_0 = \text{chol}(P_{0|0}) \end{cases} \tag{3.223}$$

Step 2. Time update

1. Calculation of the $2n$ cubature points at the current time point, as shown in Equation (3.224).

$$X_{i,k-1|k-1} = S_{k-1|k-1}\xi_i + X_{k-1|k-1} \tag{3.224}$$

2. The propagation of the cubature point, that is, the predicted value of the cubature point is calculated by the nonlinear state transfer function, as shown in Equation (3.225).

$$X^*_{i,k|k-1} = f(X_{i,k-1|k-1}, u_{k-1}) \tag{3.225}$$

3. Prediction of the state variables through the predicted value of the cubature point, as shown in Equation (3.226).

$$\hat{X}_{k|k-1} = \frac{1}{2n}\sum_{1}^{2n} X^{*}_{k|k+1} \tag{3.226}$$

4. Estimating the square-root factor of the prediction error covariance matrix through the obtained cubature point prediction value and the state variable prediction result, as shown in Equation (3.227).

$$\chi^{*}_{i,k|k-1} = \frac{1}{\sqrt{2n}}\left[X^{*}_{1|k-1} - \hat{X}_{1|k-1}, X^{*}_{2|k-1} - \hat{X}_{2|k-1}, \cdots, X^{*}_{2n|k-1} - \hat{X}_{2n|k-1}\right] \tag{3.227}$$

Step 3. Measurement update

The state prediction value at time point k is obtained through the time update, and then, the optimal state estimation value can be obtained through the update of the measurement value at this time point.

1. Calculation of the $2n$ cubature points of the updated state, as shown in Equation (3.228).

$$X_{k|k-1} = S_{k|k-1}\xi_i + \hat{X}_{k|k-1} \tag{3.228}$$

2. Spreading of cubature point, as shown in Equation (3.229).

$$Z_{k|k-1} = h\left(X_{k|k-1}\right) \tag{3.229}$$

3. Estimation of the predicted measurement value, as shown in Equation (3.230).

$$\hat{Z}_{k|k-1} = \frac{1}{2n}\sum_{1}^{2n} Z_{k|k-1} \tag{3.230}$$

4. Calculation of the square-root factor of the innovation covariance matrix, as shown in Equation (3.231).

$$S_{zz,k|k-1} = qr\left[\eta_{k|k-1}, S_{R,k}\right] \tag{3.231}$$

In Equation (3.231), $S_{R,k}$ is equal to sqrt (R_{k-1}), and the calculation equation of $\eta_{k|k-1}$ is shown in Equation (3.232).

$$\eta_{k|k-1} = \frac{1}{2n}\left[Z_{1|k-1} - \hat{Z}_{1|k-1}, Z_{2|k-1} - \hat{Z}_{2|k-1}, \cdots, Z_{2n|k-1} - \hat{Z}_{2n|k-1}\right] \tag{3.232}$$

5. Calculation of the square-root factor of the cross-covariance matrix, as shown in Equation (3.233).

$$P_{xz,k|k-1} = x_{k|k-1}\eta_{k|k-1}^{T} \tag{3.233}$$

In Equation (3.233), the calculation equation of $x_{k|k-1}$ is shown in Equation (3.234).

$$x_{k|k-1} = \frac{1}{\sqrt{2n}}\left[X_{1|k-1} - \hat{X}_{1|k-1}, X_{2|k-1} - \hat{X}_{2|k-1}, \cdots, X_{2n|k-1} - \hat{X}_{2n|k-1}\right] \tag{3.234}$$

6. Kalman gain update, as shown in (3.235).

$$K_k = P_{xz,k|k-1} \Big/ \left(S_{zz,k|k-1}^{T} S_{zz,k|k-1}\right) \tag{3.235}$$

7. Calculation of state variable, as shown in (3.236).

$$\hat{K}_{k|k} = \hat{K}_{k|k-1} + K_k\left(Z_k - \hat{Z}_{k|k-1}\right) \tag{3.236}$$

8. Update of the square-root factor of the state estimation error covariance matrix, as shown in (3.237).

$$S_{k|k} = qr\left(\chi_{k|k-1} - K_k\eta_{k|k-1}, K_k S_{R,k}\right) \tag{3.237}$$

In Equation (3.237), $S_{R,k}$ is the square-root factor of the observed noise variance matrix. So far, the state filtering at time point k is completed, set the time point $k = k + 1$ and return to step 2 to realize the state estimation at the next time. The second-RC model is used to perform the circuit equivalent of the lithium-ion battery, and the recursive least square (RLS) method is used to identify the model parameters. The SOC estimation flowchart based on SR-CKF is shown in Figure 3.44.

In Figure 3.44, the function of U_{OC} concerning SOC is obtained by polynomial fitting of HPPC experimental data. To verify the feasibility of the SR-CKF algorithm, a ternary lithium-ion battery with a rated capacity of 70 Ah is used as the research object, and a lithium-ion battery SOC estimation model is established. The SOC estimation based on SR-CKF and RLS algorithm is carried out under different working conditions such as HPPC and BBDST, and the superiority of the proposed algorithm is verified by analyzing the SOC estimation results. Setting the measurement noise variance $R = 0.02$, the process noise variance $Q = \text{diag}\left(1e^{-6}, 1e^{-6}, 1e^{-6}\right)$, and the initial value of the reference SOC is 1, the initial value of the experimental SOC is set to 0.8 to verify the convergence effect of the algorithm. The SOC estimation results under HPPC condition are shown in Figure 3.45.

In Figure 3.45a, S1 is the SOC reference value, S2 is the SOC estimation result of the CKF algorithm, and S3 is the SOC estimation value of SR-CKF. In Figure 3.45b,

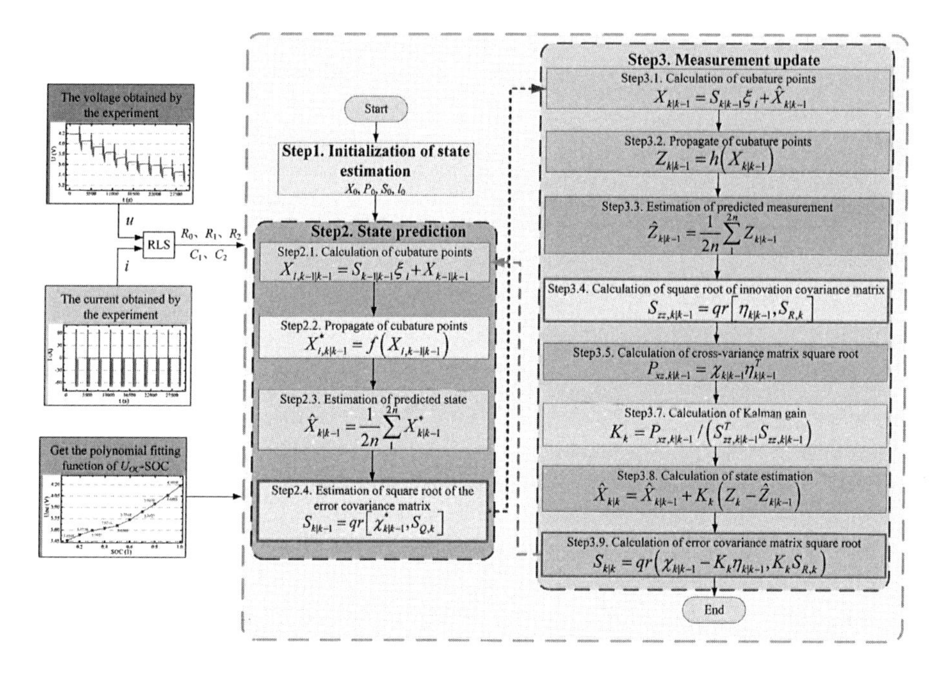

FIGURE 3.44 The SOC estimation flowchart based on the SR-CKF algorithm.

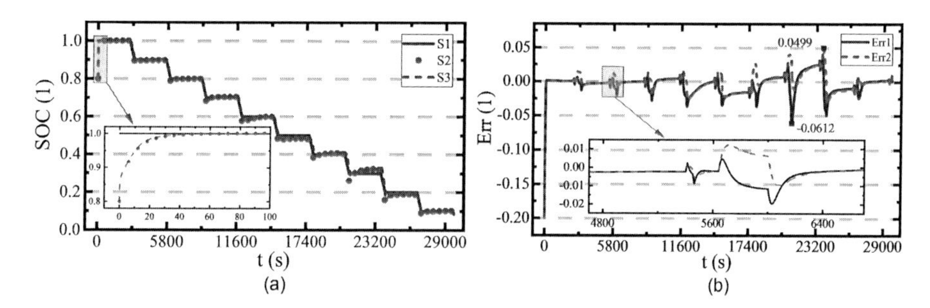

FIGURE 3.45 SOC estimation results of BBDST working condition. (a) Comparison of SOC estimation. (b) SOC estimation errors.

Err1 represents the SOC estimation error of the CKF algorithm, and Err2 is the SOC estimation error of the SR-CKF algorithm. From Figure 3.45a, it can be observed that the convergence effects of the two algorithms are not much different. When the system changes suddenly, the estimation result of the SR-CKF algorithm is more stable. It also can be observed from Figure 3.45b that the absolute maximum error of Err1 is 0.0612, and the absolute maximum error of Err2 is 0.0499, the overall fluctuation range of the SR-CKF algorithm is smaller. The results prove the superiority of the SR-CKF algorithm in SOC estimation.

Lithium-ion batteries have a wide range of applications and the working environment is always complex and changeable. To further verify the effect of the SR-CKF algorithm for lithium-ion battery SOC estimation under complex working conditions,

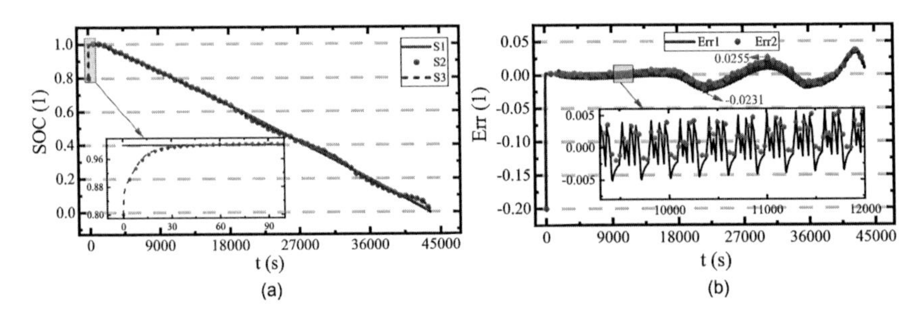

FIGURE 3.46 SOC estimation results of BBDST working condition. (a) Estimation results. (b) Error comparison.

the BBDST working condition is used for experimental verification. BBDST working conditions include starting, coasting, acceleration, and rapid acceleration of electric vehicles, which is more complicated than the HPPC working conditions. It can better simulate the actual working conditions of lithium-ion batteries. The relevant parameter settings are consistent with the verification of the HPPC operating condition, and the SOC estimation results under the BBDST operating condition are shown in Figure 3.46.

In Figure 3.46a, S1 is the SOC reference value, S2 is the SOC estimation result of the CKF algorithm, and S3 is the SOC estimation value of SR-CKF. In Figure 3.46b, Errl represents the SOC estimation error of the CKF algorithm, and Err2 is the SOC estimation error of the SR-CKF algorithm. It can be observed from Figure 3.46a that the convergence effect of the SR-CKF algorithm for SOC estimation under the BBDST condition is slightly better than that of the CKF algorithm, but the convergence effect of the two algorithms is not much different. It can be observed from Figure 3.46b that, except for the end of discharge, the overall SOC estimation error of the SR-CKF algorithm is slightly smaller than that of the CKF algorithm, but the effect is not obvious. The maximum absolute value of the SOC estimation error of the CKF algorithm and SR-CKF algorithm is 0.0231 and 0.0255 respectively, which means that some parts of the SR-CKF algorithm have worse SOC estimation results than CKF.

Synthesizing the SOC estimation results of the two working conditions, the introduction of the square-root algorithm can only guarantee the nonnegative definiteness of the covariance matrix, increase the calculation speed of the algorithm, and ensure the stability of the algorithm, but it cannot significantly improve the accuracy and convergence speed of the estimation results.

3.7.9 Adaptive H∞ Filter

The KF algorithm is an effective method to solve the state estimation problem of the lithium-ion battery system, but its premise is that the system model is accurate and the external input statistical characteristics are known, which is inconsistent with the actual situation. In addition, the traditional KF algorithm needs to set the initial noise information in advance, and the mismatch of the initial noise will cause the estimation result to diverge [233]. Also, the complexity of the lithium-ion battery condition weakens the estimation performance of the traditional EKF algorithm. Therefore, an adaptive H-infinity filtering algorithm based on a noise information covariance matching algorithm is proposed and applied to the adaptive estimation of the system to accurately estimate the SOC value. For the lithium-ion battery system, its state equation and observation equation are shown in Equation (3.238).

$$\begin{cases} x_k = \begin{bmatrix} Z(k) \\ U_P(k) \end{bmatrix} = \begin{bmatrix} 1 & 0 \\ 0 & 1-\dfrac{T}{R_P C_P} \end{bmatrix} \begin{bmatrix} Z(k-1) \\ U_P(k-1) \end{bmatrix} + \begin{bmatrix} -\dfrac{\eta T}{C_a} \\ \dfrac{T}{C_P} \end{bmatrix} i_L(k-1) + w_{k-1} \\ y_k = [U_L(k)] = [U_{OC}] - \begin{bmatrix} 0 & 1 \end{bmatrix} \begin{bmatrix} Z(k) \\ U_P(k) \end{bmatrix} - [R] i_L(k-1) + v_k \end{cases} \tag{3.238}$$

In Equation (3.238), $Z(k)$ is the SOC value at time point k, T is the sampling period, w and v are the system noise and observation noise, respectively. The general form corresponding to the above equation is shown in Equation (3.239).

$$\begin{cases} x_k = f(x_{k-1}, u_{k-1}) + w_{k-1} \approx A_{k-1} x_{k-1} + B_{k-1} u_{k-1} + w_{k-1} \\ y_k = h(x_k, u_k) + v_k \approx C_k x_k + D_k + v_k \end{cases} \tag{3.239}$$

For the above state–space model of the battery, the matrices of A, B, C, and D are the system parameter matrices, which are determined by the system structure. To realize the adaptive updating of the process noise and measurement noise, the corresponding symmetric positive definite matrices Q and R are designed by referring to the principle of adaptive EKF algorithm, which represents the covariance of the system state noise and the observation noise respectively; x represents the state variable, and matrix P is no longer the covariance of state estimation error, but a special matrix in H-infinity filtering algorithm; $\sum 0|0$ is the error covariance matrix of the state estimation; the symmetric positive definite matrix S is a weight matrix designed based on a specific problem, which is used to give different weights to each component of the state quantity; λ represents the performance boundary of the H-infinity filter, M is the window function of the noise adaptive update process, and N is the window size, which is used to determine the length of the adaptive update period, and the estimation flowchart of adaptive H-infinity filter is shown in Figure 3.47.

In Figure 3.47, SOC estimation based on an adaptive H-infinity filtering algorithm mainly includes the following steps: first, the model parameters and matrix, including $x_{0|0}$, $P_{0|0}$, $\sum_{0|0}$, Q_0, S, and λ are initialized, then a prior estimation of the state variable x and update of the matrix P and $\sum_{0|0}$ is performed. According to the real-time measured terminal voltage y_k, the residual and filter gain matrix K is calculated, and the gain matrix is then used to modify the prior estimation of the state variable and update the matrix P and $\sum_{0|0}$. Finally, the covariance matching algorithm in

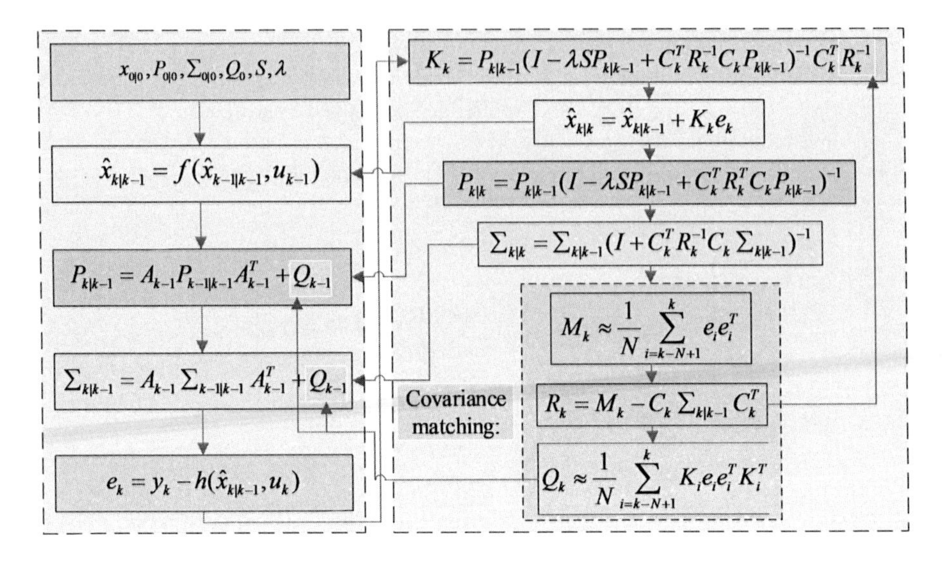

FIGURE 3.47 AHF flowchart.

Figure 3.47 is used to adaptively update the noise covariance matrix Q and R of the battery system.

3.7.10 NARX-EKF Network

The essence of the KF is a series of mathematical calculation equations to realize functions such as prediction and correction. The optimal estimation algorithm and EKF are KF-based methods. Since the NARXNN lacks stability in the SOC estimation of the charge–discharge process under complex conditions, this study uses the EKF to improve the NARXNN model to optimize the estimation performance of the network model [234–236]. The NARX-EKF model is shown in Figure 3.48.

In Figure 3.48, the KF estimation method usually uses the equivalent circuit model to establish the observation equation. The ampere–hour integration process gives the discrete state-space equation of the system. In this study, the NARXNN model is equivalent to the observation equation. The SOC value estimated by the NARXNN is input into the EKF module for filtering, thereby eliminating noise and random errors and further optimizing the SOC estimation. Equations (3.240) and (3.241) are the state equation and measurement equation of the EKF algorithm, respectively.

$$\mathrm{SOC}_{k+1} = \mathrm{SOC}_k - \left(\frac{\eta \Delta t}{Q_N}\right) i_k + w_k \tag{3.240}$$

$$E_k = \mathrm{SOC}_k + v_k \tag{3.241}$$

In Equations (3.240) and (3.241), i_k is the current value at the time point k, SOC_k is the estimated value of the SOC state time point k, η is the charge–discharge efficiency, Q_N is the rated battery capacity, w_k and v_k are process noise and measurement noise, and E_k is the estimated value of the NARXNN at time point k. The Jacobian

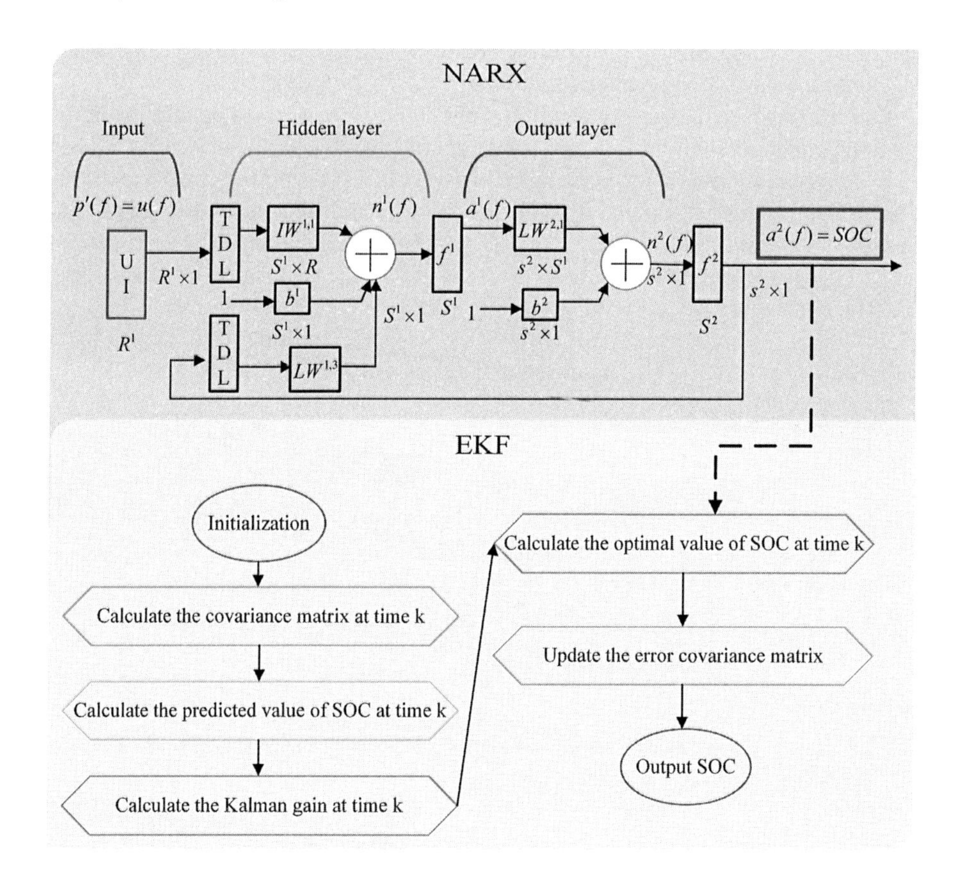

FIGURE 3.48 NARX-EKF network structure.

matrix is obtained by the Taylor series expansion to linearize the state–space equation. A more accurate result can be obtained in a nonlinear dynamic system than the classical KF, thereby reducing the cumulative error of the OCV. The KF method after the extended application is called the EKF. $x_{0|0}$ is the initial value of the state quantity, $p_{0|0}$ is the initial value of the state error covariance, $x(k \mid k-1)$ is the one-step prediction value of the state, i_{k-1} is the input of the system at $k-1$ which is the current at k in the SOC estimation, and A_{k-1} and B_{k-1} are the state transition matrix at $k-1$. $P(k \mid k-1)$ and $\hat{P}_{k-1}$ are the predicted value and estimated value of the state error covariance matrix, respectively. K_k and C_k are the Kalman gain value and state conversion matrix at the time point k. Q and R are the variance values of process noise and observation noise, respectively. The noise value is difficult to determine and therefore, usually, it is enough to keep debugging to achieve the best algorithm. The value of R and Q in this study is 0.1 and 0.001, respectively. $U_L(k)$ is the system observation value at time point k, that is, the SOC value output by NARX.

Through open-loop network training and closed-loop network testing, the improved algorithm has the characteristics of high accuracy and low delay, but this method is suitable for uncomplicated cycle conditions. To further verify the applicability of NARXNN, the closed-loop NARXNN is used for training and testing. On this basis, the complexity of the working conditions is increased. The dynamic working condition dataset is shown in Figure 3.49.

Considering that the battery is running under dynamic working conditions in the actual application environment, to verify the effect and accuracy of the NARXNN algorithm following complex working conditions, the DST working condition data is used as the training set, and the BBDST working condition data is used as the test set. In Figure 3.49, (a) is the voltage under DST working conditions, and (b) is the current under DST working conditions, both of which are the input variables of the training set. The current and voltage curves for DST conditions are shown in Figure 3.50.

In Figure 3.50, (a) and (b) are BBDST operational condition data used as input variables of the test set. The algorithm simulation is shown in Figures 3.51–3.53.

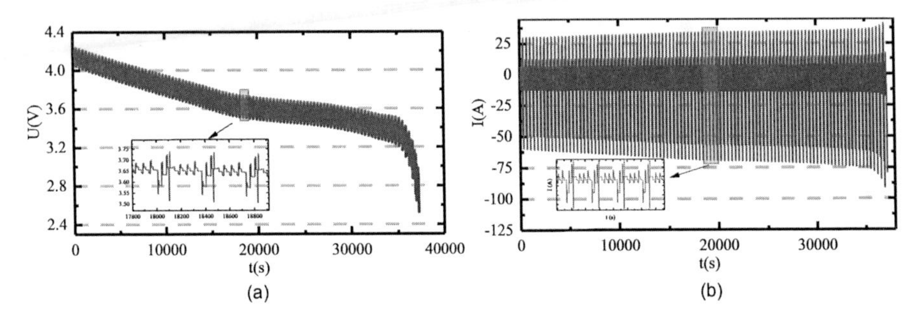

FIGURE 3.49 DST working condition dataset. (a) DST voltage curve. (b) DST current curve.

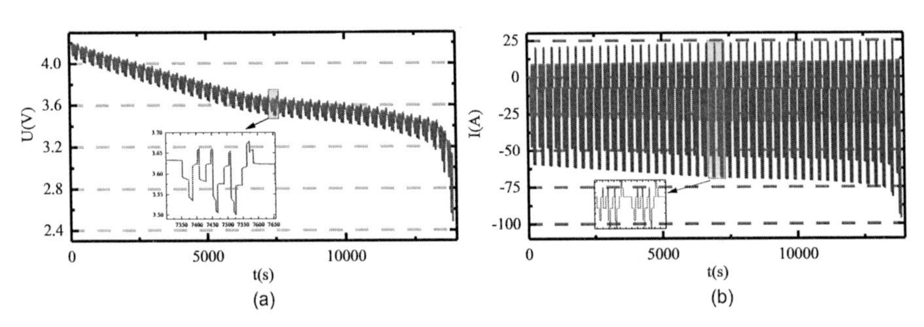

FIGURE 3.50 BBDST working condition dataset. (a) BBDST voltage curve. (b) BBDST current curve.

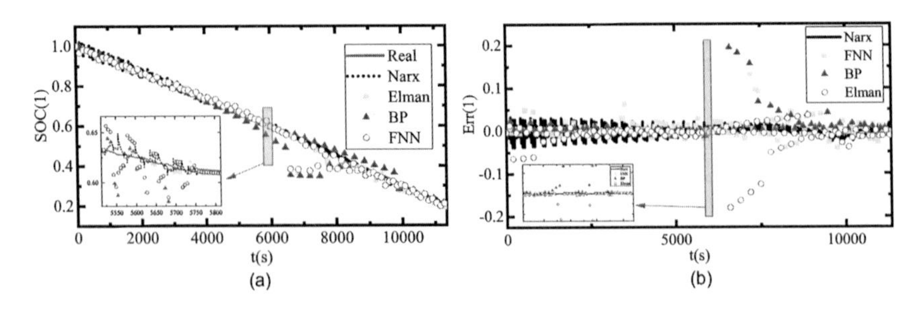

FIGURE 3.51 The SOC estimation result without EKF optimization. (a) SOC estimation without EKF optimization. (b) Estimated error without EKF optimization.

In Figure 3.51a and b, other neural networks except NARXNN have a considerable divergence during the estimation process. The NARXNN can follow the reference value steadily. The error is always less than 4%, which is much lower than other neural networks. Due to the rapid dynamic changes of the working conditions, the maximum value of the algorithm error often occurs when the voltage suddenly rises or falls, and the extreme change of data causes the network structure to be unstable. To suppress the noise caused by the data mutation, the EKF algorithm is added to reduce the tip error, as shown in Figure 3.52.

Figure 3.52 shows the various algorithms optimized by EKF have significantly reduced the tip error, and the maximum error has been reduced by about 50%. It proves that the proposed algorithm is feasible, as shown in Figure 3.53.

Figure 3.53 shows the comparison of the evaluation indicators of the estimation results of different algorithms. It can be observed more intuitively from Figure 3.53a and b that the RMSE and MAE of NARXNN are only 20% of other neural networks, proving the superiority of the NARXNN. After EKF optimization, the error of each neural network is significantly reduced, which dramatically improves the estimation accuracy of the algorithm. This shows that EKF can effectively maintain the stability of the network and improve the estimation accuracy of the network. In terms of calculation cost, as the amount of data and the complexity of working conditions increase, to obtain the best estimation effect, the training time of each neural network has to be increased relative to simple working conditions. NARXNN, FNN, and BPNN take about 90 seconds to optimize, but Elman NN takes more time.

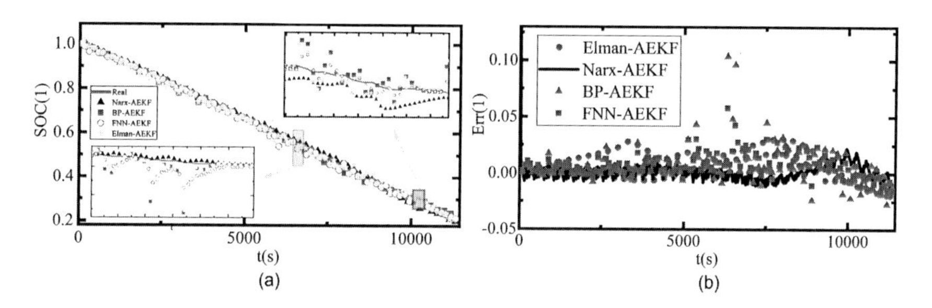

FIGURE 3.52 The SOC estimation result with EKF optimization. (a) SOC estimation optimized by EKF. (b) Estimated error optimized by EKF.

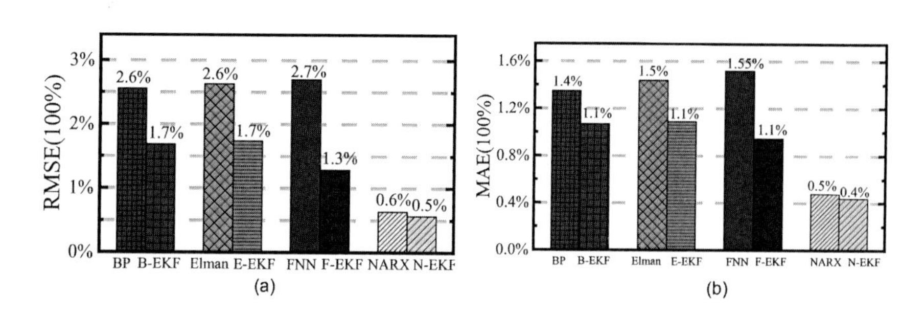

FIGURE 3.53 Evaluation indicators. (a) RMSE. (b) MAE.

3.8 SOC ESTIMATION BASED ON GA-BP ALGORITHM

The GA-BP neural network uses a GA to optimize the initial weights and thresholds of the BP neural network so that the optimized neural network can better predict the function output [237,238]. The flowchart of the GA-BP algorithm is shown in Figure 3.54.

The optimization process can be divided into three parts:

1. The network structure is determined according to the functional requirements.
2. The parameter values of the BP neural network structure are optimized by a GA. All initial weights and thresholds in the network are genetically operated to obtain the initial optimal weights and thresholds.

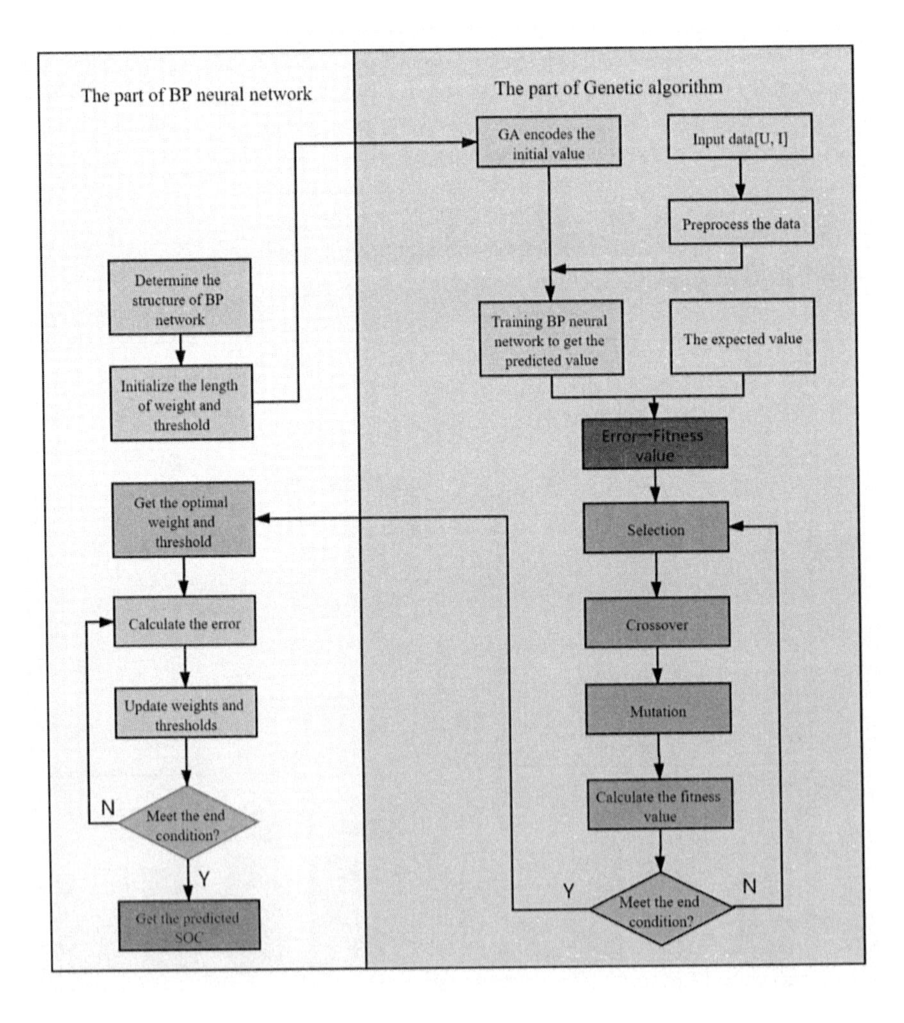

FIGURE 3.54 The flowchart of the GA-BP algorithm.

3. Maintain the topological structure of the BP neural network, assign the initial optimal weight and threshold to the network, and then use the BP neural network for training, testing, and verification.

3.8.1 Backpropagation Neural Network

The BP neural network is a multilayer network that generalizes the W-H learning rules and performs weight training on nonlinear differentiable functions [239,240]. BP network is mainly used for function approximation, pattern recognition, classification, and data compression. The flowchart of the BP algorithm is shown in Figure 3.55.

3.8.2 Genetic Algorithm

The GA is a parallel stochastic search optimization method that simulates natural genetic mechanisms and biological evolution [241]. The elements of BP neural network optimization by the GA include population initialization, fitness function, selection operation, crossover operation, and mutation operation [242].

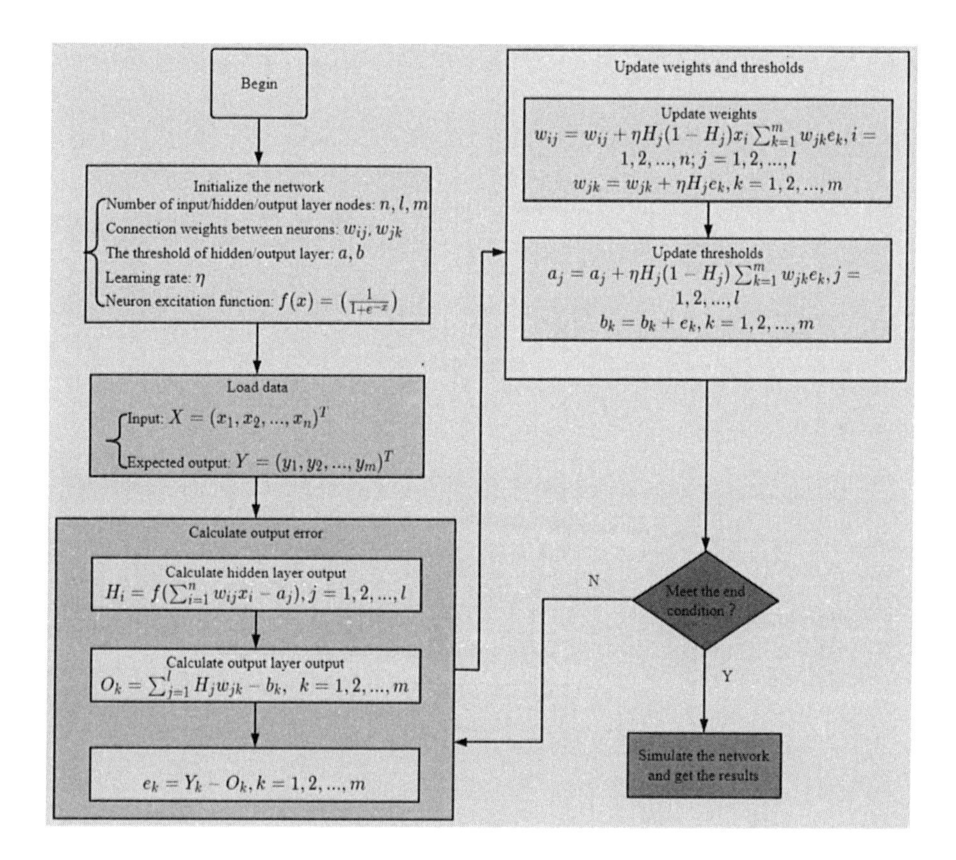

FIGURE 3.55 The flowchart of the BP algorithm.

1. Population initialization
   The individual coding method is real number coding, and each individual is a real number string, which is composed of four parts: the connection weight between the input layer and hidden layer, the threshold value of the hidden layer, the connection weight between the hidden layer and output layer, and the threshold value of output layer. The individual contains all the weights and thresholds of the neural network. If the network structure is known, a neural network with its structure, weights, and thresholds can be constituted.
2. Fitness function
   According to the initial weights and thresholds of the BP neural network obtained by individuals, the BP neural network is trained with training data to predict the system output. The absolute error value between the predicted output and the expected output is taken as the individual fitness value F. The calculation equation of F is shown in Equation (3.242).

$$F = k\left(\sum_{i=1}^{n}\left|y_i - o_i\right|\right) \tag{3.242}$$

   In Equation (3.242), n is the number of output nodes, y_i is the expected output of the ith node of the BP neural network, o_i is the predicted output of the ith node, and k is a coefficient.
3. Selection operation
   The selection operation of GA includes roulette, tournament, etc. When the roulette method is used to calculate the selection probability p_i of each individual i, the calculation equation is shown in Equation (3.243).

$$\begin{cases} f_i = \dfrac{k}{F_i} \\ p_i = \dfrac{f_i}{\sum_{j=1}^{N} f_j} \end{cases} \tag{3.243}$$

   In Equation (3.243), F_i represents the fitness value of individual i. Since the smaller, the fitness value, the better, the inverse of the fitness value is calculated before individual selection. N represents the number of individuals in the population, and k is a coefficient.
4. Crossover operation
   The crossing operation of the kth gene a_k and the l_{th} gene a_l at the j position is realized using Equation (3.244).

$$\begin{cases} a_{kj} = a_{kj}(1-b) + a_{lj}b \\ a_{lj} = a_{lj}(1-b) + a_{kj}b \end{cases} \tag{3.244}$$

   In Equation (3.244), b represents a random number in [0,1].

5. Mutation operation

The jth gene of the i-th individual factor is selected for mutation operation, as shown in Equation (3.245).

$$\begin{cases} f(g) = r_1\left(1 - \dfrac{g}{G_{\max}}\right)^2 \\ a_{ij} = \begin{cases} a_{ij} + \left(a_{ij} - a_{\max}\right) * f(g), r > 0.5 \\ a_{ij} + \left(a_{\min} - a_{ij}\right) * f(g), r \le 0.5 \end{cases} \end{cases} \tag{3.245}$$

In Equation (3.245), $a_{\max}$ and $a_{\min}$ represent the upper and lower bounds of a_{ij}, r represents the random number, g represents the current iteration, and $G_{\max}$ represents the maximum number of evolutions.

3.8.3 Experimental Analysis

To verify whether the accuracy of the BP neural network optimized by GA is better than that of the BP neural network in SOC prediction, simulation experiments are carried out. The simulation effect is shown in Figure 3.56a, and the error is shown in Figure 3.56b.

It can be observed that the GA–BP network output is closer to the expected output than the BP network, and it shows better fitting. The comparison of the experimental results of the BP and GA–BP algorithm to estimate SOC under DST conditions is shown in Table 3.11.

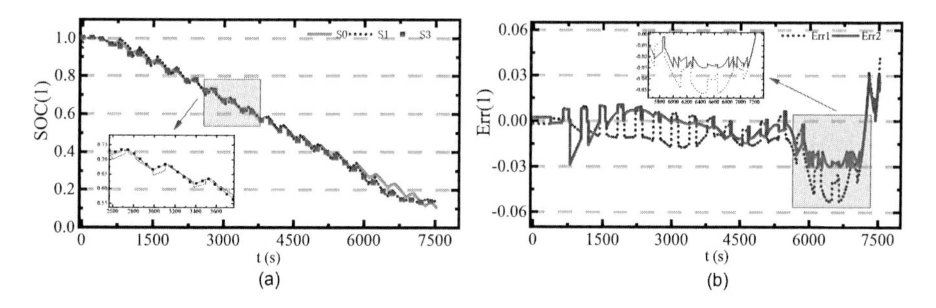

FIGURE 3.56 SOC estimation result based on BP and GA-BP neural network. (a) Comparison of SOC estimation results. (b) SOC estimation error.

TABLE 3.11
Comparison of Experimental Results of Different Algorithms

SOC Estimation Method	BP	GA-BP
Maximum error	5.2843%	3.1125%
Average error	1.3824%	1.0811%

The calculation results show that the BP neural network optimized by GA is more accurate than the BP neural network in estimating the SOC value of lithium-ion batteries.

3.9 ONLINE ESTIMATION ALGORITHM

Aiming at the complexity and working characteristics of the internal characteristics of the lithium-ion battery, an improved PNGV lithium-ion battery equivalent model is proposed, and the model parameters are identified by the RLS algorithm with the forgetting factor. Based on parameter identification, the SOC is estimated by the EKF algorithm.

3.9.1 Improved PNGV Equivalent Circuit Modeling

Based on the PNGV model of the lithium-ion battery, the RC circuit is connected in series to describe the surface effect resistance and surface effect capacitance of the lithium-ion battery, to improve the accuracy of the physical model, as shown in Figure 3.57.

In Figure 3.54, C_b is the change in the OCV due to the cumulative effect of the current. U_{OC} is the OCV. U_L is the terminal voltage. R_1 is the ohmic resistance. U_{R1} is the ohmic voltage, which is the effect of battery voltage drop at the time point of battery discharge and termination of discharge. The first RC parallel circuit is composed of a polarization resistor R_{p1} and polarization capacitor C_{p1}, which is used to characterize the polarization effect of the lithium-ion battery, and U_{p1} is the polarization voltage. The second RC parallel circuit is composed of a surface polarization resistor R_{p2} and surface polarization capacitor C_{p2}, which is used to characterize the surface effect of the lithium-ion battery, and U_{p2} is the surface effect voltage. According to Kirchhoff's law, the improved PNGV equivalent circuit model is analyzed, and the

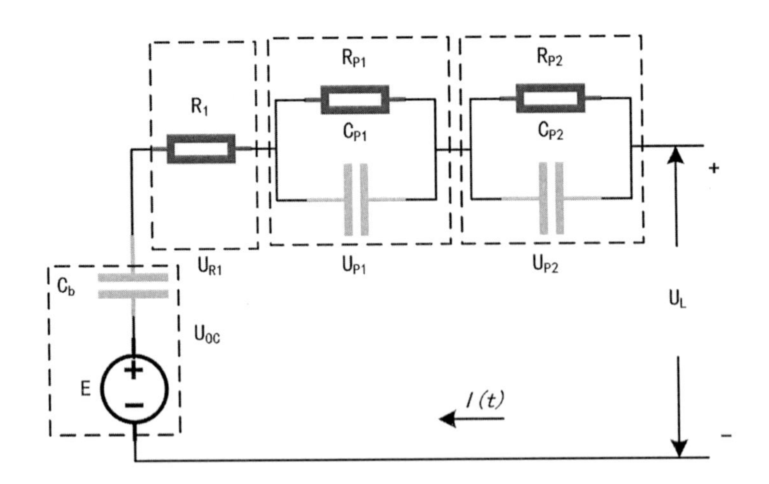

FIGURE 3.57 Improved PNGV equivalent circuit model.

voltage and current expressions of the equivalent circuit can be obtained, as shown in Equation (3.246).

$$\begin{cases} U_L = U_{OC} - U_{R1} - U_{p1} - U_{p2} \\ I = C_{p1}\dfrac{dU_{p1}}{dt} + \dfrac{U_1}{R_1} = C_{p2}\dfrac{dU_{p2}}{dt} + \dfrac{U_1}{R_1} \end{cases} \tag{3.246}$$

3.9.2 Online Identification of Forgetting Factor RLS Algorithm

Due to the phenomenon of "filtering saturation" in the RLS method, the values of gain K and P become smaller and smaller as the number of iterations of the algorithm data increases. As a result, the ability of the algorithm to correct the data is gradually weakened, and the degree of data saturation is increasing, which eventually leads to the increasing error of parameter identification [243]. Therefore, it is considered to add a forgetting factor based on least square identification, to improve the online estimation ability of the RLS algorithm. The function of the forgetting factor is to give less weight to the long-running data in the identification process, while the latest observation data occupies more weight [244]. The least square data fitting is to select a function fitting from a given function type according to a given dataset to minimize the sum of squares of residuals, as shown in Equation (3.247).

$$\begin{cases} \delta_i = y_i - y_i^* \\ \sum\limits_{i=1}^{n} \delta_i^2 = \min\limits_{\varphi \in H} \sum\limits_{i=1}^{n} \left[y_i - \varphi(x_i) \right]^2 \end{cases} \tag{3.247}$$

According to the PNGV model of the lithium-ion battery, the complex frequency domain equation can be obtained and the discrete transfer function can be obtained, as shown in Equation (3.248).

$$G\left(Z^{-1}\right) = \frac{c_3 + c_4 Z^{-1} + c_5 Z^{-2}}{1 - c_1 Z^{-1} - c_2 Z^{-2}} \tag{3.248}$$

In Equation (3.248), c_1, c_2, c_3, c_4, and c_5 are the corresponding constant coefficients, so we can know that the system studied is a single-input/single-output system. Equation (3.248) is transformed into a difference equation, as shown in Equation (3.249).

$$\begin{aligned} y(k+1) &= U_{OC} - U_L \\ &= c_1 y(k-1) + c_2 y(k-2) + c_3 I(k) + c_4 I(k-1) + c_5 I(k-2) \end{aligned} \tag{3.249}$$

In Equation (3.249), $I(k)$ is the system input and $y(k+1)$ is the system output. The relationship between lithium-ion battery parameters and the coefficient can be deduced, as shown in Equation (3.250).

$$\begin{cases} R_1 = \dfrac{c_3 - c_4 + c_5}{1 + c_1 - c_2} \\ \tau_1\tau_2 = \dfrac{T^2(1 + c_1 - c_2)}{4(1 - c_1 - c_2)} \\ \tau_1 + \tau_2 = \dfrac{T(1 + c_2)}{1 - c_1 - c_2} \\ R_1 + R_{P1} + R_{P2} = \dfrac{c_3 + c_4 + c_5}{1 - c_1 - c_2} \\ R_1\tau_1 + R_1\tau_2 + R_{P1}\tau_2 + R_{P2}\tau_1 = \dfrac{T(c_3 - c_5)}{1 - c_1 - c_2} \end{cases} \tag{3.250}$$

Input variables and parameter variable expressions are represented by matrices, as shown in Equation (3.251).

$$\begin{cases} \phi(k) = \left[y(k-1) y(k-2) I(k) I(k-1) I(k-2) \right] \\ \theta = \left[\begin{matrix} a_1 & a_2 & a_3 & a_4 & a_5 \end{matrix} \right] \end{cases} \tag{3.251}$$

The matrix expression of the system output can be obtained, as shown in Equation (3.252).

$$y(k) = \phi(k)^{\mathrm{T}}\theta + e(k) \tag{3.252}$$

The genetic factor λ ($0 < \lambda < 1$) is introduced to weaken the influence of old data and enhance the feedback effect of new data. The forgetting factor least square recursive equation is obtained, as shown in Equation (3.253).

$$\begin{cases} K(k+1) = P(k)\phi(k+1)\left[\lambda + \phi^T(k+1)P(k)\phi(k+1)\right]^{-1} \\ \hat{\theta}(k+1) = \hat{\theta}(k) + K(k+1)\left[y(k+1) - \phi^T(k+1)\hat{\theta}(k)\right] \\ P(k+1) = \dfrac{1}{\lambda}\left[I - K(k+1)\phi^T(k+1)\right]P(k) \end{cases} \tag{3.253}$$

3.9.3 Online Parameter Identification Effect

The parameters of the lithium-ion battery PNGV model are identified with the forgetting factor recursive least square method. The identification results are shown in Figure 3.58.

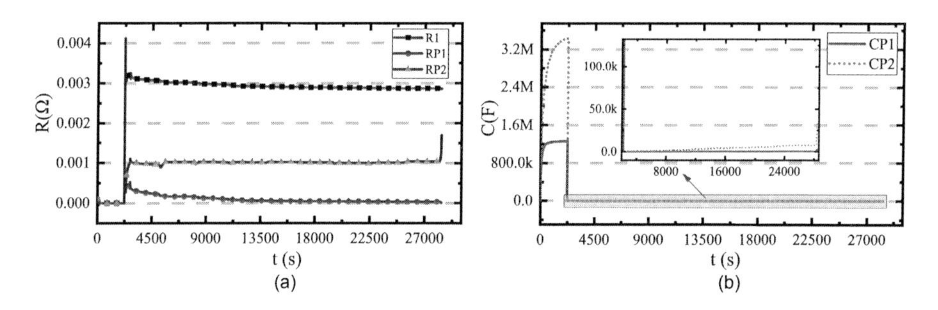

FIGURE 3.58 PNGV model fitting result. (a) Online identification resistance. (b) Online identification capacitance.

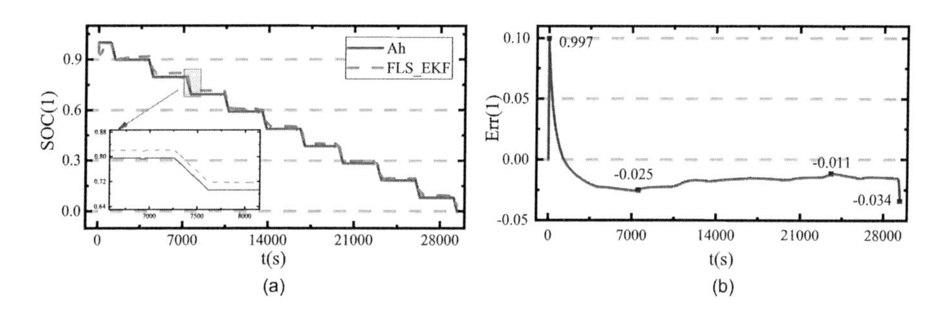

FIGURE 3.59 Online estimation effect based on improved FLS-EKF algorithm. (a) The online SOC estimation results. (b) Online SOC estimation error.

3.9.4 Online Identification and Estimation Results

The results of online identification and estimation of SOC are obtained. The initial SOC value of simulation is set as 1.0, and the SOC calculated by the ampere–hour integration method is also taken as the standard value. The error is the difference between the ampere–hour integration method and the improved fixed-lag smoothing (FLS)-EKF. It can be observed from the figure that the error of SOC estimation by online parameter identification combined with the improved FLS-EKF is less than ±5%, and the maximum estimation error is −4.32%, as shown in Figure 3.59.

3.10 CONCLUSION

This chapter mainly introduces the definition of SOC and existing estimation methods, as well as various algorithms for SOC estimation. The research on lithium-ion batteries mainly starts with the prediction of SOC value. The SOC estimation of a lithium-ion battery is defined as the ratio of the remaining capacity of the battery to the nominal capacity of the battery, which is the remaining battery capacity and represents the percentage of the current remaining capacity. Estimating the SOC estimation of the lithium-ion battery accurately can make the lithium-ion battery pack safer when it is used, and it can also improve the performance of the batteries. However, the SOC interferes with various factors during the working process of the battery, which brings difficulties to the estimation of the SOC value. The main factors affecting the SOC are temperature, charge–discharge rate, self-discharge rate, and battery aging. These factors affect the accuracy of SOC estimation by using different equivalent circuit models and algorithms and could lead to different effects on the SOC estimation of batteries. Generally speaking, the more complex the equivalent model, the more accurate it is, the more complex its calculation is, and the higher its operating cost. Also, algorithms depend on their operating environment, and the adaptation of different batteries is different, so it is very important to select the appropriate models and algorithms for the lithium-ion battery.

4 Battery State of Health Estimation

Carlos Fernandez
Robert Gordon University

Ji Wu
Hefei University of Technology

Lei Chen, Mingfang He, Peng Yu, and Pu Ren
Southwest University of Science and Technology

Shunli Wang
Sichuan University
Southwest University of Science and Technology

Xiao Yang
Southwest University of Science and Technology

Siyu Jin
Aalborg University

Yangtao Wang
Southwest University of Science and Technology

CONTENTS

DOI: 10.1201/9781003333791-4

Introduction: At present, due to the strong nonlinearity of lithium-ion batteries, the relevant research methods for the state of health (SOH) estimation are mainly divided into two types, mathematical model-driven, and data-driven. Compared with the data-driven methods, the mathematical model-driven model is matured relatively. By establishing an equivalent circuit model (ECM) to simplify the complex internal reactions of lithium-ion batteries, intelligent algorithms are combined to realize the estimation of health indicators (HIs). The established equivalent models mainly include electrochemical mechanism models, ECMs, and empirical models. In the application process, it is necessary to consider the internal factors such as electrolyte concentration, state of charge (SOC), and external factors such as temperature and SOH [245].

4.1 OVERVIEW OF BATTERY HEALTH INDICATOR

From a long-term perspective, SOH management improves the life cycle by evaluating the impact on battery aging by using the HIs, which reflects the capacity degradation of batteries. In recent studies, the estimation method of battery capacity uses the correlation between characterization parameters and aging [225]. It is a necessary issue to choose an appropriate HIs estimation method to reflect the degenerate nonlinear model. For most types of batteries, the ohmic internal resistance and polarization internal resistance of the ECM can be used as relatively suitable HIs [246].

Data-driven methods of SOH estimation are making progress. Several neural network algorithms are continually discovered, which are represented by the backpropagation (BP) neural network algorithms. The process involves the combination of the BP neural network with the simulated annealing algorithm [247]. The health factors of the battery are input into the BP neural network, and the simulated annealing algorithm is taken advantage of to optimize the weights of the BP neural network to obtain the optimal solution to the prediction model. The parameters of voltage, current, temperature, and other external battery characteristics are selected as the input variables. Then, particle swarm optimization (PSO) is introduced in the construction of the BP network to optimize the network weights, which enhances the global optimization ability of the network. Comparatively, the accuracy of SOH estimation for lithium-ion batteries has highly improved by using the data-driven method.

Although significant progress has been made in the lithium-ion battery state estimation at this stage, the diversity and various complex operating conditions make it difficult to achieve a high-precision simulation of the real operating conditions of lithium-ion batteries [248–250]. Also, there is a lack of a unified measurement standard for the SOH estimation of lithium-ion batteries. Therefore, the related research and application are still in their initial stage.

The research status of SOH estimation is introduced for lithium-ion batteries as well as some classical estimation algorithms. Based on several related studies and the improvement of the particle filtering algorithm, a new theoretical system using experimental data from the capacity attenuation of lithium-ion batteries and the cycle verification theory system has been proposed, which realizes the adaptive optimization of the SOH estimation [251–253]. The innovation of this method is to gradually improve the accuracy of SOH estimation in terms of self-adaptation conditions and comprehensively analyze the convergence and accuracy of the algorithm for multiple working conditions. The loop structure combines the experiment and algorithm into an automatic optimization system with good adaptability in various complex operating conditions. Also, it has the advantages of a simple measurement method and fast parameter identification.

Based on the improvement of the Kalman filtering (KF) algorithm, a new dual adaptive Kalman filtering (DAKF) algorithm is proposed for the co-estimation of SOC and SOH [254,255]. The innovation of this method is to adaptively co-estimate the SOC and SOH by comprehensively analyzing the convergence and accuracy of the algorithm under three working conditions. In the algorithm, both sides are mutually revised and have good applicability to several complex working conditions. This method has the advantages of a simple measurement technique and fast parameter identification. Based on the DAKF algorithm, the SOH value of the battery can be predicted for some time in the future [256]. In general, statistical methods for predicting battery status rely on historical data measured through experiments. The DAKF algorithm can make up for the lack of data in practical applications. A comprehensive model that combines DAKF and autoregressive models is used to predict SOH through statistical models when the amount of measured data is insufficient.

4.2 MODEL-BASED SOH ESTIMATION

The SOH estimation method based on mathematical modeling relies on a battery equivalent model combined with an adaptive intelligent algorithm to perform real-time filtering calculations on battery HIs such as internal resistance and polarization internal resistance [257,258]. Then, the real-time estimation of SOH is realized through the mapping relationship between HIs and battery SOH. This method resolves the shortcoming that the direct measurement method can only estimate the health status off-line, which is suitable for practical engineering applications. However, its shortcoming is that it relies heavily on the battery model. Also, the characteristic parameters of the SOH estimation, such as internal resistance, are affected by compound factors such as SOC, which increases the difficulty of estimating the SOH value. An accurate and real-time SOH estimation method should meet the requirements of high precision and low computational cost.

4.2.1 Thevenin Equivalent Modeling

The Thevenin equivalent model is widely used because it can fully characterize the internal characteristics of lithium-ion batteries, and the complexity of the model is not high. The schematic diagram of the Thevenin model is shown in Figure 4.1.

In Figure 4.1, U_{OC} is the open-circuit voltage, R_0 is the ohmic resistance, R_1 is the polarization resistance, and C_1 is the polarization capacitance. The parallel circuit of R_1 and C_1 describes the battery polarization process. As the battery ages, a series of changes occur inside the battery, which is characterized by changes in internal parameters such as capacity and ohmic resistance. Some scholars define SOH from the perspective of ohmic resistance by studying the change of internal resistance during the aging process, as shown in Equation (4.1).

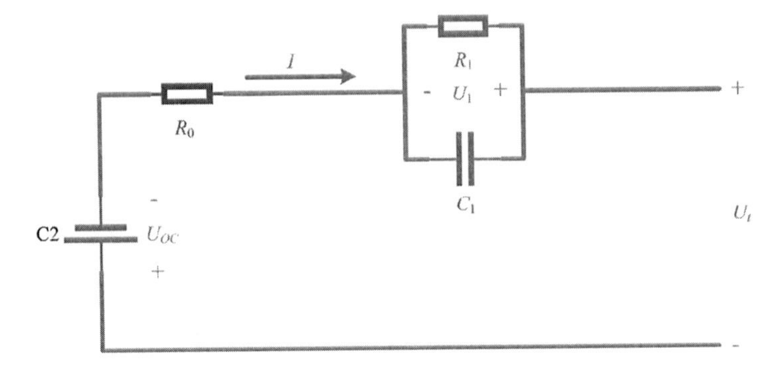

FIGURE 4.1 Thevenin model.

$$\text{SOH} = \frac{R_{\text{EOL}} - R}{R_{\text{EOL}} - R_{\text{BOL}}} * 100\% \tag{4.1}$$

In Equation (4.1), R_{EOL} is the internal resistance at the end of the battery life, R_{BOL} is the resistance of the new battery, and R is the internal resistance of the battery in the application processes. The battery characteristics are considered by the Thevenin model, and the iterative estimation algorithm is used to estimate the characteristic quantities that can reflect the battery SOH in real-time by the online approach. At present, the commonly used intelligent algorithms include the KF, particle filter (PF), and the improved algorithms derived from these two algorithms.

4.2.2 Kalman Filter and Its Extension

The KF method is a filtering theory created by the state-space theory in the time domain. It is an algorithm that uses the system state-space and input-output observation data to optimally estimate the state of the system [259,260]. The main idea of KF is to use the minimum mean square error criterion to dynamically estimate the target of linear Gaussian conditions and use the observable output error of the system to correct the unobservable state error. Thus, the noise interference in the data is reduced, and the estimation accuracy of the system is improved. The state and observation equations of the linear system are designed, as shown in Equation (4.2).

$$\begin{cases} x_{k+1} = f(x_k, u_k) + w_k \\ y_k = h(x_k, u_k) + v_k \end{cases} \tag{4.2}$$

In Equation (4.2), x_k is the state variable, u_k is the system input, y_k is the observation variable, w_k and v_k are the state and observation error, respectively. Each iteration of the KF algorithm is divided into two stages: prediction and measurement. In the prediction stage, a time-based prediction update is performed on the state variables based on the system state, which provides a prior estimation value for the posterior estimation at that time. In the measurement phase, the predicted value is corrected by the system observation value, and the deviation is corrected to update the estimated value. Equation (4.2) is then discretized as shown in Equation (4.3).

$$\begin{cases} x_{k+1} = A_k x_k + B_k u_k + w_k \\ y_k = C_k x_k + D_k u_k + v_k \end{cases} \tag{4.3}$$

In Equation (4.3), x_k and x_{k+1} are the state variable values of the system at the time points k and $k+1$, respectively. u_k is the input variable of the system, and y_k is the expression of the observed value of the system at the time point k. A_k is the state transition matrix, which predicts the system variables, B_k is the system control input matrix, and C_k is the systematic observation matrix, driving and predicting the system observations. w is the system process noise, and v is the measurement noise with covariance matrices of Q and R, respectively. Both of w and v are zero-mean Gaussian white noise.

Since the traditional KF algorithm can only be applied to linear systems, improved algorithms, such as EKF and unscented Kalman filter, are generally utilized to estimate the battery health characteristic parameters [18]. Then, the battery SOH is estimated in real time by the relationship between the health parameters and the battery SOH values. Using the functional relationship between the ohmic resistance and the battery SOH, the recursive update of the battery ohmic resistance can realize the real-time state updating. By the Thevenin equivalent model established for the battery, the internal ohmic resistance is selected as the state quantity to establish the state-space equation, as shown in Equation (4.4).

$$\begin{cases} R_k = R_{k-1} + w_k \\ U_k = U_{\mathrm{OC}}(\mathrm{SOC}) - i_k R_{0,k} - U_{1,k} - v_k \end{cases} \tag{4.4}$$

In Equation (4.4), R_k is the internal ohmic resistance of the battery at the time point k, w_{k-1} is the system process noise at the previous time point $k-1$ and U_k is the terminal voltages of the battery at the time point k, and $U_{\mathrm{OC}}(\mathrm{SOC})$ is the fitting function relationship between the battery open-circuit voltage U_{OC} and the battery SOC, which is generally a nonlinear empirical or high-order polynomial. i_k is the expression of the current flowing through the internal resistance R_0. $U_{1,k}$ is the polarization voltage in the Thevenin model at the time point k to obtain Equation (4.5):

$$U_{1,k}(k) = U_{1,k-1} e^{-\frac{t}{R_1 C_1}} + R_1 i_k \left(1 - e^{-\frac{t}{R_1 C_1}}\right) \tag{4.5}$$

In Equation (4.5), R_1 and C_1 are the polarization resistance and capacitance in the Thevenin model, respectively. Based on the iterative steps of the KF algorithm, the battery ohmic resistance is filtered and estimated. The specific steps are described as follows:

1. Initializing state vector R_0 and error covariance P_0 as shown in Equation (4.6).

$$\begin{cases} \hat{R}_0 = E[R_0] \\ P_0 = E\left[\left(R_0 - \hat{R}_0\right)\left(R_0 - \hat{R}_0\right)^T\right] \end{cases} \tag{4.6}$$

The initial value of the state and the error covariance matrix needs to be adjusted that is based on actual values. The initial value of the error covariance indicates the confidence bounds in the initial value of the state. The initialization step affects the filtering time and the error of the iterative process.

2. Time update of state variables and error covariance
In the recursive time update process of the ohmic resistance, it can be known from the established state-space equation that $A_{k-1}=1$ can obtain the value of the ohmic internal resistance R_0 and the error covariance as shown in Equation (4.7):

$$\begin{cases} R_{k|k-1} = A_{k-1}R_{k-1} + w_k = R_{k-1} + w_k \\ P_{k|k-1} = A_{k-1}P_{k-1}A_{k-1}^T + Q_k = P_{k-1} + P_r \end{cases} \tag{4.7}$$

3. Kalman gain updating
From the state-space equation, the state transition matrix $C_k = -i_k$ is obtained, and the Kalman gain K_k is updated as shown in Equation (4.8):

$$K_k = \widetilde{P_k}(C_k)^T\left[C_k\widetilde{P_k}(C_k)^T + R_k\right]^{-1} \tag{4.8}$$

In Equation (4.8), R_k is the variance matrix of the observation noise v_k. The Kalman gain is the core of the EKF algorithm, and the precise calculation directly determines the reliability and accuracy of the state estimation.

4. Measurement update
By the observation expression of the state-space equation, the state estimated value after the time update is corrected, as shown in Equation (4.9):

$$\begin{cases} R_k = \widetilde{R_k} + K_k\left[U_k - U_{OC}(\mathrm{SOC}) - U_{1,k} - i_k\widetilde{R_k}\right] \\ P_k = (I - K_kC_k)\widetilde{P_k} \end{cases} \tag{4.9}$$

In Equation (4.9), the state variable test update is an accurate estimation of the state vector at the time point k, which is performed by selecting an accurate Kalman gain value. By the output U_k, the optimal estimation value of the state vector is obtained. The main idea of the EKF algorithm is to convert the nonlinear system into a linear system first [261]. Then, the linear system is calculated according to the process of the classical KF algorithm. The EKF algorithm is an improvement of the classical KF algorithm, which makes it possible to solve the nonlinear system problem that the classical KF algorithm cannot solve. In this algorithm, after a series of multi-iterations, the estimated value of the internal resistance of the battery is finally obtained.

4.2.3 Extended Particle Filter

The PF calculation method is a Bayesian estimation algorithm based on recursion. The Monte Carlo method is taken advantage of to extract random particles from the posterior probability and assign a weight to each particle. It can be utilized in any form of the state-space model [262]. The calculation principle of the PF algorithm is divided into the following five steps.

Step 1. Initial state: A large number of particles are taken advantage of to simulate $x(t)$, and the particles are evenly distributed in the space.
Step 2. Prediction stage: Each particle is predicted by the state transition equation.
Step 3. Correction stage: The predicted particles are evaluated; the closer the particles are to the real state, the greater the weight coefficient.
Step 4. Resampling: The particles are screened according to their weight. In the process, both large and small numbers of particles with large and small weight coefficients are, respectively, retained.
Step 5. Filtering: The resampled particles are brought into the state transition equation to obtain the new prediction mentioned in Step 2.

The state-space model of the PF system is established, as shown in Equation (4.10).

$$\begin{cases} x_t = f_t\left(x_{t-1}, u_{t-1}\right) + v_t \\ Z_t = h_t\left(x_t, u_t\right) + n_t \end{cases} \tag{4.10}$$

In Equation (4.10), these two equations are called the state transition model and observation equation, respectively. x_t is the state of the system; Z_t is the output of the system; v_t and n_t are the system noise and observation noise, respectively; f_t and h_t are the nonlinear functions. Assuming that the state vector x_t follows a first-order Markov process, the prior probability equation of x_t can be derived from the Chapman–Kolmogorov equation as shown in Equation (4.11).

$$p\left(x_t \mid z_{0\sim t-1}\right) = \int p\left(x_t \mid x_{t-1}\right) p\left(x_{t-1} \mid z_{0\sim t-1}\right) dx_{t-1} \tag{4.11}$$

Then, the posterior probability of x_t is updated by z_t in Equation (4.12):

$$\begin{cases} p\left(z_t \mid z_{0\sim t-1}\right) = \int p\left(x_t \mid z_{0\sim t-1}\right) p\left(z_t \mid x_t\right) dx_t \\ p\left(x_t \mid z_{0\sim t}\right) = \dfrac{\int p\left(x_t \mid z_{0\sim t-1}\right) p\left(z_t \mid x_t\right) dx_t}{p\left(z_t \mid z_{0\sim t-1}\right)} \end{cases} \tag{4.12}$$

Recursive Bayesian filtering consists of Equations (4.11) and (4.12), and it has high adaptability in many cases. However, obtaining posterior probability $p\left(z_t \middle| z_{0\sim t}\right)$ by the analytical solution is a hard thing in most situations. Compared with the analytic solution, the PF algorithm based on the Monte Carlo algorithm is more suitable. The core idea of PF is to approximate the posterior probability density by a set of weighted particles. The prior distribution probability of the known particles is supposed to be $p(x_t \mid z_{0\sim t-1})$ and N samples are drawn from the system. The posterior distribution is approximate, as shown in Equation (4.13):

$$p(x_{0\sim t} \mid z_{0\sim t}) \approx \sum_{i=1}^{N} \omega_t^i \delta\left(x_{0\sim t} - x_{0\sim t}^i\right) \tag{4.13}$$

In Equation (4.13), for $\delta\left(x_{0\sim t} - x_{0\sim t}^i\right)$, the functional relationship is shown in Equation (4.14):

$$\delta\left(x_{0\sim t} - x_{0\sim t}^i\right) = \begin{cases} 1, & x_{0\sim t} = x_{0\sim t}^i \\ 0, & x_{0\sim t} \neq x_{0\sim t}^i \end{cases} \tag{4.14}$$

The importance of distribution $q(x_{0\sim t} | z_{0\sim t-1})$ and the importance weighting factor ω_t^i for each sample are shown in Equation (4.15), respectively.

$$\begin{cases} q(x_{0\sim t} \mid z_{0\sim t-1}) = q(x_t \mid z_{t-1}, z_t) q(x_{0\sim t-1} \mid z_{0\sim t-1}) \\ \omega_t^i = \dfrac{p(x_{0\sim t} \mid z_{0\sim t})}{q(x_{0\sim t} \mid z_{0\sim t})} \propto \omega_{t-1}^i \dfrac{p(z_t \mid x_t^i) p(x_t^i \mid x_{t-1}^i)}{q(x_t^i \mid x_{0\sim t-1}^i, z_{1\sim t})} \end{cases} \tag{4.15}$$

In Equation (4.15), the symbol $\propto$ indicates to be proportional. Next, it is necessary to get the normalized importance of weight $\tilde{\omega}_t^i$. The method is shown in Equation (4.16):

$$\tilde{\omega}_t^i = \frac{\omega_t^i}{\sum_{i=1}^{N} \omega_t^i} \tag{4.16}$$

After getting the normalized importance weight $\tilde{\omega}_t^i$, the state $\hat{x}_t$ can be estimated, as shown in Equation (4.17).

$$\hat{x}_t \approx \sum_{i=1}^{N} \tilde{\omega}_t^i x_i^t \tag{4.17}$$

The state prediction process of the PF algorithm is an iterative process of repeated prediction and update. The prediction flowchart and the schematic diagrams of algorithm implementation are shown in Figures 4.2 and 4.3, respectively.

In Figures 4.2 and 4.3, x_k^i is the state of the kth sample of the system at time point k. Unlike the KF algorithm, which has to constrain a Gaussian linear function, the PF algorithm uses the Monte Carlo sampling mechanism that can be used for any

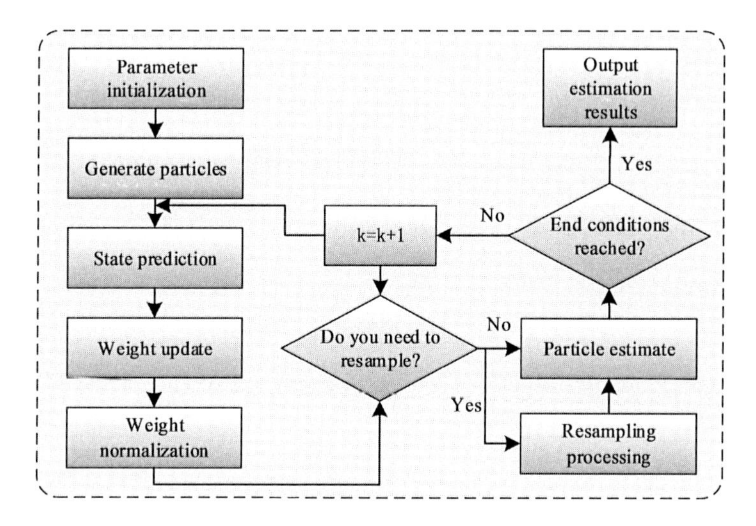

FIGURE 4.2 Flowchart of PF state estimation.

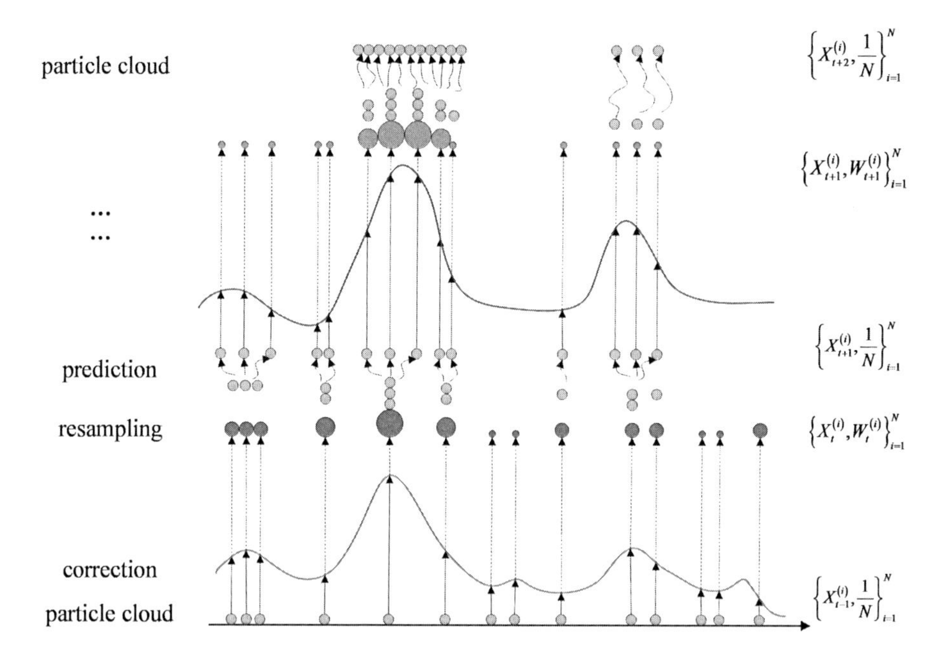

FIGURE 4.3 Schematic diagrams of the PF algorithm.

dynamic model with any form of noise. The PF-based SOH prediction is similar to the KF algorithm, which establishes a state-space based on the established ECM to iteratively estimate the internal resistance of the batteries. Then, the battery SOH can be obtained according to the corresponding relationship between internal resistance and SOH. The specific iterative steps are described as follows:

1. Setting the initialization parameters, the initial value $k=0$, the number of iterations $N=1000$, and the number of samples $M=100$.
2. Generating sampled particles from the prior probability density, and all particles have equal weights.
3. Iteratively calculate the value according to the battery ohmic resistance state.
4. Calculating the weight of each P at the time point and performing normalization processing.
5. Determining whether the number of particles is greater than the set effective number of particles, if not resample is performed, and weights are redistributed.
6. The posterior probability calculation equation is taken advantage of outputting the estimated state value.

Steps (3)–(6) should be repeated until the number of iterations is $N=1000$. The PF method uses the Monte Carlo sampling algorithm to approximate the probability density by assigning different weights to the sampled particles. It avoids the linearization of the KF algorithm and the simplified assumption of Gaussian noise. It provides a suitable prediction method by effectively calculating the uncertainty related to the estimated variable, which can achieve higher prediction accuracy.

4.2.4 Dual Adaptive Kalman Filter

The DAKF algorithm based on multi-resolution is used to analyze the nonstationary voltage signal during the battery charge–discharge processes. It simultaneously extracts the characteristics of the battery in the time and frequency domains [263]. By analyzing the voltage curve of the low-frequency part of the battery with different aging degrees, it is observed that the greater the aging degree is, the lower the voltage curve of the low-frequency part is relative to the new battery. The DAKF algorithm is a novel method for the SOH estimation, which is easy to be understood, relatively simple to use, and has a small amount of calculation time [108]. It avoids the estimation error caused by many unknown parameters in the actual BMS due to general electrochemical and ECM error.

The second-order RC ECM includes an ohmic resistor and two parallel-connected resistor–capacitor networks. These components take into account the internal hysteresis effect and the polarization phenomenon of the chemical reaction by comprehensively describing the dynamic and static characteristics of the lithium-ion batteries in the charge–discharge process. The structure of the second-order RC ECM is relatively simple, as shown in Figure 4.4.

In Figure 4.4, U_{OC} is the open-circuit voltage of the battery, and R_0 is the ohmic resistance. R_1 and R_2 are the electrochemical and concentration polarization resistances, respectively. C_1 and C_2 are the electrochemical and concentration polarized capacitors, respectively. The current I passing through the entire circuit is used as the system input. When charge–discharge, the battery terminal voltage is the observed value U_L. According to Kirchhoff's law, the state-space for discretization is shown in Equation (4.18):

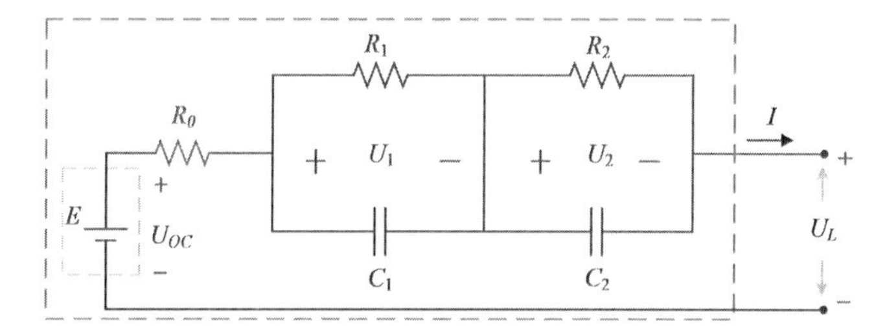

FIGURE 4.4 Second-order RC equivalent circuit model.

$$\begin{cases} \begin{bmatrix} U_{1,k+1} \\ U_{2,k+1} \\ S_{k+1} \end{bmatrix} = \begin{bmatrix} e^{-\frac{T}{(R_1C_1)}} & 0 & 0 \\ 0 & e^{-\frac{T}{(R_2C_2)}} & 0 \\ 0 & 0 & 1 \end{bmatrix} * \begin{pmatrix} U_{1,k} \\ U_{2,k} \\ S_k \end{pmatrix} + I_k \begin{bmatrix} R_1\left(1-e^{-\frac{T}{(R_1C_1)}}\right) \\ R_2\left(1-e^{-\frac{T}{(R_2C_2)}}\right) \\ -\frac{\eta T}{C} \end{bmatrix} \\ U_{L,k} = E_k - U_{1,k} - U_{2,k} - R_0 I_k \end{cases} \tag{4.18}$$

In Equation (4.18), $U_{1,k+1}$ and $U_{2,k+1}$ are the terminal voltages of the two RC circuits, S_{k+1} is the estimated SOC value of the lithium-ion battery. T is the sampling time, η is the Coulomb efficiency, C is the available battery capacity, U_k is the terminal voltage, and E_k is the power supply voltage. The equation for calculating the SOC and internal resistance of the battery is directly derived from the above equation.

1. Research on the state of charge and state of health
   The SOC estimation of a lithium-ion battery refers to the ratio of the remaining capacity to the rated capacity under certain conditions, and it ranges from 0% to 100%. The SOC cannot be measured directly, which is only estimated through other internal parameters of the batteries. According to the definition of SOC, its expression is shown in Equation (4.19):

$$\text{SOC}_t = \frac{Q_t}{Q_0} * 100\% \tag{4.19}$$

   In Equation (4.19), Q_t is the remaining battery capacity at time point t, and Q_0 is the rated capacity. Q_t is obtained by discharging, and the battery is fully charged and discharged to the cut-off voltage under various working conditions. Therefore, in theory, the SOC is only obtained by discharging electricity. However, the type of method only calculates the SOC under off-line conditions and cannot estimate it in real time. The discharge capacity is expressed as the integral of current over time. As long as the sampling period is short enough, it is considered that the current does not change to realize the real-time estimation of SOC. Therefore, the SOC can be estimated in real-time according to Equation (4.20):

$$SOC_t = SOC_{t0} - \frac{\int_{t0}^{t} I(t)\eta \, dt}{Q_0} \tag{4.20}$$

In Equation (4.20), η is the Coulomb efficiency, which is directly used as $\eta = 1$ and SOC_{t0} is the SOC of the battery at the time point t_0. The filtering algorithm discretizes the equation and divides it into adequate intervals. Then, the iterative nature of SOC is used to estimate the changing law in the whole process. The SOH of the power battery is usually analyzed from the perspective of an increase in internal resistance. If the internal resistance increases to a threshold level, the power becomes limited, and the battery reaches the end of life (EOL) state at the time. Therefore, the definition of battery SOH in terms of resistance is shown in Equation (4.21).

$$SOH = \frac{R_{EOL} - R}{R_{EOL} - R_{BOL}} * 100\% \tag{4.21}$$

In Equation (4.21), R_{EOL} is the internal resistance when the battery reaches the EOL state, which is 10 times that of the new battery for lithium-ion batteries. R is the internal ohmic resistance of the battery at the current time point, and R_{BOL} is the internal resistance of the new batteries. When the SOH drops to 80%, the battery reaches the EOL state and needs to be replaced or scrapped.

2. Research on dual adaptive Kalman filter
The classical KF algorithm is only suitable for linear systems, and a linearization step needs to be added to obtain the EKF algorithm. It is necessary to improve the traditional EKF algorithm by adding an adaptive factor to estimate the statistical characteristics of the system to realize the joint estimation of internal resistance and SOC more accurately. First, the state-space expression of the system is obtained as shown in Equation (4.22):

$$\begin{cases} x_{k+1} = A_k x_k + B_k u_k + w_k \\ z_k = C_k x_k + D_k u_k + v_k \\ R_{k+1} = R_k + m_k \\ U_k = U_{OC}(SOC) - U_{1,k} - U_{2,k} - i_k R_k - d_k + r \end{cases} \tag{4.22}$$

In Equation (4.22), w_k is the process noise, and v_k is the observation noise. Because the variation of the internal resistance is very low, a small amount of noise m_k is used to express its change. d_k is the forgetting factor, and r

is the mean value of v_k. $U_{OC}(SOC)$ is the functional relationship of open-circuit voltage concerning SOC value, and $U_{1,k}$ and $U_{2,k}$ are the polarization voltages. The matrices of A_k, B_k, C_k, and D_k are all determined by the model state-space expression, as shown in Equation (4.23).

$$\begin{cases} A_k = \begin{bmatrix} 1 & 0 & 0 \\ 0 & e^{-T/(R_1C_1)} & 0 \\ 0 & 0 & e^{-T/(R_2C_2)} \end{bmatrix} & C_k = \begin{bmatrix} U'_{OC} \\ -1 \\ -1 \end{bmatrix} \\ B_k = \begin{bmatrix} -\eta T / C \\ R_1\left(1-e^{-T/(R_1C_1)}\right) \\ R_2\left(1-e^{-T/(R_2C_2)}\right) \end{bmatrix} & D_k = R_k \end{cases} \tag{4.23}$$

In Equation (4.23), R_k is the internal resistance at time point k. The estimation process of the DAKF algorithm mainly includes five parts: initialization, time update, Kalman gain calculation, filter update, and noise update. The flowchart of the DAKF algorithm is shown in Figure 4.5.

1. Initialization: The initial values are set for the variables and error covariance matrix at time point $k = 0$. The first element of the state vector x_k is SOC, and the remaining elements are zeros. The initial value of the internal resistance is identified by the parameter.

$$K_{x,k} = \tilde{P}_{x,k}\left(C_{x,k}\right)^T\left[C_{x,k}\tilde{P}_{x,k}\left(C_{x,k}\right)^T + P_v\right]^{-1} \tag{4.24}$$

   Taking $\tilde{x}_0$ and x_0 as the initial value of the state vector filter, $\hat{R}_0$ and R_0 as the initial value of the internal resistance filter, the initialization is realized as shown in Equation (4.24):

$$\begin{cases} \hat{x}_0 = E[x_0] \\ \hat{P}_{x,0} = E\left[\left(x_0 - \hat{x}_0\right)\left(x_0 - \hat{x}_0\right)^T\right] \\ \hat{R}_0 = E[R_0] \\ \hat{P}_{r,0} = E\left[\left(R_0 - \hat{R}_0\right)\left(R_0 - \hat{R}_0\right)^T\right] \end{cases} \tag{4.25}$$

2. Time update: $\tilde{x}_k$ is the time update value of the state vector, x_{k-1} is the last filter update value, q is the mean value of w_k, F is the noise driving matrix, P_w is the w_k covariance, and P_r is the error covariance. $\tilde{P}_{x,k}$ is the time updating of the state vector error covariance, and $P_{x,k-1}$ is the last filter update value. Let $\tilde{R}_k$ be the internal resistance time update value,

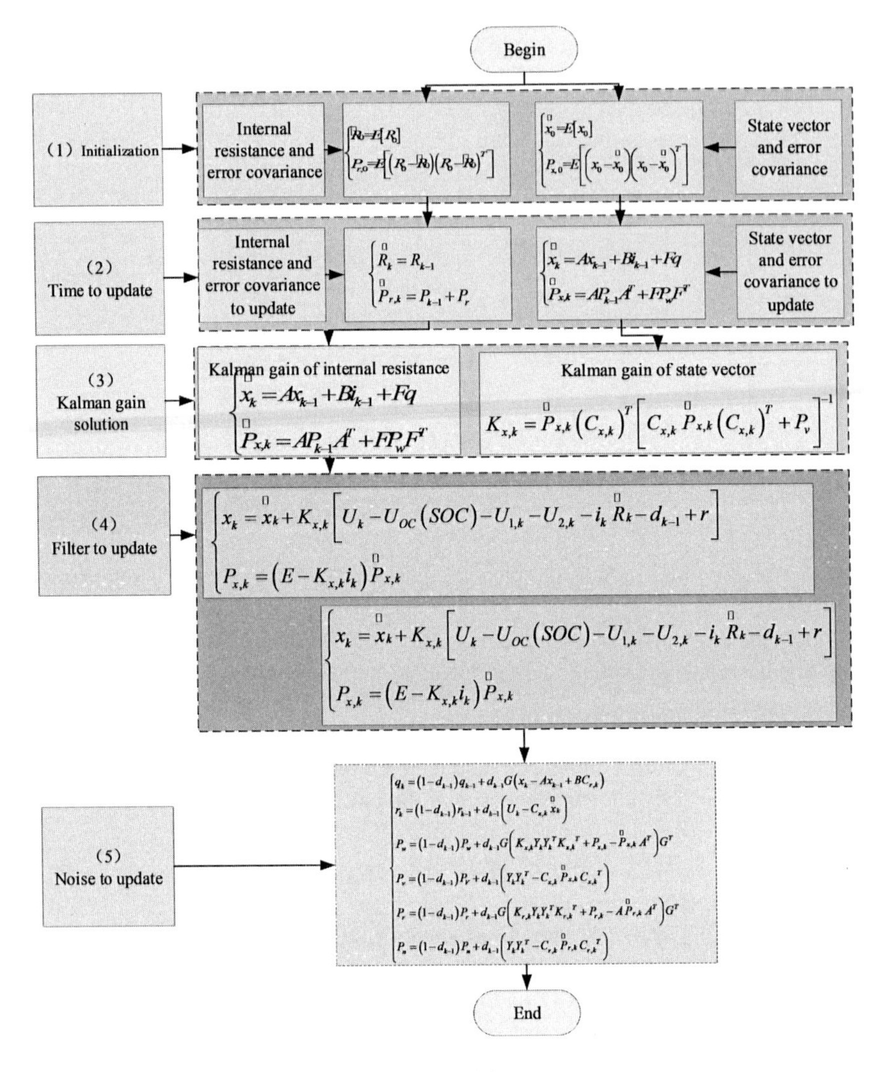

FIGURE 4.5 The flowchart of the DAKF algorithm.

R_{k-1} be the last filter update value, $\tilde{P}_{r,k}$ be the internal resistance error covariance time update value, and $P_{r,k-1}$ be the last filter update value as shown in Equation (4.26):

$$\begin{cases} \tilde{x}_k = A_{k-1}x_{k-1} + B_{k-1}i_{k-1} + Fq \\ \tilde{P}_{x,k} = A_{k-1}P_{x,k-1}A^T + FP_wF^T \\ \tilde{R}_k = R_{k-1} \\ \tilde{P}_{r,k} = P_{r,k-1} + P_r \end{cases} \tag{4.26}$$

3. Kalman gain calculation: Kalman gain is used to update variables and error covariance. Taking $K_{x,k}$ as the state vector Kalman gain, $K_{r,k}$ as the internal resistance Kalman gain, the expression is shown in Equation (4.27):

$$\begin{cases} K_{x,k} = \tilde{P}_{x,k} C_k \left(C_k \tilde{P}_{x,k} C_k^T + P_v \right) \\ K_{r,k} = \tilde{P}_{r,k} i_k \left(i_k \tilde{P}_{x,k} i_k^T + P_r \right) \end{cases} \tag{4.27}$$

4. Measurement update: Using the Kalman gain obtained in the previous step, and the time updated value of the variable and the covariance, U_k is the terminal voltage experimentally measured, and Y_k is the residual. The Kalman gain is used to filter and update the state variables and error covariance to obtain the measurement update value at time point k as shown in Equation (4.28):

$$\begin{cases} x_k = \tilde{x}_k + K_{x,k} Y_k \\ P_{x,k} = \left(E - K_{x,k} C_k \right) \tilde{P}_{x,k} \\ R_k = \tilde{R}_k + K_{r,k} Y_k \\ P_{r,k} = \left(1 - K_{r,k} i_k \right) \tilde{P}_{r,k} \end{cases} \tag{4.28}$$

5. Noise update: The AKF algorithm adds an adaptive step based on the EKF algorithm and updates the noise to make it more in line with the changing situation of the interference under actual conditions. The estimated mean noise and covariance values are updated by the state vector, and the noise covariance is estimated by the internal resistance as shown in Equation (4.29):

$$\begin{cases} q_k = \left(1 - d_{k-1} \right) q_{k-1} + d_{k-1} G \left(x_k - A x_{k-1} + B C_{r,k} \right) \\ r_k = \left(1 - d_{k-1} \right) r_{k-1} + d_{k-1} \left(U_k - C_{x,k} \tilde{x}_k \right) \\ P_w = \left(1 - d_{k-1} \right) P_w + d_{k-1} G \left(K_{x,k} Y_k Y_k^T K_{x,k}^T + P_{x,k} - \tilde{P}_{x,k} A^T \right) G^T \\ P_v = \left(1 - d_{k-1} \right) P_V + d_{k-1} \left(Y_k Y_k^T - C_{x,k} \tilde{P}_{x,k} C_{x,k}^T \right) \\ P_r = \left(1 - d_{k-1} \right) P_r + d_{k-1} G \left(K_{r,k} Y_k Y_k^T K_{r,k}^T + P_{r,k} - A \tilde{P}_{r,k} A^T \right) G^T \\ P_n = \left(1 - d_{k-1} \right) P_n + d_{k-1} \left(Y_k Y_k^T - C_{r,k} \tilde{P}_{r,k} C_{r,k}^T \right) \end{cases} \tag{4.29}$$

The accuracy of the algorithm depends on two factors. The first part is the initial value of the internal resistance and the noise variance.

The initial value of the internal resistance is the internal resistance of the new battery, but the noise variance is adjusted in real time. The second part is the current and voltage of the imported algorithm. If the data are noisy, the algorithm convergence performance may be limited.

3. Experiments and results
   a. Parameter identification
   Based on the second-order RC model, the off-line method is used for parameter identification, and the parametric results are shown in Table 4.1.
   From Table 4.1, the data are selected from 30% to 80% SOC. Because the battery has a high power at high SOC, the internal resistance is low. When the battery is about to be fully discharged, the internal resistance rises rapidly, and the estimation result of SOH is low, which is inconsistent with the actual health status. Also, the open-circuit voltage is proportional to the SOC value. The curve-fitting method is used to obtain a sixth-order linear fit between the OCV and SOC parameters, and the same operation is performed on the remaining parameters.
   b. State estimation under HPPC operating conditions
   Using the DAKF algorithm, the experimental current and voltage obtained by the HPPC test are introduced to obtain the change process of the battery SOC and internal resistance. The DAKF algorithm estimates the SOH value of the battery with the forgetting factor least square method. The obtained result is used as a reference, and the SOC estimation results of the lithium-ion battery are shown in Figure 4.6.
   In Figure 4.6a, SOC1 is the change process of the SOC obtained by the ampere–hour integration method, and SOC2 and SOC3 are the results of SOC obtained by EKF and DAKF algorithms, respectively. In Figure 4.6b, Err1 and Err2 are, respectively, the errors between SOC2and SOC3 compared with SOC1. It can be seen that the error of DAKF is more stable, and the mean value is smaller than that of EKF, and the maximum error is reduced by 2.163%. The SOH estimation results under HPPC working conditions are shown in Figure 4.7.
   In Figure 4.7a, R1 is the battery internal resistance change process, R2 is the internal resistance of the forgetting factor least square method

TABLE 4.1
Parameter Identification Result of Second-Order RC Equivalent Circuit Model

SOC	U_{OC}(V)	R_0(Ω)	R_2(Ω)	C_2(F)	R_1(Ω)	C_1(F)
0.8	3.9368	0.00215142	0.00093170	29108.1340	0.00032683	1240.7267
0.7	3.8323	0.00210102	0.00113284	28547.7525	0.00032246	1263.1096
0.6	3.7252	0.00204085	0.00085208	30102.8295	0.00031913	1285.6905
0.5	3.6527	0.00200420	0.00074835	40529.3208	0.00031561	1275.3141
0.4	3.6185	0.00198594	0.00028976	1465.0150	0.00066627	40299.0720
0.3	3.5907	0.00197412	0.00080352	39849.8270	0.00029968	1418.5330

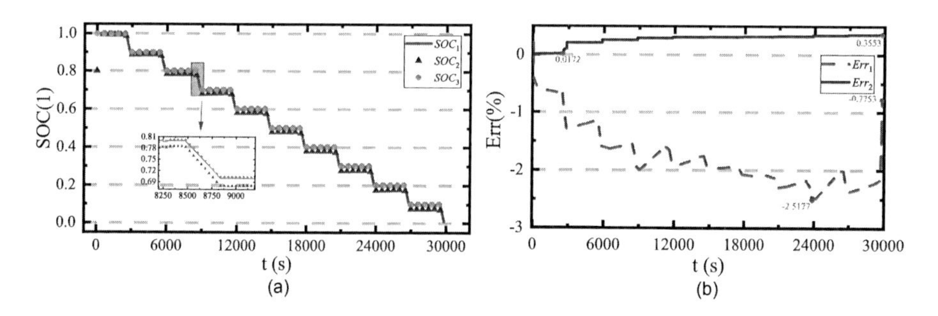

FIGURE 4.6 State of charge estimation under HPPC condition. (a) SOC estimation. (b) SOC estimation error.

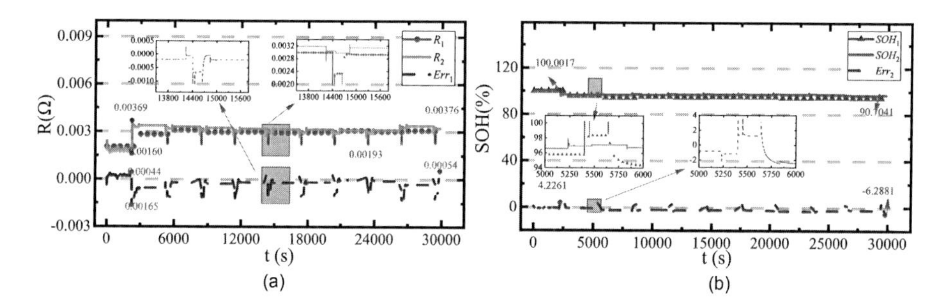

FIGURE 4.7 State of health estimation under HPPC condition. (a) Internal resistance change process. (b) SOH estimation.

as a result, the maximum deviation is 0.00165 Ω. In Figure 4.7b, SOH1 and SOH2 are estimation results of the SOH by the DAKF algorithm and the forgetting factor least square method, respective. It can be seen that SOH1 fluctuates roughly around SOH2, with a maximum deviation of 6.288%, and the overall SOH changes little.

c. Changes in SOC and SOH estimation under BBDST condition
Under the BBDST condition, the battery imitates the working state of the bus, and the working condition is divided into morning peak, evening peak, and suitable period. The battery is charged and discharged with constant power. The test finally discharges the battery to a cut-off voltage of 2.75 V. The estimated SOC of the battery is shown in Figure 4.8.

In Figure 4.8, the SOC estimation results of DAKF have better convergence. Compared with the estimation results of EKF, the error is lower. The maximum error is 1.30%, which is less than 4.88% of the former and a reduction of 3.58%. The SOH estimation results under BBDST working conditions are shown in Figure 4.9.

In Figure 4.9, the SOH1 and SOH2 are estimation results of the state of heaith by DAKF algorithm and theforgetting factor least square method respective the internal resistance estimation result fluctuates

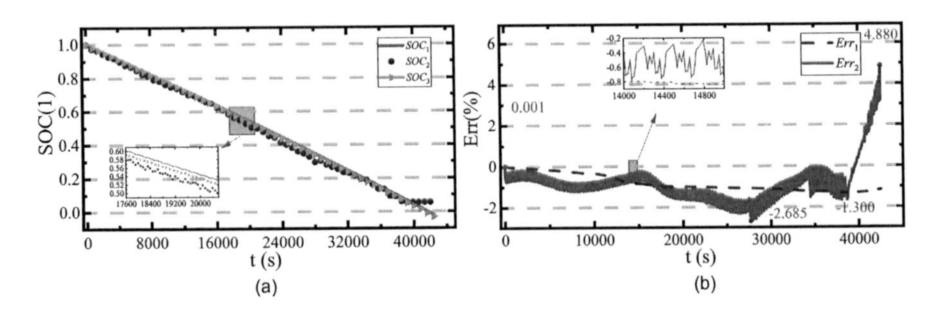

FIGURE 4.8 Changes in SOC under BBDST working conditions. (a) SOC estimation. (b) SOC estimation error.

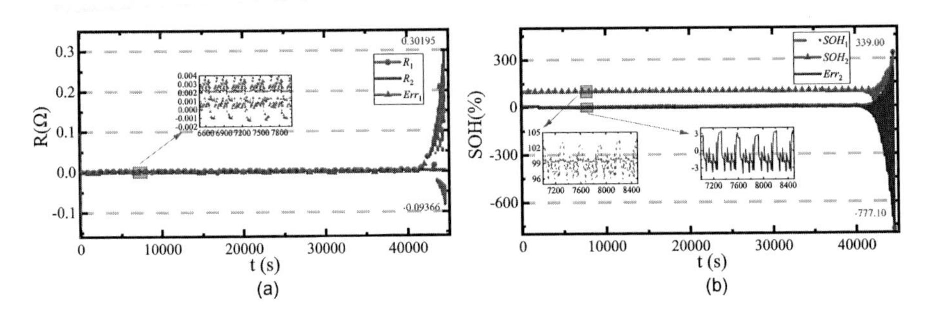

FIGURE 4.9 Changes in SOH under BBDST working conditions. (a) Internal resistance change process. (b) Changes in SOH under BBDST working conditions.

around the results of the forgetting factor least square method. The internal resistance rises rapidly closer to the cut-off voltage. The reason is that the internal polarization reaction rapidly increases when the battery is about to be fully discharged. The SOH estimation results show that the DAKF algorithm has good convergence. Comparatively, the results of the forgetting factor least square method have a divergence of less than 3%, which have a better evaluation effect on the battery SOH estimation.

d. Estimation of battery SOC and SOH under DST working condition
The dynamic stress test (DST), that is, the cyclic constant current charge–discharge test, was performed on the ternary lithium-ion battery. The battery is discharged to a cut-off voltage of 2.75 V. The voltage and current data of the battery obtained from the test are imported into the DAKF algorithm. The SOC and internal resistance change of the battery are obtained under the working condition; the SOC estimation result is shown in Figure 4.10.

In Figure 4.10, it can be observed that the proposed DAKF has a better effect in estimating SOC than the EKF. When the battery is about to be fully discharged, the error of the EKF estimation of SOC suddenly increases, but the DAKF algorithm does not, indicating that the

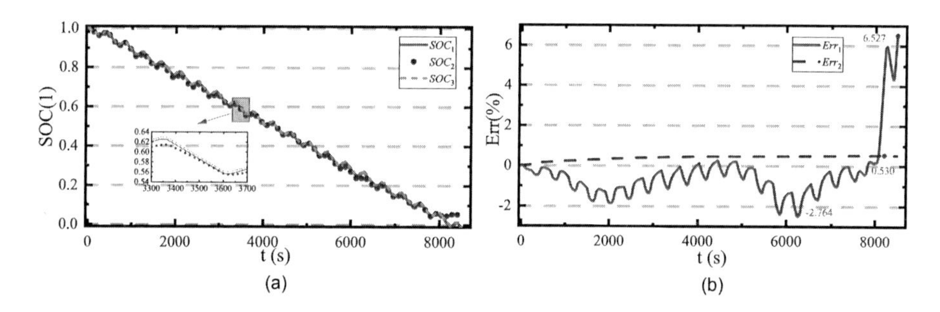

FIGURE 4.10 Estimation of SOC under DST conditions. (a) SOC estimation. (b) SOC estimation error.

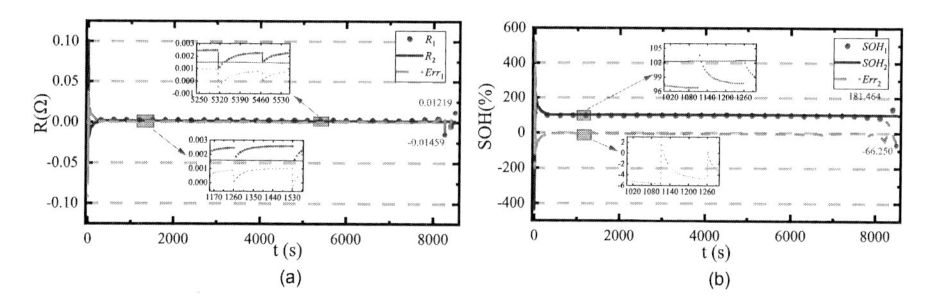

FIGURE 4.11 Estimation of SOH under DST conditions. (a) Internal resistance change process. (b) SOH estimation.

algorithm has high convergence. Under DST conditions, the error of the SOC estimation result is less than 1.00%, and the error is reduced by 5.997% compared with the 6.527% of EKF. The SOC estimation result is shown in Figure 4.11.

In Figure 4.11, the SOH1 and SOH2 are estimation results of the SOH by the DAKF algorithm and the forgetting factor least square method respective. The internal resistance estimation result is slightly higher than the forgetting factor least square method result. Meanwhile, the SOH estimation result is slightly lower, and the divergence is less than 6%.

4.3 DATA-DRIVEN SOH ESTIMATION

The entire vehicle battery system uses multiple lithium-ion batteries in series and parallel to form a battery pack to meet the power output and ensure the high-power output of the power supply system [264]. However, it leads to more complex structures and brings hidden dangers to the safety of new energy vehicles. During the actual application, the aging of the lithium-ion batteries is observed by the declination in the remaining capacity and the increase of the internal resistance leading to the continuous degradation of the SOH. The aging battery brings potential safety hazards to the operation of electric vehicles. Hence, it is necessary to know the SOH of the battery in real-time applications [262,265,266]. Also, the lithium-ion battery is a complex electrochemical system with many factors affecting the remaining capacity and has a high degree of nonlinearity, which makes the SOH extremely difficult to be predicted.

The battery state prediction methods mainly include physical degradation models and data-driven methods. Advanced degradation models for lithium-ion batteries are highly accurate and adaptable, but they are too complex and have high calculation costs, while simple degradation models have low accuracy and poor adaptability in predicting battery status. Compared with the physical degradation model method, the data-driven method does not require the establishment of a complicated battery model. The internal information of the battery can be fully displayed through the collected sample data so that the performance of the battery can be predicted through the collected sample data. In the context of current big data, data-driven methods have good use value in the field of battery health monitoring. The health status information of the lithium-ion battery can be obtained through the analysis of the experimental data, and the battery status can be monitored, which can effectively improve the safety of new energy vehicles and eliminate user "mileage anxiety".

4.3.1 Support Vector Machine

A support vector machine (SVM) is a concrete realization of the dimensional theory of statistical learning theory and the principle of structural risk minimization [267]. The SVM method uses a small number of support vectors to represent the entire sample set and seeks the optimal balance between the complexity of the model and the learning ability to obtain the best promotion ability. The SVM algorithm is good at dealing with the linear inseparability of sample data, which is mainly realized by penalty and kernel function technology. At present, the SVM algorithm has applications in pattern recognition, regression estimation, probability density function estimation, and so on [268–270]. There are n training sample data (x_i, z_i), where x_i is the ith input sample and z_i is the corresponding output value. The functional expression considering linear regression is shown in Equation (4.30):

$$f(x) = (\omega x) + b, \ \omega \in R^n, b \in R \tag{4.30}$$

In the numerical calculation, a margin of error is allowed between $f(x)$ and the true value z, and a suitable loss function needs to be introduced as shown in Equation (4.31):

$$\left| f(x_i) - z_i \right| \le \xi^* + \varepsilon \tag{4.31}$$

In the numerical calculation, a margin of error is allowed between $f(x)$ and the true value z, and a suitable loss function needs to be introduced, as shown in Equations (4.32)–(4.35):

$$\min_{\omega,b,\xi,\xi^*} \frac{1}{2}\omega^T\omega + C\sum_{i=1}^{l}\xi_i + C\sum_{i=1}^{l}\xi_i^* \tag{4.32}$$

$$\omega^T \phi(x_i) + b - z_i \le \varepsilon + \xi_i \tag{4.33}$$

$$z_i - \omega^T \phi(x_i) - b \le \varepsilon + \xi_i^* \tag{4.34}$$

$$\xi_i, \xi_i^* \ge 0,\ i = 1, \ldots, l \tag{4.35}$$

In Equations (4.32)–(4.35), $\omega^T \omega$ determines the confidence range; ξ_i and ξ_i^* are slack variables. The constant C is the penalty factor, which reflects the tolerance of the model to errors in the learning process. Since the optimization problem is a convex quadratic optimization problem, the Lagrange function is introduced, as shown in Equations (4.36) and (4.37):

$$\min_{\alpha,\alpha^*} \frac{1}{2}\left(\alpha_i^* - \alpha_i\right)^T \left(x_i, x_j\right)\left(\alpha_j^* - \alpha_j\right) + \varepsilon \sum_{i=1}^{l} \left(\alpha_i + \alpha_i^*\right) + \sum_{i=1}^{l} z_i \left(\alpha_i - \alpha_i^*\right) \tag{4.36}$$

$$e^T \left(\alpha - \alpha^*\right) = 0 \tag{4.37}$$

The final regression function is shown in Equation (4.38):

$$f(x) = \left(-\alpha_i + \alpha_i^*\right) K\left(x, x_i\right) + b \tag{4.38}$$

The analysis shows that the health state estimation model of the lithium-ion battery is a multivariate nonlinear model related to monitoring parameters, which is described by the mathematical model shown in Equation (4.39):

$$C = f(i, \upsilon, T, \text{SOC}) \tag{4.39}$$

In Equation (4.39), i is the discharge current of the battery, and υ is the terminal voltage of the battery. T is the battery temperature, SOC is the electrical state of the battery, and C is the battery capacity in the present SOH situation. For nonlinear regression models, the basic idea of the support vector regression machine is to use the inner product function to map the training data to a high-dimensional space and use the function to perform linear regression in the high-dimensional space. Then, return to the original space to obtain the original input space. The inner product function is called the kernel function. In the training process of the model, the kernel function is the key factor that determines the performance of the SVM algorithm. The Gaussian radial basis kernel function is selected, and its expression is shown in Equation (4.40):

$$K\left(x, x_i\right) = \exp\left(-\gamma \left\| x - x_i \right\|^2\right), \gamma > 0 \tag{4.40}$$

The SVM-based SOH estimation is mainly divided into two parts: support vector regression machine model training and lithium-ion battery SOH estimation. (1) Support vector regression machine model training: conduct aging tests on batteries under dynamic working conditions, monitor and record operating parameters,

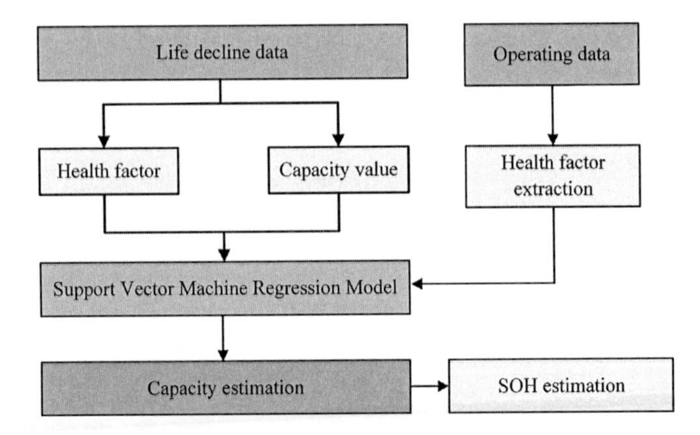

FIGURE 4.12 SVM-based SOH estimation flowchart.

and extract health factors. At regular intervals, the battery capacity is calibrated and used as the output. Voltage, current, temperature, and SOC are used as input factors. And then, a regression model is established. (2) Lithium-ion battery SOH estimation: extract HIs based on battery operating data and input them into the SVM model to estimate the current lithium-ion battery capacity. The SOH value is obtained by the SOH definition expression. The flowchart of the SVM-based SOH estimation is shown in Figure 4.12.

The SVM method of estimating the SOH value requires a large amount of life decline data to establish a regression model [271]. When selecting data, it is necessary to ensure the comprehensiveness of the data and try to include battery life data under various environmental conditions. For data with large individual errors, the reasons should be analyzed to eliminate individual interference data.

4.3.2 Deep Convolutional Network

The convolutional neural network (CNN) is a feedforward neural network (FNN), which not only includes a convolutional calculation function but also has a certain depth structure. It is often used in computer vision, natural language processing, and other fields. Similar to BP neural network, CNN has an input layer, hidden layer, and output layer. But each layer is different from the BP neural network. Compared with the BP neural network, the hidden layer of CNN is no longer a single neuron combination. It does not contain a convolutional layer with a weighted convolution kernel but also a pooling layer for filtering. There is a tiled layer that plays a role in connection called a fully connected layer [272,273].

1. Convolutional layer: The convolutional layer uses the internal convolution kernel to extract features from the input samples. It has three important parameters, namely the size, the step size, and the padding. These parameters determine the size of the output feature map of the convolution layer. In the convolutional layer, the activation function is used to express the complex features of the input. The CNN algorithm is used in the remaining

useful life (RUL) function. Generally, the activation function operation is performed after the convolution kernel but is not fixed. There are many types of activation functions, but their application can be changed.

2. Pooling layer: After the input samples are passed through the convolutional layer, the feature map is obtained. Then, the pooling layer is used to perform feature selection and information filtering on the feature map. The pooling layer also includes pooling size, step size, and fill in the control parameter index. The difference is that the pooling layer only adjusts these parameters to extract the maximum or average value in the feature map. The pooling layer can also reduce the number of nodes in the fully connected layer and prevent overfitting.
3. Full connect layer: The fully connected layer is similar to the hidden layer in the BP neural network. Therefore, its structure and corresponding working principle are the same as the hidden layer in BP neural network. Its function is to perform a series of nonlinear combinations on the features extracted after the input sample is convolutional to obtain the corresponding output.

The difference between the CNN and BP neural network is that the hidden layer of the CNN is not only fully connected but also a network layer for extracting features. The hidden layer is interactively connected between the convolutional layer and the pooling layer. Before trying the method, it is assumed that the ability to automatically extract features would be more beneficial than discovering feature parameters by analyzing the data in advance.

4.3.3 Machine Learning

Machine learning (ML) is a data analysis method that automates the construction of analysis models. The ML system learns from data and makes decisions or predictions with minimal human intervention. The basic workflow required for ML applications for online SOH estimation is divided into four steps. The first step is data collection [274]. The measurable battery parameters, such as temperature, current, and voltage data, are recorded by the BMS during operation and used as input for the training model. However, not all data are related to battery aging. The second step is to extract characteristic values that represent the aging process. The third step is to train an ML model to describe the relationship between the battery SOH and the extracted feature values. After training the model, the last step is to implement it in BMS and apply it online. Among them, feature value extraction is a key step, which seriously affects the performance of SOH estimation [275,276]. A reasonable eigenvalue extraction makes the input data more meaningful and accurate, which also makes the prediction more targeted and accurate.

From the perspective of extracting feature values, ML-based SOH estimation methods are divided into three categories. The first type is model fitting eigenvalues. Thus, eigenvalues are calculated through basic electrical models and online parameter/state estimation algorithms. This type of eigenvalues cannot be directly accessed from BMS sensors, so they are not suitable for real-time applications. The second type is the indirect characteristic value, which usually extracts the values from the differential charging curve at a constant current rate. The change of characteristic value is closely related to the aging process. Due to the limited computing power of the BMS, it is difficult to obtain many external features in operation. Therefore, ML models with variables that can be measured online are more suitable for implementation in more complex devices such as electric vehicles (EVs).

For the third type, direct feature value is obtained by directly accessing the BMS sensor without using models and differential curves for processing data, such as terminal voltage, current, and temperature, to reduce the computation and shorten the SOH estimation time. After collecting and properly representing the dataset, selecting the appropriate ML model is necessary. The ML models are divided into supervised and unsupervised learning models. For the supervised learning model, the training data consist of a collection of inputs and their corresponding output values. The goal of the algorithm is to learn a mapping from input to output with acceptable fidelity. The form of the output value can be in a discrete set to classify a battery as faulty or nonfaulty or in a continuous set of capacitance or resistance values. When the output is classified, the problem is called classification. But when it is a real value, it is called regression. All battery health estimation and prediction problems fall into the regression category because they generate SOH or service life values.

Contrary to a supervised learning model, an unsupervised learning model determines the variables or feature values to train the model and make predictions. The values are imported into a given input to find the trend or clustering of the data without other inputs. So far, a supervised model is the most mature and powerful method. Most research on machine learning methods for battery health diagnosis and prediction uses supervised learning models. Therefore, the ML method described here is centered on supervised learning models. Several supervised ML techniques have been used for battery SOH estimation, including SVM, artificial neural networks, correlation vector machine, Gaussian process regression (GPR), and random forest regression.

Although the ML method is used for SOH estimation and RUL prediction, there are differences between the two applications in terms of input and output characteristics. In the SOH estimation, the input feature values for estimation are extracted from the BMS, and the output is the capacity estimation at a given time point. However, the ML method used for RUL prediction usually requires estimated or measured SOH information, which means the capacity value is used as an input to predict the remaining lifetime or period. The supervised ML model can be a probabilistic model or a non-probabilistic model. For the non-probabilistic, the result is determined by the known relationship between the state and the event, without the need to model the potential probability distribution, including artificial neural networks, SVM, etc. [277,278] However, the important aspect of RUL diagnosis is not only to predict the RUL value but also to propose the uncertainty level of the prediction. Therefore, GPR, RVM, and other probabilistic models derived from the Bayesian framework have attracted more attention due to the level of uncertainty. In addition, GPR is a machine learning method based on kernel function, which can be combined with shell-based prior knowledge of the Yees framework used to realize the prediction. The relevant uncertainty is described by providing the confidence interval of the predicted value. For accurate SOH and RUL estimation, in addition to choosing a better modeling method, it is also very important to select the health factors that can characterize the degree of battery aging.

4.3.4 Fusion Modeling

With the introduction of battery health assessment methods, the battery model and data-driven SOH assessment methods have advantages and disadvantages in terms of accuracy and feasibility. Therefore, the latest research hotspot is the strategy of estimating the SOH by combining the fusion model with any of these two methods. The main idea of the fusion method is to combine, correlate, and merge multiple types of data, models, or algorithms to take full advantage of data information to achieve a more accurate and reliable SOH collaborative estimation of lithium-ion batteries. The co-estimation methods make up for the shortcomings of low estimation accuracy, poor reliability, or misjudgment of a single method and have attracted more research attention [279].

There are two practical methods in fusion modeling. One is to fuse different data-driven algorithms, which combine PSO and SVM algorithms. The PSO algorithm improves the speed of parameter optimization to enhance the SVM algorithm and realize accurate SOH prediction. With the same idea, multiple data-driven methods can be weighted and merged. Combining the data-driven algorithm with the battery model-based method to optimize the error detection of the battery model, the error of the observed state through the data-driven method improves the accuracy of SOH estimation. Using PSO and SVM algorithms to model historical data, predict future measurement data, and establish battery capacity decline and model. The PSO algorithm estimates the battery health status to update the forecast data in real-time applications [280–282]. Also, through a large number of off-line tests, the gray neural network algorithm is used to train the off-line parameters to obtain the battery aging model. Then, the battery equivalent model is established, and the internal impedance parameters are obtained. The gray neural network algorithm identifies the online parameters to obtain the battery SOH value.

Combining the voltage and current data with the ECM in the real-time operation, it is observed that the ohmic resistance and capacitance of the model change monotonously with the attenuation of battery performance to realize the SOH estimation of the lithium-ion batteries. SVM is used to predict the future residual of lithium-ion batteries [283,284]. Then, the future residual combined with the unscented Kalman filter recursively estimates the remaining life and available capacity of the battery to ensure the reliability and accuracy of the estimation method. Also, combining PF and SVM to estimate the capacity and impedance of lithium-ion batteries can obtain high-precision SOH diagnostic results.

Meanwhile, to solve this problem, a dual extended Kalman filter is proposed to estimate the battery state. The variation law of model parameters with the premature aging of the battery is studied. A method for online identification of model parameters based on an autoregressive external source model based on an accurate parameter model is proposed. The regression external source model provides accurate battery open-circuit voltage information, realizes the collaborative estimation of battery state, and improves the accuracy of state estimation. The fusion method gradually combines different state estimation methods to take respective application advantages to achieve a coordinated SOH estimation of lithium-ion batteries for different actual needs. The realization process of the fusion method is simple and fast, showing a good application prospect.

4.3.5 IC-Based SOH Estimation

Aside from the classical methods mentioned above, other methods used to estimate the SOH of the battery are the gray correlation analysis method and wavelet transform method. The gray correlation analysis is also combined with other methods to estimate the SOH value of the batteries [285]. Considering the six main factors that affect SOH, experts score them and determine weights and then use the gray correlation analysis to optimize the parameters that affect these factors. Due to the influence of many factors on SOH, the method is more reasonable and accurate in estimating SOH than the traditional method that only considers a specific factor.

1. Battery incremental capacity curve
   The incremental capacity analysis is an important method to examine the degradation mechanism of batteries, which is centered on the battery incremental capacity curve (IC curve) [286]. The research takes a ternary lithium-ion battery with 25 Ah produced by a domestic manufacturer as the experimental object. The battery terminal voltage versus charging capacity voltage curve is shown in Figure 4.10a. It is observed that the voltage rises rapidly in the initial and final stages of the battery charging period, but in the middle of the charging curve, the voltage rises gradually. The part of the curve where the voltage rises gradually has the highest charged capacity. The battery incremental capacity analysis method is to process the original charging data to obtain the dQ/dV data (charging capacity derived from the battery terminal voltage), then the $dQ/dV - V$ curve (IC curve) is obtained.

   The advantage is that the voltage curve rises gradually, and the relatively flat area of the battery internal reaction is transformed into a dQ/dV peak, which is easily observed and can only be analyzed on the incremental capacity curve, making it difficult to be identified on the voltage curve. A slight change is reflected in the IC curve. Taking a ternary lithium-ion battery as an example, during the entire charging process, the reaction of the positive electrode is mainly nickel ions (Ni^{+}) transforming from divalent ions (Ni^{2+}) to trivalent ions (Ni^{3+}). These two reactions are converted into tetravalent ions (Ni^{4+}), while the negative electrode reaction is the four-phase transitions between the five states (C6, C72, C36, C18, and C12) of graphite (carbon). The current $V - Q$ and $dQ/dV - V$ curves of the ternary lithium-ion battery can be obtained accordingly.

   The analysis of the characteristics of the variation curve of different aging degrees during the battery cycle aging process, including the height and both sides of the IC peak, are studied. The area under the peak is analyzed to determine the mode and mechanism of battery aging and degradation, which affects the diagnosis of the health status of the battery.

2. Characteristic parameters and change characteristics of battery health status
   a. Characteristic parameters in the incremental capacity curve
   The battery capacity model is used to estimate the actual battery capacity. The first step is to select and screen the independent variables of the model. The incremental capacity curve of a lithium-ion battery is composed of several IC peaks [287]. In the incremental capacity curve, the characteristic parameters that are easily extracted include the position of each peak with the corresponding voltage value, the height of each peak, the area of the peak, and the slope on both sides of each peak. The schematic diagram of each parameter in the ternary lithium-ion battery incremental capacity curve is shown in Figure 4.11. The preliminary selected characteristic parameters are used, including the height of peak I, the area under peak I, the left and right slopes of peak I, the position of peak I, and the height of peak II. There are eleven parameters including the area of the peaks, in which the left and right slopes of peak II, the position of peak II, and the area of peak III, as shown in Figure 4.13.

   In Figure 4.13, all the initial parameters extracted are used for model training, which is too redundant, and the task of data preprocessing is large. There is a strong linear relationship between multiple independent variables, but multicollinearity problems occur in the regression process, so the characteristic parameters need to be simplified. First, from the aspect of data source acquisition in practical applications in

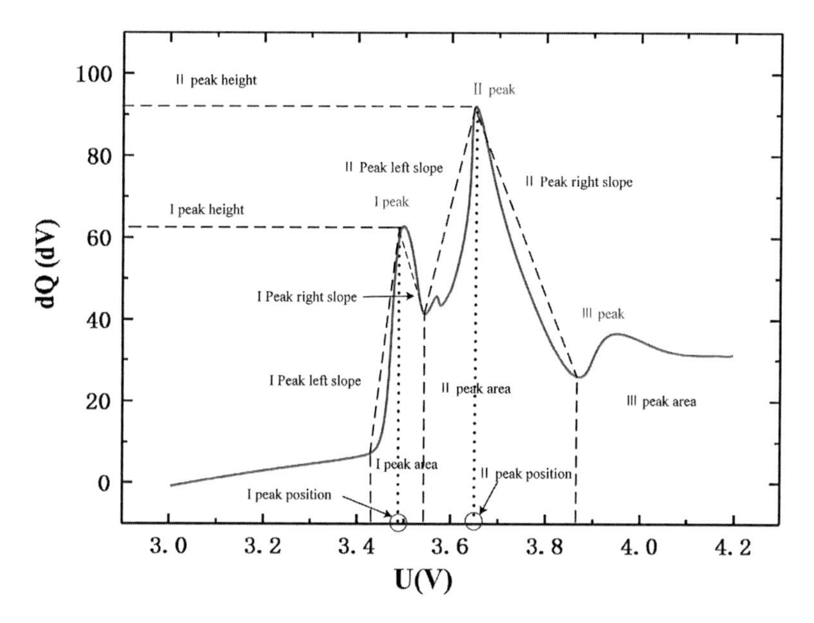

FIGURE 4.13 Schematic diagram of characteristic parameters.

energy storage battery packs of EVs, there are usually inconsistencies in single cells due to capacity, internal resistance, and SOC value.

Due to the inconsistency of the SOC estimation of each cell in the battery pack, most batteries are not fully charged. If the parameters related to the highest or lowest point of the SOC, such as the area covered by peaks I and III, are included in the independent variables, the single cells in the module may not be fully charged. If the parameter data cannot be obtained due to the large-scale operation, the area covered by peak I, peak III, and the slope on the left side of peak I, which depend on the full SOC operation of the battery, are excluded from the characteristic slope. Furthermore, it can be observed from the curves that the shift between peak I and peak II positions is almost the same due to the increase in internal resistance and polarization with aging. According to a mathematical relationship, the slopes to the right of peak I and the left of peak II are determined by the height of the two peaks and the minimum point between the two peaks.

If the heights of the two peaks are selected as the two characteristic parameters, the slope on the right side of peak I and the left side of peak II depend on the minimum value between the two peaks and both sides. The two have a negative correlation, so one of the two is selected because the main peak of the variation curve for peak II can better reflect the thermal characteristics of the batteries. The slope of peak II reflects the speed at which the battery enters the transition phase. Therefore, the SOC on the left side of peak II is selected with the same expression. Therefore, the position of peak I can be eliminated, and that of peak II is retained. In summary, the five characteristic parameters include the area of peaks I and III and the slopes on both sides of peak I, but the position of peak I can be excluded. The remaining six characteristic parameters are the height of peaks I and peak II, the area of peak II, the position of peak II, and the left and right slopes of peak II.

b. Recognition of battery aging and decay patterns

The aging of ternary lithium-ion batteries is generally divided into the decline of thermal characteristics and the decline of dynamic characteristics. Different aging modes reflect different changes in the incremental capacity curve. Thermal loss is usually studied by using the variation curve drawn by charge–discharge data of a small current, such as 0.05 C. Battery thermal loss can be divided into the loss of lithium ions and the loss of positive and negative active materials: The loss of lithium ions usually leads to a decrease in the area of the peak III. However, the loss of negative electrode material usually leads to the reduction of areas of peak I and peak II. Kinetic loss is usually studied by using the curves drawn from high current charge–discharge data, such as 1 C. The aging of battery dynamic characteristics is mainly reflected in the increase in the impedance of lithium-ion batteries, which is reflected in the overall deviation of the curve on the incremental capacity curve. For the incremental capacity curve during charging, if the impedance increases, the curve moves to the right, which is used as the high voltage direction. The experimental program is a 1 C constant current and constant voltage charge-discharge cycle. The battery has undergone a total of 600 charge–discharge cycles, and its capacity has declined by approximately 1.34 Ah for 5.28%.

The variation curve cannot observe peak III. It is because the charge–discharge current is relatively large, and the variation cannot be reflected in the phase change process. From 0 to 600 cycles, in terms of thermal characteristics, according to the low current variation curve analysis method, due to the significant decrease in the height and area of the peak II, the loss of the negative electrode active material is obvious, but the loss of lithium ions cannot be reflected by peak III. Therefore, the loss of lithium ions cannot be determined. For the dynamic characteristics, all the curves shift significantly to the right, indicating that the battery is aging, and the internal resistance is increasing.

c. Parameter correlation analysis

As the number of battery cycles increases, the battery aging and decay intensifies, and the parameters in the variation curve in Figure 4.12 also change. However, the pattern of change for each parameter is different. Some parameters change greatly, and the law is obvious, such as voltage and current, while some parameters remain unchanged or irregular, including internal resistance and SOH. Therefore, it is necessary to screen out the parameters closely related to battery aging and degradation from others. The most direct feature reflecting the degree of aging and degradation is the decrease in the maximum usable capacity of the batteries. Therefore, it can be concluded whether the parameters are closely related to the aging and degradation of the batteries by exploring the correlation between the parameters and the maximum usable capacity.

Further quantitative analysis of each parameter and the maximum usable capacity, as well as the correlation between each parameter, are performed to determine whether there is multicollinearity between the characteristic parameters. The Pearson correlation coefficient is often used as an indicator to measure the correlation between the two parameters. The calculation expression of the Pearson correlation coefficient is shown in Equation (4.41):

$$\rho_{x,y} = \frac{\mathrm{COV}(X,Y)}{\sigma_X \sigma_Y} \tag{4.41}$$

In Equation (4.41), X and Y are two variables, $\mathrm{COV}(X,Y)$ is the covariance of the variables X and Y, and σ_X and σ_Y are the standard deviations of the variables X and Y, respectively. The value of the Pearson correlation coefficient is between −1 and 1. If it is greater than 0, the two variables are positively correlated. If less than 0, the two variables are negatively correlated. If the correlation coefficient is 0, these two variables are irrelevant. The greater the absolute value of the Pearson correlation coefficient between two variables, the higher the accuracy of predicting another variable by the other. It is because the higher the correlation coefficient, the more the variations between the variables. The Pearson correlation coefficients of the seven variables are extracted from the variation curves and the maximum usable capacity of the ternary lithium-ion battery. The results are shown in Table 4.2.

Table 4.2 shows that there is a strong correlation between multiple variables and the maximum usable capacity of the batteries. Also, the correlation between the characteristic parameters is analyzed with the correlation for the other six selected variables. The Pearson correlation coefficients between the variables are shown in Table 4.3.

Table 4.3 shows that there is a strong correlation between multiple variables. For example, the correlation coefficient between the height of peak I and peak II is 0.944, and the correlation between the right slope and the height of peak II is –0.914, indicating that the two correlations

TABLE 4.2
Correlation Coefficient of Parameters and Maximum Useable Capacity

Parameter	**Height of Peak I**	**Height of Peak II**	**Area of Peak II**
Correlation coefficient	0.9483	0.9842	0.2033
Parameter	**Position of Peak II**	**Left slope of Peak II**	**Right slope of Peak II**
Correlation coefficient	–0.9380	0.9497	–0.9013

TABLE 4.3
Pearson Correlation Coefficient for the 6 Characteristic Parameters

Parameters	**Height of Peak I**	**Height of Peak II**	**Area of Peak II**
Height of peak I	1.000	0.944	0.096
Height of peak II	0.944	1.000	0.181
Area of peak II	0.096	0.181	1.000
Position of peak II	–0.905	–0.927	–0.173
Left slope of peak II	0.918	0.967	0.103
Right slope of peak II	–0.883	–0.914	0.179
Parameters	**Position of Peak II**	**Left Slope of Peak II**	**Right Slope of Peak II**
Height of peak I	–0.905	0.918	–0.883
Height of peak II	–0.927	0.967	–0.914
Area of peak II	–0.173	0.103	0.179
Position of peak II	1.000	–0.927	0.833
Left slope of peak II	–0.927	1.000	–0.869
Right slope of peak II	0.833	–0.869	1.000

between these parameters are high. In linear regression, the parameters with higher correlation as the independent variables have the problem of multicollinearity. Multicollinearity means that due to the high correlation between the explanatory variables in the linear regression model, the model estimation is distorted or difficult to estimate accurately. The challenge is that it increases the number of unnecessary calculations and makes the regression unsuitable.

The coefficients of the variables are too sensitive to the sample extraction. Also, the signs of some regression coefficients are opposite to the actual meaning, and it is difficult to distinguish the individual influence of each explanatory variable on the dependent variable. Therefore, the significance of some independent variables is hidden in the regression, which fails the coefficients in the significance test. The variance inflation factor of VIF can quantitatively characterize the severity of multicollinearity, and the calculation formula is determined, as shown in Equation (4.42).

$$\mathrm{VIF} = \frac{1}{1 - R^2} \tag{4.42}$$

In Equation (4.42), R^2 is the multiple determination coefficient of regression on other independent variables when taken as the dependent variables. The empirical judgment method shows that there is no multicollinearity when $0 < \mathrm{VIF} < 10$. Also, for $10 \leq \mathrm{VIF} < 100$, there is strong multicollinearity, but for $\mathrm{VIF} \geq 100$, there is severe multicollinearity. The VIF value of each characteristic parameter is calculated, as shown in Table 4.4.

The experimental results show that the VIF of multiple characteristic parameters is greater than 100, and the maximum value is 375. Therefore, there is serious collinearity between variables, which cannot be directly estimated by simple multiple linear regression. A regression method is used to solve multicollinearity-principal component regression (PCR) to establish a ternary lithium-ion battery capacity estimation model.

TABLE 4.4
The Variance Inflation Factor Value of the Characteristic Parameter

Parameter	Variance Inflation Factor
Height of peak I	32.586
Height of peak II	374.339
Area of peak II	21.129
Position of peak II	130.195
Left slope of peak II	173.787
Right slope of peak II	207.902

The regression estimation results are compared with the most commonly used multiple linear regression results to prove the superiority of the proposed model.

1. Capacity estimation based on PCR
    a. The principle of the PCR
       The PCR is a typical regression method to solve multicollinearity problems, including the principal component analysis (PCA) and regression analysis. The PCA is a typical dimensionality reduction statistical method. Its basic principle is to recombine the original multiple correlated variables x_1, x_2, …, and x_p to generate a few uncorrelated variables F_1, F_2, …, and F_N, causing them to reflect the original variable information as much as possible. Among them, the converted F_1, F_2, …, and F_N are called principal components, which are called the first principal component, and they are the main ingredient. It is geometrically expressed as transforming the original nonorthogonal coordinate system into a new orthogonal coordinate system, making it point to the N orthogonal directions where the sample points are most dispersed. The regression part of principal component regression is ordinary least square regression. Its independent variables are N principal components obtained after PCA, and the dependent variables are standardized cause variables. Since the principal components obtained after PCA are orthogonal, there is no problem with multicollinearity. The main steps of principal component regression are shown in Figure 4.13, and the data standardization method in Step a is shown in Equation (4.43):

$$Z_x = \frac{x - u}{\sigma} \tag{4.43}$$

       In Equation (4.43), x is the original variable, u is the mean value of the variable data, and σ is the standard deviation of the variable data. The principal component regression steps are shown in Figure 4.14
    b. Realization of capacity estimation model based on principal component regression
       Specific to the experimental data, the PCA method is performed on six parameters. The Kaiser–Meyer–Olkin value is a parameter that characterizes whether the data are suitable for PCA between 0 and 1. The closer its value is to 1, the stronger the correlation between variables and the more suitable the data for PCA. The Kaiser–Meyer–Olkin value of the characteristic parameter data of the ternary lithium-ion battery variation curve is 0.716, indicating that PCA is more suitable for dimensionality reduction processing of the data.

(a) Standardize the data of the original independent variable and dependent variable

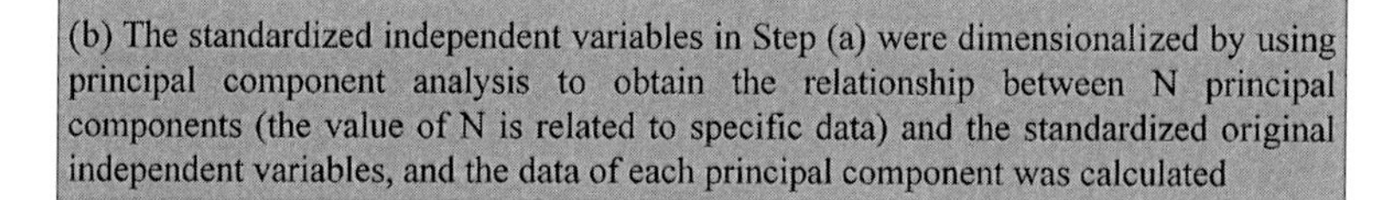

(c) Taking N principal components in Step (b) as independent variables and the standardized dependent variables obtained in Step (a) as dependent variables, the least squares regression was carried out to obtain the relationship between the standardized dependent variables and N principal components

(d) Substituting the relationship between the independent variables of standardization of N principal components in Step (b) into the relationship between the dependent variables of standardization and standardization in Step (e). The relationship between the standardized dependent variable and the standardized original independent variable is obtained

(e) All independent variables and dependent variables in the relationship between standardized dependent variables and standardized independent variables in Step (d) are restored to a non-standardized form to obtain the relationship between the cause variable and the original independent variable

FIGURE 4.14 Principal component regression steps.

The dimensionality reduction results show that the variance of the two principal components is greater than 1. Also, the cumulative variance percentage of the two principal components is 95.025%, indicating that the two principal components contain 95.025% of the information of the original six variables, so the two principal components can be identified. The dimensionality reduction results of the PCA are named Z_{F_1} and Z_{F_2}, and their expressions are obtained as shown in Equation (4.44):

$$\begin{cases} Z_{F_1} = 0.488Z_{x_1} + 0.459Z_{x_2} + 0.047Z_{x_3} - 0.443Z_{x_4} + 0.451Z_{x_5} - 0.431Z_{x_6} \\ Z_{F_2} = -0.001Z_{x_1} + 0.071Z_{x_2} + 0.953Z_{x_3} - 0.086Z_{x_4} + 0.0126Z_{x_5} - 0.282Z_{x_6} \end{cases} \tag{4.44}$$

Equation (4.44) is the data standardization method, Z_{x_1} to Z_{x_6} are the standardized data of parameters, respectively. According to Equations (4.43) and (4.44), the data of principal component variables Z_{F_1} and Z_{F_2} are calculated. Taking Z_{F_1} and Z_{F_2} as independent variables and the standardized capacity Z_Q as the dependent variable, the regression model is established using the least square method and the relationship between

Z_{F_1} and Z_{F_2}. The expression for the capacity Z_Q is obtained as shown in Equation (4.45):

$$Z_Q = 0.449Z_{F_1} + 0.107Z_{F_2} \tag{4.45}$$

Substituting Equation (4.44) into (4.45), the relationship between the standardized capacity and the standardized independent variables to obtain Equation (4.46):

$$Z_Q = 0.201Z_{x1} + 0.214Z_{x2} + 0.123Z_{x3} - 0.208Z_{x4} + 0.204Z_{x5} - 0.163Z_{x6} \tag{4.46}$$

Reversing all the variables in Equation (4.45) to nonstandardized original variables according to Equation (4.46), the final capacity estimation model equation can be obtained as shown in Equation (4.47):

$$Q = 0.0503x_1 + 0.0128x_2 + 0.131x_3 - 8.847x_4 + 0.001x_5 + 51.765 \tag{4.47}$$

In Equation (4.47), x_1 to x_5 are height of peak I, height of peak II, area of peak II, the position of peak II, and left and right slopes of peak II in the ternary lithium-ion battery incremental capacity curve, respectively. The original data are used to test the accuracy of the regression model, and error results show that most of the cycle errors are controlled within 0.2 Ah (1%). Taking into account the measurement error and the outliers that may occur during the measurement process, the model error is less than 2%, indicating the accuracy of the capacity estimation is high. Also, the sign of the coefficient for each parameter is consistent with the correlation between the capacity and each characteristic parameter in Table 4.2, which shows that the results are well interpretable and there is no multicollinearity problem.

By the definition of SOH based on battery capacity, the ratio of the actual capacity estimated by the model to the rated capacity when it leaves the factory is the SOH. A ternary lithium-ion battery is selected to test the above modeling process to verify the applicability of the method using different batteries. The initial capacity of the battery is 36 Ah, and the capacity decline shows 3.026 Ah (8.55%) after 490 cycles. Using the principal component regression to establish a capacity estimation model, the equation is obtained as shown in Equation (4.48):

$$Q = 0.01072x_1 + 0.05891x_2 + 0.38991x_3 - 46.4756x_4 + 0.00019x_5 - 0.06694x_6 + 187.7522 \tag{4.48}$$

The sign of the model expression coefficients is reasonable, and the capacity estimation result is obtained. The residual error values of all cycles are less than 0.2 Ah (2%), which verifies the accuracy and applicability of the model.

2. Comparison of capacity estimation model based on principal component regression and multiple linear regression
The most commonly used multiple linear regression method is used to perform regression analysis on the data. The results are compared with that of the above-mentioned SOH estimation model based on the principal component regression. Using the original six characteristic parameters as the independent variables of the multiple linear regression model and the actual capacity as the model-dependent variable, train the multiple linear regression model to obtain the regression model expression as shown in Equation (4.49):

$$Q = 0.063x_1 - 0.045x_2 + 0.43x_3 - 3.657x_4 + 0.003x_5 - 0.007x_6 + 30.109 \quad (4.49)$$

In Equation (4.49), $x_1 - x_6$ are the height of peak I, the height of peak II, the area of peak II, the position of peak II, and the left and right slopes of peak II, respectively. The results of the principal component regression are compared with the multiple linear regression model. The residual errors for the two methods are the same, less than 0.20 Ah (2%). For interpretability, the result expression of multiple linear regression is unreasonable. The coefficient of x_2 representing the height of peak II is negative. From Table 4.4, the capacity and height of peak II are positively correlated. So, the sign of the coefficients of these two characteristic parameters is unreasonable, indicating that multiple linear regression does not solve the multicollinearity problem. The method causes model distortion and cannot explain the physical meaning of the battery. Conversely, the result expression of the principal component regression is interpretable, and the sign of each parameter corresponds to the correlation, so the collinearity problem is solved with high accuracy.

For the SOH estimation of lithium-ion batteries, there are mainly two methods: mathematical-driven and data-driven models. The research is introduced for state estimation of lithium-ion batteries using classical algorithms and proposes a DAKF algorithm to adaptively co-estimate SOC and SOH. In the lithium-ion battery SOH estimation based on mathematical modeling, the nonlinear system is transformed into a linear system by the EKF algorithm. Then, the nonlinear system is calculated according to the process of the classical KF algorithm.

The PF algorithm avoids the linearization of the KF algorithm and the simplified assumption of Gaussian noise to achieve high prediction accuracy by effectively calculating the uncertainty related to the estimation variables. In the data-driven SOH estimation model, the SVM method needs a large amount of life decay data to establish a regression model [288,289]. The fusion modeling method uses dual extended Kalman filter based on accurate model parameters to regress the external model and provide the battery open-circuit voltage information to realize the collaborative estimation of battery state, which has a high application prospect. The DAKF algorithm is better than the EKF algorithm in SOC estimation, but the SOH estimation result is slightly lower than the forgetting factor least square method, and the error is less than 6%.

4.4 CONCLUSION

This chapter mainly introduces the definition of SOH estimation of batteries and existing estimation methods, as well as several algorithms for co-estimation of SOC and SOH. Generally, for pure EVs, only capacity decay needs to be considered. The SOH of lithium-ion batteries is expressed as the ratio of the current maximum practical capacity to the initial capacity. Mathematical model-driven is a relatively mature mathematical method, which mainly establishes the equivalent model, simplifies the complicated internal reaction, and realizes the estimation of HIs combined with other intelligent algorithms. The established models mainly include the electrochemical mechanism model, an ECM, together with the empirical model. The KF and PF algorithms are introduced to estimate the battery SOH. The SOH estimation model based on mathematical modeling relies on a battery equivalent model combined with an adaptive intelligent algorithm to perform real-time filtering. It helps to calculate battery HIs, such as battery internal resistance and polarization internal resistance, and time SOH through the mapping relationship between HIs and battery SOH estimation. This method solves the shortcoming that the direct measurement method uses to estimate the SOH off-line and is suitable for practical engineering applications. The disadvantage is that this relies heavily on the battery model. The internal resistance and other SOH parameters are affected by compound factors such as SOC and internal resistance, making it difficult to be estimated. An accurate and real-time SOH estimation model depends on meeting high accuracy and low model requirements in computing by SVM, DCN, ML, and other new algorithms.

5 Battery State of Power Estimation

Bowen Li and Jiawei Peng
Southwest University of Science and Technology

Jinhao Meng
Sichuan University

Lili Xia, Li Zhang, Liying Xiong, and Wen Cao
Southwest University of Science and Technology

Shunli Wang
Sichuan University
Southwest University of Science and Technology

Xiao Yang and Yuyang Liu
Southwest University of Science and Technology

CONTENTS

Introduction: In the battery management system (BMS) of the power pack, state estimation is an important feature. The accuracy of battery state estimation directly affects the user's judgment about the electric vehicle, which turns out to determine the vehicle's drivability. However, most of the current research just focuses on the estimation methods of the state of charge (SOC) and the state of health (SOH) of

DOI: 10.1201/9781003333791-5

lithium-ion batteries [290,291]. Although these two state indicators are essential to reflect the remaining power and life of the battery, they all lack the expression of instantaneous response to the change of internal load in the study of the continuous state of the battery. Therefore, it is necessary to carry out the SOP estimation of lithium-ion batteries.

5.1 OVERVIEW OF SOP ESTIMATION

In an in-depth study, SOP is used to describe the peak charge–discharge power of the battery in the current state. The peak power of the battery is estimated to evaluate the charge–discharge limit capability of the batteries. The SOP guides the lithium-ion battery pack to optimally match the power requirement of the equipment, which provides an effective safety guarantee and power output for the equipment and battery pack within a safe working range. Also, it helps to maximize the braking energy recovery function of the motor in the electric vehicle (EV) or hybrid electric vehicle (HEV). The estimation of the peak power can avoid the overcharge and overdischarge of the battery in advance, which has important theoretical and practical value for prolonging the service life of the batteries. Therefore, the accuracy of the SOP estimation for lithium-ion batteries has great significance to electric vehicles [292].

At present, there are several studies on the battery SOC and SOH estimation, and further research on the SOP estimation is needed. The SOP estimation of batteries has received more attention from manufacturers. Some manufacturers have already written about the SOP estimation in technical documents and even some on a commercial basis. However, most of the BMSs of batteries do not involve the SOP estimation. The SOP estimation of the lithium-ion battery provides a reference for the power distribution strategy in electric vehicle systems. As an important reference index of energy output, it can meet the acceleration and climbing performance of the vehicle by providing the needed maximum energy [293,294]. Also, it provides a reference for braking energy recovery so that the energy of electric vehicles is fully utilized. Accurate SOP estimation effectively prevents the occurrence of battery pack overcharge and overdischarge, as well as preventing accidents and prolonging battery cycle life. Accurate acquisition of peak power is indispensable for the overall performance and comfort in the usage of the vehicle. In the energy storage device, the power state of the battery is switched based on the SOP value of the battery to make the energy fully utilized.

The test methods of battery peak power mainly include the United States Advanced Battery Consortium LLC (USABC) test method in the United States, the hybrid pulse power characterization (HPPC) test method proposed by the Freedom CAR project, the power density test method, and the power test method proposed by the Chinese Ministry of Science and Technology in the battery test specification. The core of the USABC test method is the peak power test of the battery. The battery is discharged to a different depth of discharge (DOD), and then, the battery is continuously discharged for t_s to reach the battery discharge cutoff voltage U_{min}. In the HPPC test, when the battery is continuously discharged at a constant current for t_{ss}, the rear voltage drops to the battery discharge cutoff voltage U_{min}, and the constant current value is the peak current of the battery in that state.

The objects tested by the Japan electric vehicle association standard (JEVS) power density test method are mainly lithium-ion batteries and nickel-metal hydride batteries. This method tests the peak power at different power levels from 0% to 100%. The batteries are charged alternately at different rates. After discharging, the voltage–current fitting curve is used to get the highest charging voltage and the lowest discharging voltage of the battery at different DODs, and then, the peak power of the battery is obtained by substituting the corresponding calculation equation. Based on sufficient experiments, the power test method has been formulated as appropriate. The peak power of the battery refers to the product of the voltage value and the current value collected by the BMS every 0.1 seconds. The cumulative value of the battery peak power is divided by the discharge time to calculate the average power of the battery at 1, 3, 5, and 10 seconds. According to the various working conditions of the battery, different continuous discharge times are defined, respectively. For hybrid vehicles, high-current continuous discharge is required for 10 seconds for pure electric vehicles, and it is required for continuous discharge for 30 seconds. For hybrid passenger vehicles, it is continuous discharge for 60 seconds.

At present, the research on the battery power state focuses on the impact of a single factor on the peak power of the battery and the maximum power during the entire battery usable period. Finally, it can be concluded that temperature has the highest impact on the estimation of SOP. Some literature proposes multidimensional dynamic control algorithms to predict the battery voltage to calculate the available battery power. Some researchers believe that the estimated variables, such as SOP and aging, are not accurate. So they directly use the parameters of the model to estimate the online peak power of the batteries. Later studies proposed an electrochemical model considering the effect of temperature on internal resistance and capacity. The extended Kalman filter (EKF) algorithm is used to estimate the power state of the lithium-ion battery. Finally, the maximum peak power of the charged state and the minimum discharge state is obtained through multi-parameter constraints.

5.2 INTERPOLATION-BASED SOP ESTIMATION

The interpolation method is one of the most basic methods for estimating the SOP of lithium-ion batteries. The method utilizes the known static functional relationship between battery power capability and parameter status [295,296]. The method obtains SOP values under different conditions through experiments, constructs an interpolation table for query through different SOP values, and calculates the required data by querying the corresponding SOP values available in the interpolation table. Generally, the battery-related parameters and conditions such as SOC, SOH, and temperature are used for testing. Through the HPPC experiment under different temperatures, SOC, SOH, and electrochemical impedance spectroscopy test conditions, the SOP of the battery in different states is obtained to construct an interpolation table to realize its estimation.

The advantage of interpolation is that it is simple and does not need complex models and algorithms. However, this method requires a large amount of experimental data for training. Also, the interpolation method is the data obtained under static conditions, which is not suitable for SOP estimation under dynamic conditions.

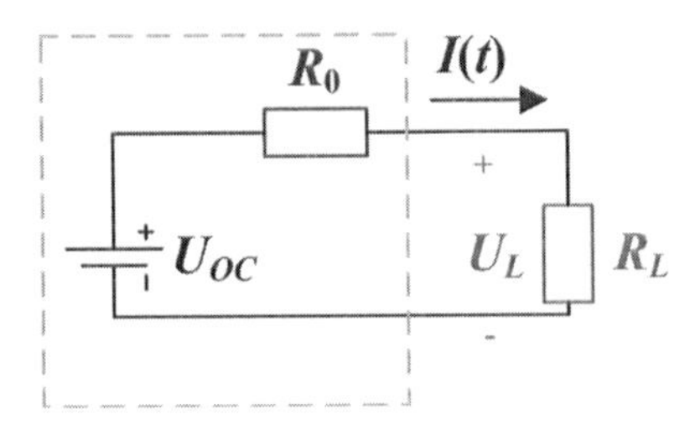

FIGURE 5.1 The linear model.

5.3 MODEL-BASED SOP ESTIMATION

5.3.1 Linear Modeling

A linear modeling method is an approach to estimating SOP using a linear model as an equivalent circuit model. The linear model is simple, and it contains only one resistor. It can only reflect the internal resistance characteristics but cannot effectively characterize the polarization effect of the battery [297]. The linear model of the lithium-ion battery is shown in Figure 5.1.

In the figure, R_0 is used to characterize the internal resistance, U_{OC} is used to characterize the open circuit voltage (OCV), and R_L is the load of the lithium-ion battery.

The partnership for new generation of vehicles (PNGV) compound pulse method uses the charge–discharge cutoff voltage in the battery design as a limiting condition to calculate the peak value of charge–discharge when estimating SOP. With the current model, the peak current and SOP of the battery can be obtained. The calculation equation of SOP is shown in Equation (5.1).

$$\text{SOP} = U_{\text{lim}} * \frac{U_{\text{OC}} - U_{\text{lim}}}{R_0} \tag{5.1}$$

In Equation (5.1), U_{lim} is the limit voltage of the lithium-ion battery. When estimating the SOP during charging, it is the maximum cutoff voltage, and during discharge, it is the minimum cutoff voltage.

5.3.2 RC Loop Modeling

Since the linear equivalent circuit model cannot characterize the dynamic characteristics of the lithium-ion battery, later research improves it by adding an resistor-capacitor (RC) loop. The RC loop can effectively characterize the polarization effect of the lithium-ion battery. The Thevenin model is also called the first-order RC model [298,299]. The Thevenin model is one of the most commonly used equivalent circuit models in battery engineering applications. It has the advantages of simple modeling, few model parameters, higher accuracy, and convenient model parameter identification. It is commonly used for battery power state prediction, fault diagnosis, etc. The added RC loop circuit connected parallel to the linear model characterizes the polarization reaction inside the battery, which can effectively describe the dynamic and static operating characteristics.

In this model, R_0 is used to describe the ohmic characteristics of the battery. R_p and C_p are the polarization resistance and capacitance, respectively, which are used to describe the polarization effect of the battery. U_{OC} is an ideal voltage source, which is used to characterize the open-circuit voltage of lithium-ion batteries. Since there is a relationship between the open-circuit voltage and SOC, it is usually used to characterize the open-circuit voltage U_{OC}. The electrical behavior of the Thevenin model is shown in Equation (5.2).

$$\begin{cases} \dot{U}_p = -\dfrac{1}{R_p C_p} U_p + \dfrac{1}{C_p} I_L \\ U_L = U_P + R_0 \cdot I_L + U_{OC} \end{cases} \tag{5.2}$$

In Equation (5.2), U_p is used to represent the terminal voltage of the RC parallel circuit in the model, and I_L is the load current. The relationship between U_{OC} and SOC is usually obtained through experiments to estimate the OCV in real time based on the SOC variation. Therefore, the model has only three parameters to be identified, which is one of the reasons why the model is widely used.

During the time of battery operation, the terminal voltage needs to be limited, including the battery charge and discharge cutoff voltages, which can be expressed as U_{max} and U_{min}, respectively. U_{lim} is taken as U_{max} when charging, and U_{lim} is taken as U_{min} when discharging. When the terminal voltage is low, the peak discharge power is carefully adjusted to avoid exceeding the lower voltage limit. Under voltage constraints, a detailed calculation method of peak current is given. When estimating the continuous peak current of the battery and the terminal voltage reaches the cutoff voltage, it can be expressed as shown in Equation (5.3).

$$U_L(k+N) = U_{OC}(\mathrm{SOC}(k+N)) + U_P(k+N) + I_{k+N}(k)R_0 \tag{5.3}$$

In Equation (5.3), during the time of discharge, $U_L(k+N)$ is U_{min}, which is the discharge cutoff voltage of the lithium-ion battery. I_{k+N} is the peak discharge current of the lithium-ion battery. During charging, $U_L(k+N)$ is U_{max}, which is the charging cutoff voltage of a lithium-ion battery. The open-circuit voltage is a variable coupled with the SOC, so decoupling is required to accurately estimate the peak current. The most commonly used method for decoupling SOC and open-circuit voltage is the Taylor series expansion. According to the first-order Taylor series expansion, the state-space expression of the open-circuit voltage U_{OC} is shown in Equation (5.4).

$$\begin{aligned} U_{OC}\left(\mathrm{SOC}(k+1), Q_0\right) &= U_{OC}\left(\mathrm{SOC}(k+1) + \frac{\eta i(k+1)T}{Q_0}, Q_0\right) \\ &= U_{OC}\left(\mathrm{SOC}(k+1), Q_0\right) + \frac{\eta i(k+1)T}{Q_0} \frac{\partial U_{OC}}{\partial \mathrm{SOC}}\bigg|_{\mathrm{SOC}=\mathrm{SOC}(k+1)} \\ &\quad + \delta\left(\mathrm{SOC}(k+1), \frac{\eta i(k+1)T}{Q_0}\right) \end{aligned} \tag{5.4}$$

In Equation (5.4), Q_0 is the capacity of the lithium-ion battery, δ is the high-order residual term obtained by the Taylor series expansion, and T is the sampling time of the voltage and current of the BMS. Ignoring the high-order terms, the final OCV calculation equation is obtained, as shown in Equation (5.5).

$$U_{OC}\left(\text{SOC}(k+1),Q_0\right) \approx U_{OC}\left(\text{SOC}(k+1),Q_0\right) + \frac{\eta i(k+1)T}{Q_0}\frac{\partial U_{OC}}{\partial \text{SOC}}\bigg|_{\text{SOC}=\text{SOC}(k+1)} \tag{5.5}$$

In Equation (5.5), $U_{OC}\left(\text{SOC}(k+1),Q_0\right)$ is the OCV after decoupling. According to the state-space expression and observation of the Thevenin model, the lithium-ion battery system can be obtained, as shown in Equation (5.6).

$$\begin{cases} X(k+L) = A \cdot X(k) + B \cdot I(k) \\ U_L(k+L) = U_{OC}(k+L) + [\begin{matrix} 0 & 1 \end{matrix}]X(k+L) + R_0(k)I(k) \end{cases} \tag{5.6}$$

In Equation (5.6), X is the state matrix, $X = \left[\text{SOC}(k);\, U_p(k)\right]$. A and B are the state transition and control matrices of the lithium-ion battery system, $A = \left[1 \;\; 0;\, 0\; e^{-T/\tau}\right]$, $B = \left[-\eta T/Q_0;\, R_p\left(1 - e^{-\frac{T}{\tau}}\right)\right]$. The peak current calculation equation of the lithium-ion battery of the Thevenin model can be obtained, as shown in Equation (5.7).

$$\begin{cases} I_{\max}^{\text{dis}} = \dfrac{U_{L,\min} - U_{OC}\left(\text{SOC}_{k+N}, Q_0\right) - \left[\begin{matrix} 0 & 1 \end{matrix}\right] \cdot A^N(k) \cdot X(k)}{\dfrac{\eta T \cdot N}{Q_0}\dfrac{\partial U_{OCV}}{\partial \text{SOC}}\bigg|_{\text{SOC}=\text{SOC}_K} + \left[\begin{matrix} 0 & 1 \end{matrix}\right]\sum\limits_{j=0}^{N-1} A^{N-1-j}(k) \cdot B(k) + R_0(k)} \\ I_{\min}^{\text{chg}} = \dfrac{U_{L,\max} - U_{OC}\left(\text{SOC}_{k+N}, Q_0\right) - \left[\begin{matrix} 0 & 1 \end{matrix}\right] \cdot A^N(k) \cdot X(k)}{\dfrac{\eta T \cdot N}{Q_0}\dfrac{\partial U_{OCV}}{\partial \text{SOC}}\bigg|_{\text{SOC}=\text{SOC}_k} + \left[\begin{matrix} 0 & 1 \end{matrix}\right]\sum\limits_{j=0}^{N-1} A^{N-1-j}(k) \cdot B(k) + R_0(k)} \end{cases} \tag{5.7}$$

According to the obtained peak current and cutoff voltage, the SOP of continuous L sampling points can be obtained, and the SOP calculation is shown in Equation (5.8).

$$\begin{cases} \text{SOP}_{\max}^{\text{dis},V} = I_{\max}^{\text{dis},V} \cdot U_{L,\min} \\ \text{SOP}_{\min}^{\text{chg},V} = I_{\min}^{\text{chg},V} \cdot U_{L,\max} \end{cases} \tag{5.8}$$

As an improvement of the Thevenin model, the second-order Thevenin model uses two parallel RC loops to characterize the electrochemical and concentration polarizations of the lithium-ion battery, which is shown in Figure 5.2.

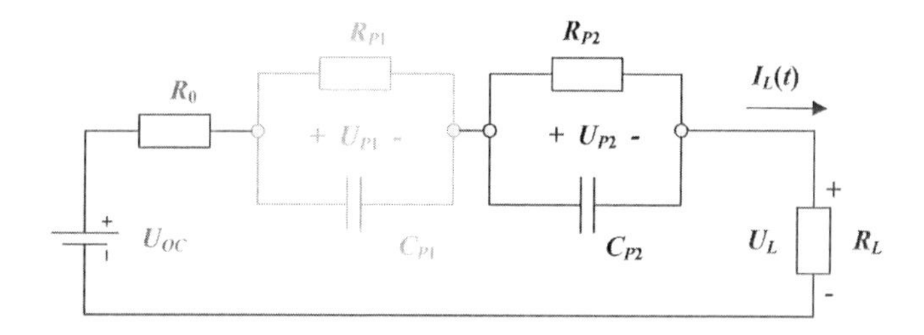

FIGURE 5.2 Second-order Thevenin model.

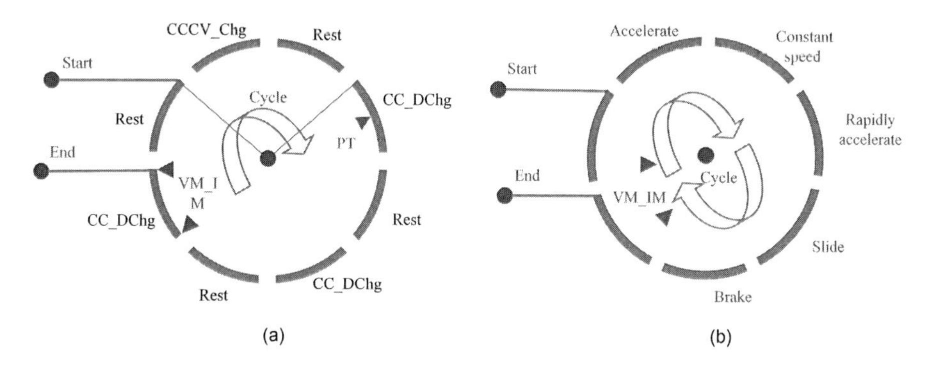

FIGURE 5.3 The experimental data test flow of the research object. (a) HPPC test condition flowchart. (b) BBDST test condition flowchart.

In Figure 5.2, R_{p1} and C_{p1} are used to characterize the electrochemical polarization of the lithium-ion battery. R_{p2} and C_{p2} are used to characterize the concentration polarization of the lithium-ion battery. When the model is used to estimate the SOP value, an estimation method is formed similar to the Thevenin model that can realize SOP estimation based on the second-order Thevenin model. The following is a validation analysis of the first-order model. The continuous response of the lithium-ion battery changes the state of the battery, thus changing the estimation accuracy. In this research, discharge conditions with different durations were constructed, as shown in Figure 5.3.

By experimenting with the discharge state of lithium-ion batteries under different continuous excitation, a comparative analysis under different conditions is constructed. A lithium-ion battery with a capacity of 74.28 Ah is used, and the experiment is carried out in temperature test equipment at 25°C. Figure 5.3a shows the HPPC flowchart to describe the different discharge conditions of the battery; Figure 5.3b shows the Beijing bus dynamic stress test (BBDST) flowchart, used to describe the power output conditions of the vehicle in different states. The BBDST operating currents for different continuous peak discharges are shown in Figure 5.4.

Figure 5.4a–c shows the current state of the BBDST under different continuous peak discharges. PT is the power discharge time, which changes the continuous excitation duration; VM_IM is the current and voltage cutoff judgment.

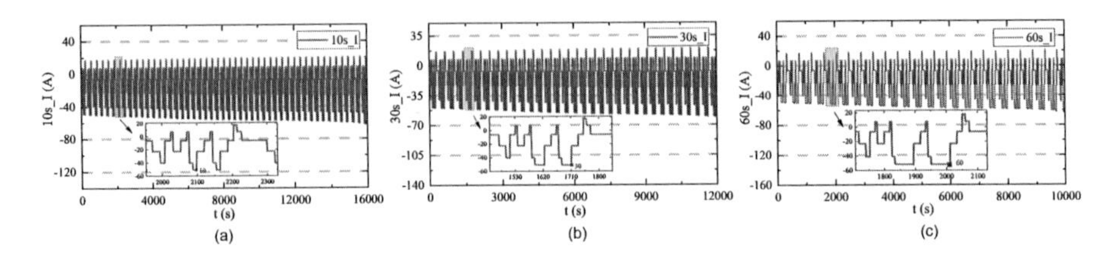

FIGURE 5.4 The experimental condition curves. (a) Peak discharge current curve for 10 seconds. (b) Peak discharge current curve for 30 seconds. (c) Peak discharge current curve for 60 seconds.

When testing a lithium-ion battery under a constant BBDST working condition, as the voltage decreases, the battery cannot maintain power output, which causes the working current of the battery to fluctuate highly. When the battery can provide sufficient voltage, it can capture the above-mentioned current change characteristics. However, when the battery is in trickle charge or discharge, the voltage change is not significant for a short time. But a single factor is difficult to track and correct, which causes system estimation errors. The SOP of a lithium-ion battery is mainly related to the battery voltage, current, and internal state parameters. The objective of this study is to analyze the discharge-based experiment. The length of the continuous discharge time causes a large variation in the voltage; that is, the state of the lithium-ion battery under the excitation of different durations changes highly. The relationship is analyzed by constructing HPPC experiments with three excitation duration of 10, 30, and 60 seconds, as shown in Figure 5.5.

Figure 5.5 shows the P-SOC fitting curve under different excitation duration. The P-SOC polynomial expression is shown in Equation (5.9).

$$\begin{cases} 10\text{ seconds\_}P_k = 3.098e+4\times \text{SOC}_k^5 - 9.741e+4\times \text{SOC}_k^4 + 1.17e \\ \quad +5\times \text{SOC}_k^3 - 6.515e+4\times \text{SOC}_k^2 + 1.727e+4\times \text{SOC}_k \\ 30\text{ seconds\_}P_k = 2.36e+4\times \text{SOC}_k^5 - 7.091e+4\times \text{SOC}_k^4 + 8.581e \\ \quad +4\times \text{SOC}_k^3 - 5.036e+4\times \text{SOC}_k^2 + 1.447e+4\times \text{SOC}_k \\ 60\text{ seconds\_}P_k = 3.618e+3\times \text{SOC}_k^5 - 1.678e+4\times \text{SOC}_k^4 + 3.242e \\ \quad +4\times \text{SOC}_k^3 - 2.77e+4\times \text{SOC}_k^2 + 1.07e+4\times \text{SOC}_k \end{cases} \tag{5.9}$$

The power variation under the HPPC test is assumed to be like the voltage variation trend. The estimation error of SOP mainly comes from internal parameters and characteristics of working conditions. The operating characteristics of EVs and HEVs have significant changes under different working conditions. By analyzing the online identification of the same battery under the excitation of different continuous peak discharges, a battery peak state estimation system under the compound model is constructed. Taking into account the influence of various factors, the SOP estimation and error curves of the lithium-ion battery, as shown in Figure 5.6.

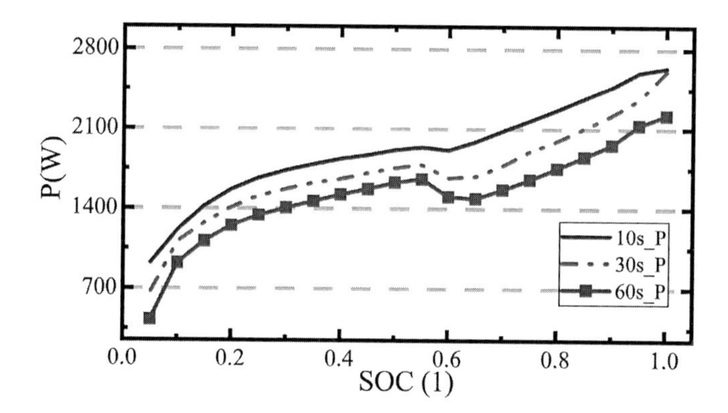

FIGURE 5.5 Power curve under different discharge continuous excitation.

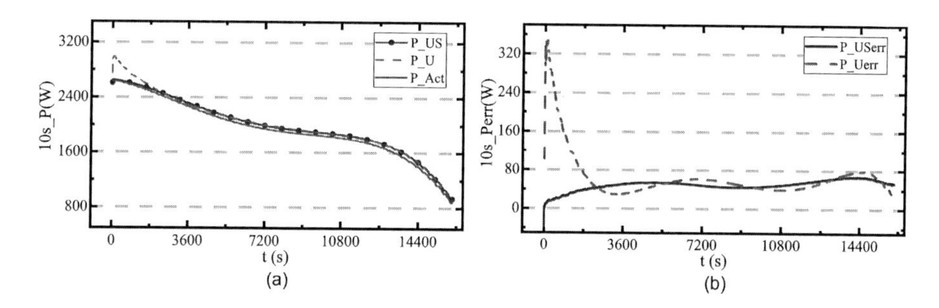

FIGURE 5.6 SOP estimation result at 10 seconds peak discharge. (a) SOP estimation curve based on fusion model at 10 seconds peak discharge. (b) SOP error curve at 10 seconds peak discharge.

As shown in Figure 5.6, *P_US* is the power estimation under the composite model of the lithium-ion battery, and the peak current is introduced to the limit; *P_U* is the estimated state of the lithium-ion battery under the voltage limit; *P_Act* is the reference state obtained through the HPPC test; P_U_{err} is the power error under the voltage limit; P_US_{err} is the power error under the composite model. In Figure 5.6a and b, the composite model can provide front-end power optimization with an overall error of less than 100 W when discharging at a sustained peak for a short time, as shown in Figure 5.7.

Figure 5.7a and b is the SOP estimation result at 30 seconds peak discharge, the legend of it is the same as Figure 5.6. The composite mode performance is sufficient to provide better back-end optimization capability, and the overall error remains within 130 W after the estimation model is stable, as shown in Figure 5.8.

Figure 5.8a and b is the SOP estimation result at 60 seconds peak discharge. As shown in the figures, with the increase in discharge time, the peak power in the latter half fluctuates highly. The composite mode performance is sufficient to provide better back-end optimization capability, and the overall error remains within 250 W.

The influence of different excitation duration under complex conditions increases along with the time extension. Analyzing the estimation error between each impact

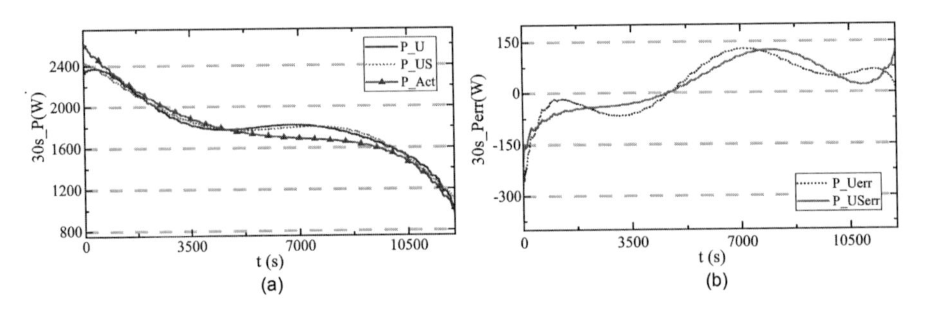

FIGURE 5.7 SOP estimation result at 30 seconds peak discharge. (a) SOP estimation curve based on fusion model at 30 seconds peak discharge. (b) SOP error curve at 30 seconds peak discharge.

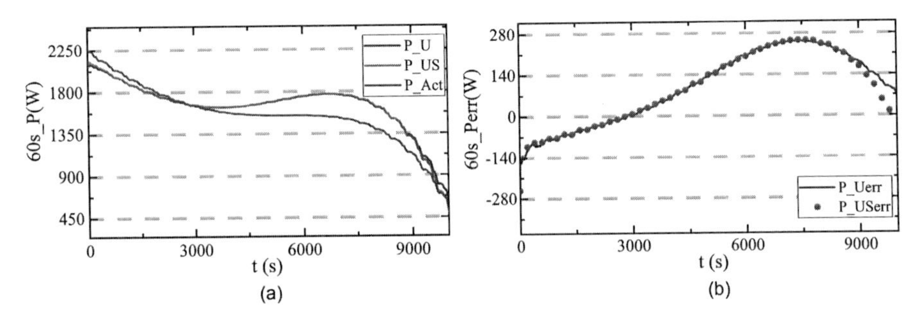

FIGURE 5.8 SOP estimation result at 60 seconds peak discharge. (a) SOP estimation curve based on fusion model at 60 seconds peak discharge. (b) SOP error curve at 60 seconds peak discharge.

factor helps to analyze the inherent problems of the algorithm. The evaluation of the algorithm's overall identification parameters is shown in Table 5.1.

From Table 5.1, it can be observed that the estimation errors of voltage and peak power show a slow upward trend with the excitation time. The fusion algorithm has a strong tracking effect on short-term excitation, but the long-term excitation receives more interference. Effective estimation of balanced time pulses can provide a higher power state tracking effect.

5.3.3 SOC Limit Estimation

SOC is a state parameter used to characterize the remaining capacity, which is closely related to the current of the lithium-ion batteries. Therefore, it can be used for peak current estimation when performing SOP estimation. The SOC calculation equation is shown in Equation (5.10).

$$\mathrm{SOC}(k+1) = \mathrm{SOC}(k) - \frac{\int_{k}^{k+1} \eta I(k)\,dk}{Q_0} \tag{5.10}$$

TABLE 5.1
Full State Performance Estimation

Time (seconds)	Performance	Error Observation			
		U	SOC	*P_U*	*P_US*
10	Mean absolute error (MAE)	0.002	0.0092	65.4179	49.0931
	Mean squared error (MSE)	1.75e−05	9.27e−05	7.05e+03	3.03e+03
	Root mean square error (RMSE)	0.0042	0.0096	83.9513	55.0342
30	MAE	0.0025	0.0082	65.3559	60.8035
	MSE	2.07e−05	7.97e−05	6.05e+03	5.89e+03
	RMSE	0.0045	0.0089	77.7608	76.9086
60	MAE	0.0032	0.0075	133.2087	128.8349
	MSE	3.02e−05	7.29e−05	2.56e+04	2.51e+04
	RMSE	0.0055	0.0085	159.8933	158.4401

The current calculation equation based on the SOC can be obtained, as shown in Equation (5.11).

$$I(k) = Q_0 \cdot \frac{\text{SOC}(k+1) - \text{SOC}(k)}{\eta T} \tag{5.11}$$

The SOC estimation needs to be controlled within an appropriate range to ensure the safe and suitable operation of the electric vehicle battery pack. The SOC at any time point k should satisfy $\text{SOC}_{\min} < \text{SOC} < \text{SOC}_{\max}$. The maximum charge–discharge current in N sampling intervals can be calculated, as shown in Equation (5.12).

$$\begin{cases} I_{\max}^{\text{dis,SOC}} = Q(k) \cdot {\left(\text{SOC}(k) - \text{SOC}_{\min}\right)} \Big/ {\eta T \cdot N} \\ I_{\min}^{\text{chg,SOC}} = Q(k) \cdot {\left(\text{SOC}(k) - \text{SOC}_{\max}\right)} \Big/ {\eta T \cdot N} \end{cases} \tag{5.12}$$

In Equation (5.12), $I_{\max}^{\text{dis,SOC}}$ is used to characterize the maximum discharge current, and $I_{\min}^{\text{chg,SOC}}$ is used to characterize the maximum charge current estimated based on SOC value. The calculation equation of SOP is shown in Equation (5.13).

$$\begin{cases} \text{SOP}_{\max}^{\text{dis,SOC}} = I_{\max}^{\text{dis,SOC}} \cdot U_L\left(I_{\max}^{\text{dis,SOC}}\right) \\ \text{SOP}_{\min}^{\text{chg,SOC}} = I_{\min}^{\text{chg,SOC}} \cdot U_L\left(I_{\min}^{\text{chg,SOC}}\right) \end{cases} \tag{5.13}$$

5.3.4 Temperature Limit Estimation

The temperature of the battery affects the internal electrochemical reaction inside it. The heat generated by the battery during the charge–discharge process causes the temperature to rise, which damages the elements of the electrolytes inside the battery and shortens its life span. Therefore, when estimating the SOP of a lithium-ion battery, the temperature limits need to be considered [300]. When the battery operating temperature is close to the limit, the power should be controlled. The temperature-based SOP estimation method uses the battery thermal characteristic and the energy balance to obtain the energy balance of the battery for the time interval Δt, as shown in Equation (5.14).

$$I^2 R + T_{\text{avg,bat}} \frac{dU}{dT} I + hS\left(T_{\text{amb}} - T_{\text{avg,bat}}\right) - C_p m \frac{T_{k+L} - T_k}{L\Delta T} = 0 \tag{5.14}$$

In Equation (5.14), R is the total internal resistance of the battery. $T_{\text{avg,bat}}$ and T_{amb} are the average temperature and ambient temperature of the battery, respectively, and C_p is the specific heat capacity of the battery. h is the heat transfer coefficient, m is the mass of the lithium-ion battery, and S is the heat transfer surface area. During the charge–discharge process, the maximum temperature threshold of the battery is set, and then, the peak current value under the temperature limit is obtained. Then, the peak power of charge–discharge can be calculated based on the terminal voltage. Therefore, based on the energy balance, the peak current is obtained to estimate the SOP of the battery. The method is usually combined with other SOP estimation methods. Also, the temperature causes changes in the parameters of the lithium-ion battery model. Therefore, most researchers consider the temperature factor in the equivalent circuit model of the battery, thereby reducing the SOP estimation error. In addition to considering the SOC and temperature limits, the voltage limit can also be used as a factor when calculating the SOP of lithium-ion batteries if necessary.

5.4 DATA-DRIVEN NON-PARAMETRIC MODELING

The non-parametric model for the SOP estimation method shifts the focus on the internal mechanism of the battery of the data itself. The method regards the battery as a black box and usually adopts data-driven algorithms as a support, using machine learning methods to estimate the battery state [301]. Each parameter is used as the input of the model, and the output is the SOP to be estimated. Through a large amount of testing data and training model, the SOP is estimated.

5.4.1 BP Neural Network

The backpropagation (BP) neural network is composed of an input layer, an output layer, and a hidden layer. It has the network model characteristics of signal forward transmission and error BP, and it is a feedforward multi-layer perceptron [302]. The input layer is the basic parameters of the battery, such as voltage, temperature, and SOC values. The output layer is the estimated SOP of the battery. In the signal transmission process, if the actual value of the output layer is different from the expected value, it goes to the BP stage. Then, it adjusts the threshold and weight of the network according to the prediction error so that the output layer of the BP neural network

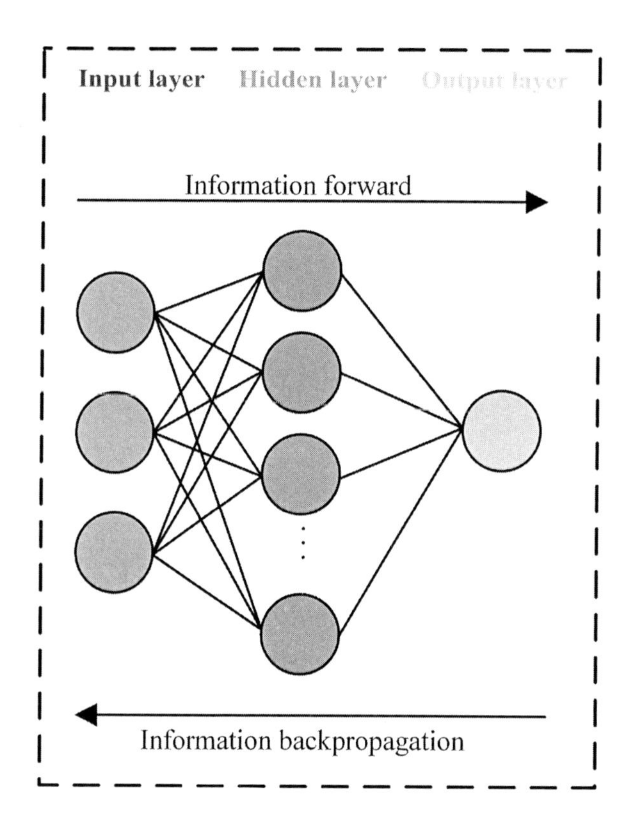

FIGURE 5.9 BP neural network structure.

constantly approaches the expected value. The structure of the BP neural network is shown in Figure 5.9.

According to theoretical research, in the two sets of nonlinearly related input and output data, if other conditions are selected reasonably, a three-layer BP neural network can meet the design accuracy requirements [303]. Also, the choice of input and output parameters directly affects the number of nodes in the input layer and output layer of the neural network and the type of transformation function. Selecting reasonable input parameters can obtain good network performance and improve the training speed and accuracy of the network. The choice of input parameter basic principle generally followed is to select parameters that have a large impact on the network output and are easy to measure. Consequently, there is no correlation or low correlation between input variables. The flowchart of battery SOP estimation is shown in Figure 5.10.

BP neural network selects the appropriate number of the input layer, hidden layer, and output layer according to the modeling requirements. It uses the hidden layer to connect the input and output layers and establishes the node weight between layers [304]. Then, the normalized training samples are sent to the output layer to train the network. Finally, the network weights are adjusted until the error reaches the acceptance range.

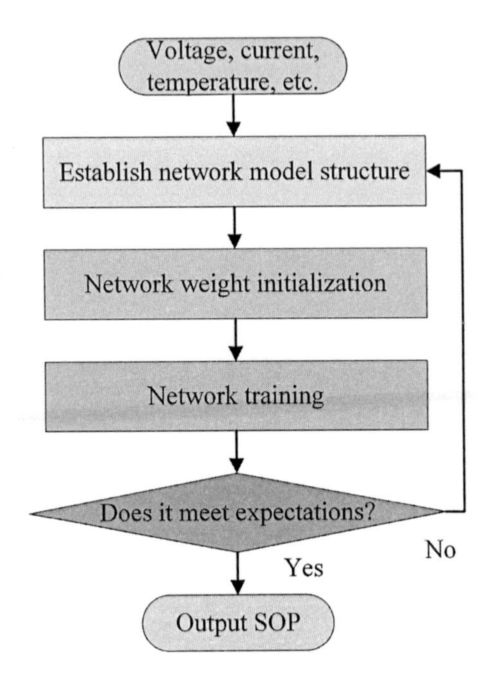

FIGURE 5.10 The flowchart of battery SOP estimation.

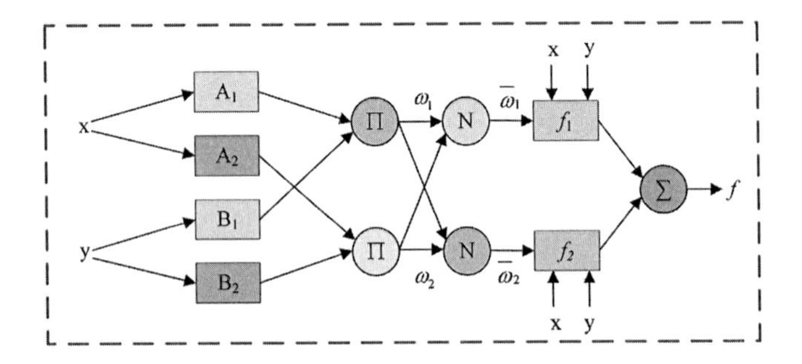

FIGURE 5.11 Typical ANFIS structure diagram.

5.4.2 Adaptive Neuro-Fuzzy Inference Modeling

The adaptive neuro-fuzzy inference system (ANFIS) directly utilizes the optimization function of the neural network to control the parameters of the fuzzy controller and improve the function of the fuzzy control system. The neural network learning mechanism is introduced into the fuzzy system, which does not only learn the training data but automatically generates and modifies the membership functions of input and output variables [305]. Also, it summarizes the optimal fuzzy rules to form an adaptive neuro-fuzzy controller. The ANFIS structure is shown in Figure 5.11.

The ANFIS SOP estimation is mainly divided into three steps: data acquisition, model training, and simulation verification. First, the experimental method is determined to obtain the peak power and ohmic resistance at different temperatures and

SOC conditions. Then, the fuzzy interval of variables can be divided by the grid generation method and subtractive clustering method [306]. After that, a single BP neural network or comprehensive training method is used for the training model, which reduces the number of iterations and increases the accuracy. Finally, simulation verification is carried out. This method has a strong adaptive ability and reduces the complexity for the premise of ensuring accuracy.

5.4.3 Support Vector Machine

Support vector machine (SVM) is a method based on the Vapnik–Chervonenkis dimensional theory of statistical learning theory and the principle of structural risk minimization. It has important advantages in solving small samples, and nonlinear and high-dimensional pattern recognition, which can be extended and applied to other machine learning problems such as function fitting with high robustness [307]. As a supervised learning model in machine learning, SVM can analyze data, recognize patterns, and be used for classification and regression analysis. The experimental procedure of SVM to estimate SOP consists of three steps: First train and learn the data with the SVM, construct a regression estimation function through a suitable kernel function, and then substitute the data into the calculation procedure to obtain the corresponding SOP predicted value. Finally, compare the experimental and predicted values to perform a comparison match.

The data-driven non-parametric estimation method has wide applicability and can estimate the SOP of any battery [308]. The more the input parameters and the larger the amount of data, the more accurate the SOP estimation. However, the method requires a large amount of experimental data. The estimation accuracy is strongly dependent on the training data and the training method, which requires high hardware and takes a long time. The comparison of the advantages and disadvantages of the data-driven estimation methods is shown in Table 5.2.

5.5 OTHER ESTIMATION METHODS

In recent years, new SOP estimation methods have also been proposed, such as the SOP estimation method based on MSP430, which is a new method that combines

TABLE 5.2
Advantages and Disadvantages of SOP Data-Driven Estimation Methods

	Estimation Method	Advantage	Disadvantage
Data-driven estimation approach	BP neural network	It has a self-learning ability without an accurate battery model.	It is easy to produce the phenomenon of local optimum, and the amount of calculation is very large
	ANFIS method	The local optimal problem is overcome and high accuracy.	It requires a lot of experiments and complicated calculations
	SVM method	It has good robustness.	

parametric and non-parametric models for the online SOP estimation. Most of these new methods are used in the theoretical or experimental simulation stage, and it takes a long time to develop before putting into practical application. The improvement of the accuracy in the battery SOP estimation requires not only mature algorithms and battery models with large data sets and integration of various methods.

Interpolation is one of the most basic methods for estimating the SOP of lithium-ion batteries, which utilizes the known static relationship between battery power capability, parameters, and status [309–311]. This method mainly obtains SOP values under different conditions through experiments, constructs an interpolation table for query through the different SOP values, and calculates the required data by querying the corresponding SOP values in the interpolation table. The advantage of constructing the interpolation table to realize the battery's SOP estimation is that it is simple and does not require complex models and algorithms.

However, this type of method requires a large amount of experimental data as support and takes up a lot of memory. Also, these data are obtained by the interpolation method under static conditions, which is not suitable for SOP estimation under dynamic conditions. SOP estimation based on parameter modeling mainly uses linear, RC loop, SOC estimation limit, and temperature estimation limit modeling. Data-driven non-parametric modeling is implemented using BP neural network, SVM, ANFIS modeling, and other new estimation algorithms.

5.6 CONCLUSION

This chapter mainly introduces the battery SOP estimation, the new and existing methods, and several algorithms. As the SOP describes the transient response to the external load changes, it is necessary to carry out the SOP estimation of lithium-ion batteries. There are two methods to estimate battery SOP: One is based on interpolation, and the other on parameter modeling. Interpolation has the advantages of simplicity and no complex algorithm. It requires a lot of memory and is not suitable for SOP estimation under dynamic conditions. For different models, the difficulty and accuracy of parameter estimation are different. In addition, the model changes along with the parameter variation, so the temperature in the battery equivalent circuit model should be considered when using this method. In the non-parametric model SOP estimation method, the data-driven method shifts the focus from the internal mechanism of the battery to the data itself. Each parameter is used as the input of the model, and the output is SOP, which is estimated through a large amount of testing data and training model. The battery state estimation is an important link in the BMS of the power battery pack. The SOP of lithium-ion batteries is estimated to describe the peak charge–discharge power in the current state. The charge–discharge limit capacity is evaluated by estimating the peak power of the batteries, which has great significance to electric vehicles.

6 Battery State of Energy and Cycle Life Estimation

Josep M. Guerrero
Aalborg University

Junhan Huang
Southwest University of Science and Technology

Kailong Liu
University of Warwick

Long Zhou
University of Shanghai for Science and Technology

Ran Xiong
Southwest University of Science and Technology

Shunli Wang
Sichuan University
Southwest University of Science and Technology

Xiao Yang
Southwest University of Science and Technology

Siyu Jin
Aalborg University

Wu Tang
University of Electronic Science and Technology of China

Weihao Shi
Southwest University of Science and Technology

DOI: 10.1201/9781003333791-6

CONTENTS

Introduction: The state of energy (SOE) is a parameter required by the lithium-ion battery management system (BMS), which can reflect the remaining energy of the battery. It is an important index for predicting the remaining mileage of electric vehicles. Common prediction methods include power integration method, open-circuit voltage method, model-based method, and data-driven method. The remaining useful life (RUL) is generally defined as the number of charge–discharge cycles remaining until the failure threshold. The RUL prediction method is designed to use historical and current performance data to predict future conditions and provide warnings before battery failures occur. Reliable and accurate RUL prediction is of great significance to ensure the stability and safety of battery-powered systems.

6.1 OVERVIEW OF SOE ESTIMATION

6.1.1 SOE Definition

SOE is an important parameter required by the lithium-ion BMS, which can reflect the size of the remaining battery energy [312]. The SOE value can be calculated with the following formula obtained by dividing the remaining electric energy by the total electric energy. It is generally considered that when the battery is charged to the cutoff voltage, the SOE value is 100%, and when the battery is discharged to the cutoff voltage, the SOE value is zero. The calculation equation of SOE is defined for lithium-ion batteries, as shown in Equation (6.1).

$$\text{SOE} = \frac{\text{surplus power}}{\text{total power}} = \frac{W_n - W_d}{W_n} * 100\% \tag{6.1}$$

In Equation (6.1), W_n is the total electric energy of the battery; W_d is the accumulated electric energy of the batteries.

6.1.2 Traditional SOE Estimation Method

Traditional SOE estimation methods are divided into static initial estimation methods and dynamic cycle estimation methods.

1. Static initial estimation process: The system collects the open-circuit voltage information of the battery and corrects the relationship curve between open circuit voltage (OCV) and SOE during the battery resting time and the state before resting. Then, the corresponding battery SOE value is calculated according to the current OCV value of the battery. Finally, after the battery is shelved, the estimated SOE values in different periods are averaged to obtain the current static SOE value corresponding to the battery. It can be used as the initial value of the dynamic loop SOE estimation.
2. Dynamic cycle estimation process: After entering the cycle estimation process, the system corrects the total rated energy of the battery in the current state according to the ambient temperature and the health state (cycle use times) of the battery; at the same time, it passes the SOE-EMF curve model and the initial SOE value (or the last recorded SOE value) is used to calculate the electromotive force (EMF) of the current state of the battery. Then, according to the principle of energy conservation, through the EMF and current at the time, the amount of change in the battery energy at the specific time point can be calculated. Finally, the self-discharge energy loss of the battery is corrected according to the time interval between the two dynamic estimations. In this way, the SOE value at the previous time point t can be used to calculate the SOE value at the time point $t+1$. Based on the continuous loop calculation, the state parameters are updated in real time to obtain the accurate SOE value of the battery.

6.1.3 SOE Integral Calculation Expression

The battery outputs electrical energy during the discharge process, while heat energy is generated inside the battery and dissipated to the outside. These two parts of energy are stored inside the battery in the form of chemical energy. The amount of chemical energy is related to the structure and activity of the internal materials of the battery [313]. SOE reflects the state of remaining capacity inside the battery, so electrical and thermal energy should be considered in the calculation process, as shown in Equation (6.2).

$$W_d = \int_0^t (U_b \cdot i)dt \tag{6.2}$$

In Equation (6.2), the electric energy output is considered without the heat loss, and the result cannot reflect the amount of remaining energy inside the battery. According to the working principle of the battery, when the battery is charged, the external electric energy is converted into chemical energy, and the heat energy is radiated outward at the same time [314]. During the discharging process, the internal chemical energy is converted into electrical energy, and heat energy is dissipated outward at the same time point. The released thermal energy includes two parts: the heat of the chemical reaction and the Joule heat generated by electrical resistance. The calculation of thermal energy is shown in Equation (6.3).

$$W_r = \int_0^t \left(i^2 \cdot \left(R_0 + R_p\right) + 1000 \cdot \frac{Q}{F} \cdot i \right) dt \tag{6.3}$$

In Equation (6.3), i is the discharge current, in which the discharge direction is positive and the charging direction is negative; Q is the thermal coefficient of reaction; F is Faraday constant; R_0 is the ohmic resistance; R_p is the polarization internal resistance. The total energy released is shown in Equation (6.4).

$$W = W_d + W_r \tag{6.4}$$

To accurately reflect the state of remaining energy in the battery, the SOE definition is modified based on the law of conservation of energy, as shown in Equation (6.5).

$$\text{SOE}_{(t)} = \frac{\text{surplus power}}{\text{total power}} = \frac{W - \left(W_{d(t)} + W_{r(t)}\right)}{W} * 100\% \tag{6.5}$$

In Equation (6.5), $W_{d(t)}$ is the electric energy released by the battery; $W_{r(t)}$ is the heat energy lost; W is the total internal energy of the battery, which represents the maximum storable energy of the battery in the current state. Whether the accuracy of the W value is accurate or not has a great influence on the calculation accuracy of the SOE. By analyzing the working characteristics of the battery, the ambient temperature, discharge rate, cycle life, and self-discharge have an impact on the total internal energy W of the battery. To make the SOE calculation result more accurate, it is necessary to determine the relationship between the influencing factors and W.

The monthly self-discharge rate of lithium-ion batteries is low (the value $<10\%$), and many factors affect the self-discharge process. In the SOE calculation process, the influence of self-discharge on the estimation accuracy is usually ignored, and the impact on environmental temperature, charge–discharge current, and battery health status is mainly analyzed. The influence of the total energy inside the battery and the SOE calculation is corrected.

According to relevant information, the best ambient temperature for lithium-ion batteries is 30°C. Therefore, the total discharge energy at 30°C is used as a benchmark to analyze the impact of other ambient temperatures on the total discharge energy. The temperature influencing coefficient η_T is used to describe the influence of temperature on the discharge energy, and the calculation equation is shown in Equation (6.6).

$$\eta_T = \frac{E_{\text{ttv}}}{E_{\text{tt30}}} * 100\% \tag{6.6}$$

In Equation (6.6), E_{ttv} is the total discharge energy at different temperatures, and E_{tt30} is the total discharge energy at 30°C. The energy that the battery can discharge under different discharge currents is very different. The total energy W that the battery can discharge under different working conditions is different, which brings errors to the SOE estimation. Therefore, the influence of the discharge current on the value needs to be considered in the SOE estimation. Based on the discharge rate of $C/3$, the influence of different discharge currents on the total discharge energy is analyzed, and the current influence coefficient is used to describe the influence of the discharge current rate on the discharge energy. The formula for this calculation is shown in Equation (6.7).

$$\eta_I = \frac{E_{\text{tcv}}}{E_{\text{tc/3}}} * 100\% \tag{6.7}$$

In Equation (6.7), E_{tcv} is the total discharge energy in the first stage of different rates, and $E_{\text{tc/3}}$ is the total discharge energy at the discharge rate of $C/3$. Through the analysis of the battery cycle characteristics, it can be observed that the maximum dischargeable energy of the battery decreases as the number of cycles increases. The adaptive voltage curve fitting method is used to obtain the health status of the battery, and then, the current maximum discharge capacity and energy are calculated. The corrected expression is shown in Equation (6.8).

$$\text{SOE}_t = \text{SOE}_0 - \frac{\left(\int_0^t (U_b \cdot i)dt + \int_0^t \left(i^2 \cdot (R_0 + R_p) + 1000 \cdot \frac{Q}{F} \cdot i\right)dt\right) * \eta_T * \eta_I}{W_e} \tag{6.8}$$

The influence of ambient temperature, discharge rate, and battery health status on the total dischargeable energy W_e is comprehensively analyzed, and the SOE calculation expression is corrected.

6.2 UKF-BASED SOE ESTIMATION

Based on the SOE calculation expression and the battery terminal voltage expression established by the simplified electrochemical combination model, the unscented Kalman filtering (UKF) principle is used to realize the SOE estimation, including the establishment of the discrete state-space equation of the system, determining the initial SOE value, and carrying out iterative estimation of SOE.

6.2.1 SOE Estimation Model Structure Establishment

In the UKF algorithm, the estimation model is established based on the state-space expression of the system. In the battery SOE estimation model, SOE is a real-time estimator that can be used as a state variable for real-time observation [315]. The SOE calculation method is given in Equation (6.8) [316]. The terminal voltage of the battery is a known quantity, which is used as an observed variable to correct the SOE estimation result. Two parameters are combined to obtain the state-space model of the system, and the discretized expression is shown in Equation (6.9).

$$\begin{cases} \mathrm{SOE}_{k+1} = \mathrm{SOE}_k \\ -\left[U_k i_k \Delta t + i_k^2 \cdot \left(R_{0\_k} + R_{p\_k} \right) \Delta t + 1000 \dfrac{Q}{F} i_k \Delta t \eta_T \eta_I \right] \Big/ W_e \\ U_{k+1} = U_{\mathrm{ocv}\_k+1} - \left(R_{0\_k} + R_{p\_k} \right) \cdot i_{k+1} \end{cases} \tag{6.9}$$

In Equation (6.9), SOE_{k+1} is the SOE value at time point $k+1$; SOE_k is the energy state value at time point k; U_k is the terminal voltage at time point k; i_k is the discharge current; R_{0_k} is the ohmic resistance at time point k; the polarization internal resistance R_{p_k} is obtained at the time point k accordingly; Q is the reaction heat coefficient; F is the Faraday constant; η_I is the current influence coefficient; η_T is the temperature influence coefficient; U_{OCV_k+1} is the open-circuit voltage at the time point $k+1$.

6.2.2 SOE Estimation Procedure Design

Before the battery starts to work, the initial SOE determination method is obtained by judging the static time of the battery. After the initial SOE is given, the BMS is used to collect the voltage, current, and temperature data of each cell in the battery pack in real time to estimate the SOE value at the current time point [317,318]. The estimation of SOE value includes two processes, SOE prediction and SOE correction. The estimation process is described as follows:

1. SOE prediction
   The SOE value as well as covariance at time point $k-1$ and the sigma point near the state at that time can be obtained using unscented transformation (UT), as shown in Equation (6.10).

$$\begin{cases} [\mathrm{SOE}_{k-1}]_0 = \mathrm{SOE}_{k-1} \\ [\mathrm{SOE}_{k-1}]_i = \mathrm{SOE}_{k-1} + \left(\sqrt{(n+\lambda P_{k-1})}\right)_i, \quad i = 1,2,\ldots,n \\ [\mathrm{SOE}_{k-1}]_{n+i} = \mathrm{SOE}_{k-1} - \left(\sqrt{(n+\lambda P_{k-1})}\right)_i, i = 1,2,\ldots,n \end{cases} \tag{6.10}$$

The sigma point at time point $k-1$ is propagated backward nonlinearly using the state to obtain the sigma point set of the state variable at time point k, as shown in Equation (6.11).

$$[\mathrm{SOE}_k] = [\mathrm{SOE}_{k-1}] - \frac{U_{k-1} \cdot i_{k-1} \cdot \Delta t + i_{k-1}^2 \cdot R_{k-1} \cdot \Delta t + 1000 \cdot \frac{Q}{F} \cdot i_{k-1} \cdot \Delta t * \eta_T * \eta_I}{W_e} \tag{6.11}$$

For the sigma point set of the state variable at the time point k, the predicted value of the state variable at time point k can be obtained through the weighted average of the UT change, so the SOE predicted value SOE_k at the time point k and the covariance P_k of SOE at the time point k are obtained, as shown in Equation (6.12).

$$\begin{cases} \mathrm{SOE}_k = \sum_{i=0}^{2n} W_i^m \cdot [\mathrm{SOE}_k]_i \\ P_k = \sum_{i=0}^{2n} W_i^c \cdot \left(\mathrm{SOE}_k - [\mathrm{SOE}_k]_i\right)\left(\mathrm{SOE}_k - [\mathrm{SOE}_k]_i\right)^T + Q_k \end{cases} \tag{6.12}$$

The sigma point of the state variable at time point k is propagated backward nonlinearly using the observation to obtain the sigma point set of the output variable at time point k, as shown in Equation (6.13).

$$[U_k] = k_0 - k_1/[\mathrm{SOE}_k] - k_2 \cdot [\mathrm{SOE}_k] + k_3 \cdot Ln([\mathrm{SOE}_k]) + k_4 \cdot Ln(1 - [\mathrm{SOE}_k]) - R_k \cdot i_k \tag{6.13}$$

For the output sigma point set at time point k, the predicted value of the battery terminal voltage U_k of the output variable at time point k can be obtained through the weighted average of the UT change, and the covariance P_{yy} of the battery terminal voltage at time point k is obtained, as shown in Equation (6.14).

$$\begin{cases} U_k = \sum_{i=0}^{2n} W_i^m \cdot [U_k]_i \\ P_{yy} = \sum_{i=0}^{2n} W_i^c \left(U_k - [U_k]_i\right)\left(U_k - [U_k]_i\right)^T + R_k \end{cases} \tag{6.14}$$

2. SOE correction

The covariance matrix P_{xy} between the state variable and the output variable at time point k is calculated, and the calculation equation is shown in Equation (6.15).

$$P_{xy} = \sum_{i=0}^{2n} W_i^c \cdot \left(\text{SOE}_k - [\text{SOE}_k]_i\right)\left(U_k - [U_k]_i\right)^T \tag{6.15}$$

The updated Kalman gain K_e is calculated, and the calculation equation is shown in Equation (6.16).

$$K_e = P_{xy} \cdot P_{yy}^{-1} \tag{6.16}$$

The filter value after the status update is obtained, so the SOE correction value SOE_k at time point k is obtained, and the calculation equation is shown in Equation (6.17).

$$\begin{cases} \text{SOE}_k = \text{SOE}_k + K_e \cdot \Delta U \\ \Delta U = U_{b\_k} - U_k \end{cases} \tag{6.17}$$

The updated covariance P_k of the state variable at time point k is obtained, and the calculation equation is shown in Equation (6.18).

$$P_k = P_k - K_e \cdot P_{yy} \cdot (K_e)^T \tag{6.18}$$

During the battery working process, the information collected includes terminal voltage, current, and temperature, which are taken as the input variables of the estimation model, and the SOE is set as the state variable to obtain the SOE estimation value of the battery in real time. The SOE estimation process based on the UKF algorithm is shown in Figure 6.1.

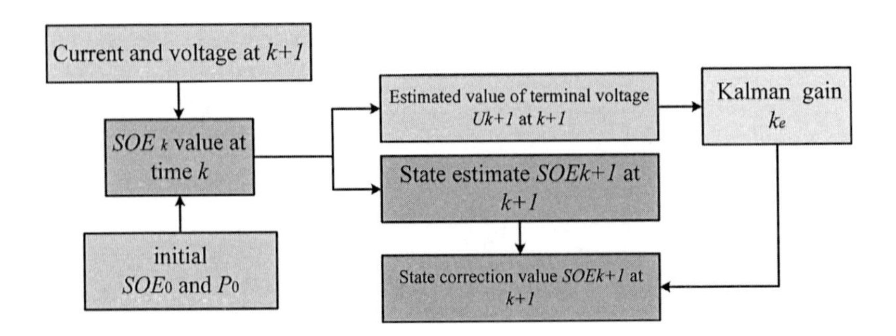

FIGURE 6.1 Flowchart of SOE estimation based on UKF algorithm.

6.2.3 AUKF-Based SOE Estimation

The control model established by the UKF algorithm requires prior knowledge of the mathematical model and noise statistics. However, in the actual system, due to the unknown battery model and noise statistics, the statistical characteristics of noise are usually set to use inaccurate noise in the filter calculation process, which leads to an increase in the error of the state variable estimation value and even causes the filtering to diverge. The Sage–Husa adaptive algorithm is introduced into the UKF algorithm to predict and correct the noise in real time, thereby improving the accuracy of SOE estimation [319]. The adaptive unscented Kalman filtering (AUKF) algorithm estimates the process noise and the measurement noise through a maximal posterior estimator, obtaining a suboptimal unbiased estimate of the noise. The estimation principle is described as follows [320]:

1. The unscented transformation is used to obtain the SOE value and covariance at the time point $k-1$, and the calculation equation is shown in Equation (6.19).

$$\begin{cases} \left[\mathrm{SOE}_{k-1}\right]_0 = \mathrm{SOE}_{k-1} \\ \left[\mathrm{SOE}_{k-1}\right]_i = \mathrm{SOE}_{k-1} + \left(\sqrt{(n+\lambda P_{k-1})}\right)_i, \quad i=1,2,\ldots,n \\ \left[\mathrm{SOE}_{k-1}\right]_{n+i} = \mathrm{SOE}_{k-1} - \left(\sqrt{(n+\lambda P_{k-1})}\right)_i, i=1,2,\ldots,n \end{cases} \tag{6.19}$$

2. Given the calculation weighting coefficient d_k, the calculation equation is further computed, as shown in Equation (6.20).

$$d_k = \frac{(1-b)}{\left(1-b^k\right)} \tag{6.20}$$

   In Equation (6.20), b is the forgetting factor, with $0 < b < 1$, d_k are calculated and the general value range is 0.95-0.99.
3. The process noise w_k and the process noise covariance Q_k are calculated, and the calculation equation is shown in Equation (6.21).

$$\begin{cases} w_k = (1-d_k)w_{k-1} + d_k\left(\mathrm{SOE}_k - \widehat{\mathrm{SOE}_k}\right) \\ Q_k = (1-d_k)Q_{k-1} + dk\left(K_e\alpha_k\alpha_k^T + P_k - A_kP_{k-1}A_k^T\right) \end{cases} \tag{6.21}$$

   In Equation (6.21), α_k is the innovation coefficient.
4. The measurement noise v_k and the measurement noise covariance R_k are calculated, and the calculation equation is shown in Equation (6.22).

$$\begin{cases} v_k = (1-d_k)v_{k-1} + d_k\alpha_k \\ R_k = (1-d_k)R_{k-1} + dk\left(\alpha_k\alpha_k^T - C_kP_{k-1}C_k^T\right) \end{cases} \tag{6.22}$$

From the above analysis, it can be seen that, depending on the new information formed by the difference between the battery terminal voltage and the battery model output terminal voltage, the Sage–Husa adaptive filtering theory can estimate process noise and measurement noise and realize online adjustment.

6.2.4 Initial State Determination

By analyzing the factors that affect the SOE estimation accuracy in the UKF algorithm, it can be known that the accuracy of the initial state of the estimation model has an impact on the SOE tracking time and accuracy. Therefore, accurately obtaining the initial SOE value helps to improve the accuracy of SOE estimation [321]. The analysis shows that the open-circuit voltage of the battery is suitable for different SOE conditions, as shown in Figure 6.2.

Thus, the current SOE can be obtained through the open-circuit voltage; that is, the SOE value of the initial state can be corrected by the open-circuit voltage method. However, under different SOE_s conditions, the time required for the battery to charge the working state to a suitable state is unequal. The resting time required for the battery to recover to a suitable state at different SOE_s is obtained through experiments, as shown in Figure 6.3.

When using the UKF algorithm to estimate the SOE value, firstly, the SOE value is obtained at the end of the last discharge, which is used as the current initial state SOE_0. Then, the rest time T of the battery pack is obtained and compared with the rest time in the current state to determine whether the initial state SOE_0 can be corrected by the open-circuit voltage method. The initial state determination method is shown in Figure 6.4.

6.2.5 DEKF-Based SOE Estimation

In the above estimation methods, experiments are needed to calculate the relationship between SOE and open-circuit voltage. When estimating the state of a lithium-ion battery, it is often necessary to obtain the relationship between the state of charge (SOC) and open-circuit voltage. Therefore, the cooperative estimation method of

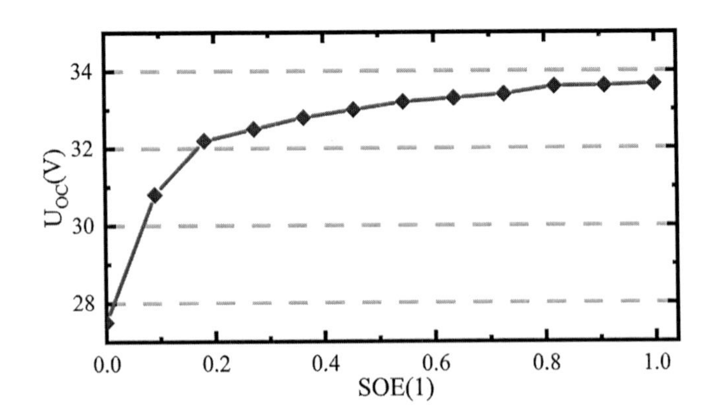

FIGURE 6.2 SOE and OCV diagram.

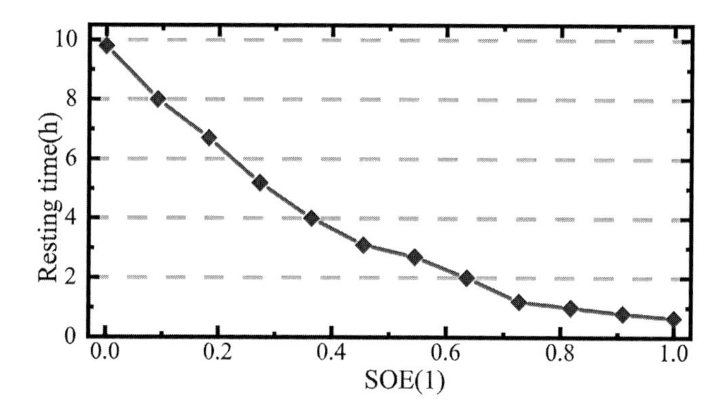

FIGURE 6.3 SOE resting time chart.

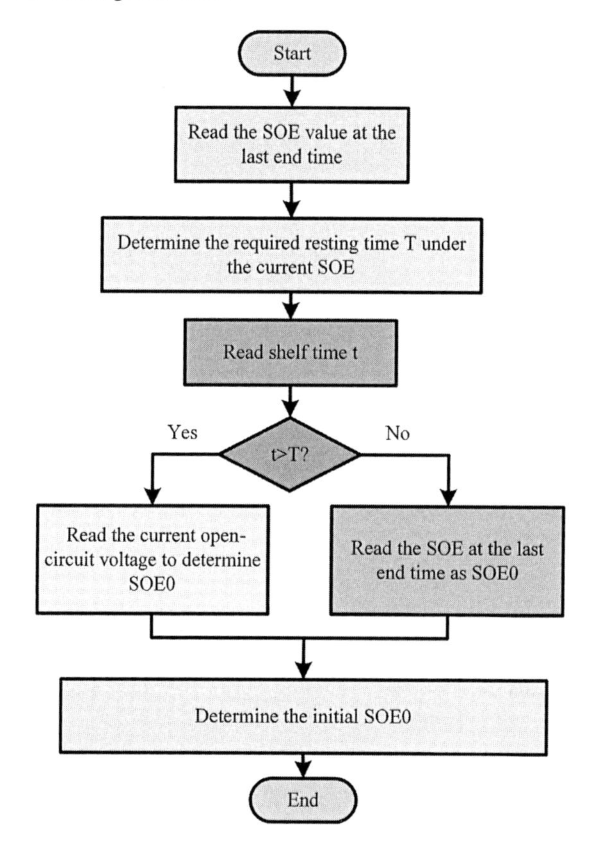

FIGURE 6.4 Initial SOE determination method diagram.

SOC and SOE based on a dual extended Kalman filter (DEKF) is adopted to realize the SOE estimation [322]. The DEKF is a method of predicting the system state by merging the previous state and measurement value. In the case of using the extended Kalman filter (EKF) method, a simpler mechanism and lower computational burden can be its advantages. Differently from a fast-varying SOC characteristic, the battery capacity has a relatively slow-varying characteristic. Due to the difference in the characteristics, a single EKF method hardly assures the stability and accuracy of the result. Therefore, the DEKF method is proposed to simultaneously identify the SOC and capacity of the battery and improve the stability of an observer at the same time. The EKF observer can run in parallel with estimating the SOC and the SOE at the same time point. For SOC and SOE co-estimation based on the DEKF algorithm, the first part is SOC estimation. The steps are shown as follows:

1. State prediction and error covariance matrix estimation

$$\begin{cases} x(k+1|k) = A_x x(k) + B_x x(k) + \omega_x(k) \\ P(k+1|k) = A_x P(x) A_x^T + Q_x(k) \end{cases} \tag{6.23}$$

   In Equation (6.23), the state transition matrix A_x, control matrix B_x, feedback matrix C_x, and measurement matrix D_x are obtained by the Taylor expansion of the lithium-ion battery system.
2. Kalman gain calculation and system residual error prediction

$$\begin{cases} K_x(k+1) = P(k+1|k)C^T\left(CP(k+1|k)C^T + R_x(k)\right)^{-1} \\ \zeta(k+1) = y(k+1) - \left(Cx(k+1|k) + Du(k+1)\right) + r_x(k) \end{cases} \tag{6.24}$$

3. State correction and error covariance matrix update

$$\begin{cases} x(k+1) = x(k+1|k) + K_x(k+1)\zeta(k+1) \\ P(k+1) = \left(I - K_x(k+1)C\right)P(k+1|k) \end{cases} \tag{6.25}$$

In Equations (6.23), (6.24), and (6.25), P is the error covariance, Q_x is the process noise error covariance, K_x is Kalman gain, R_x is measurement noise error covariance, and ζ is system residual error. For the SOE estimation, the prediction equation is obtained, as shown in Equation (6.26).

$$\begin{cases} \phi(k+1|k) = A_\phi \phi(k) + B_\phi u(k) \\ S(k+1|k) = S(k) + Q_\phi(k) \end{cases} \tag{6.26}$$

In Equation (6.26), $\phi(k) = \mathrm{SOE}(k)$, $A_\phi = 1$, $B_\phi = \eta y(k)T/E_T B_\phi$. According to the system, residual error in Equation (6.24), the gain, the SOE value, and the error covariance matrix can be updated. Then, the SOE can be estimated, as shown in Equation (6.27).

$$\begin{cases} \phi(k+1) = \phi(k+1|k) + K_\phi(k+1)\zeta(k+1) \\ S(k+1) = \left(I - K_\phi(k+1)C\right)S(k+1|k) \\ K_\phi(k+1) = S(k+1|k)C^T\left(CS(k+1|k)\right)C^T + R_\phi(k)^{-1} \end{cases} \tag{6.27}$$

6.2.6 Comparative SOE Estimation Results

To verify the estimation effect of SOE, the Beijing bus dynamic stress test (BBDST) and dynamic stress test (DST) conditions are analyzed. The SOE estimation results are shown in Figure 6.5.

In Figure 6.5, the SOE_1 and SOE_2 are the actual SOE and the SOE estimated by the DEKF algorithm, respectively. Err is the estimated error using the DEKF algorithm. For the BBDST, the maximum error of the DEKF algorithm is –3.61%. It can be seen from the error that the DEKF algorithm can estimate the SOE value accurately. The SOE estimation results under DST conditions are shown in Figure 6.6.

In Figure 6.6, the SOE_1 and SOE_2 are the actual SOE, and the SOE is estimated by the DEKF algorithm, respectively. Err is the estimated error using the DEKF algorithm. For the DST conditions, the maximum error of the DEKF algorithm is –2.54%. The definition of the energy state of lithium-ion batteries, the existing estimation methods, and various algorithms for SOE estimation are introduced. The SOE is a parameter required by the lithium-ion BMS, which can reflect the remaining energy of the battery. Traditional SOE estimation methods are divided into static initial estimation methods and dynamic period estimation methods.

Based on the SOE calculation expression and the battery terminal voltage expression established by the simplified electrochemical combination model, the UKF filtering principle is used to realize the SOE estimation, including the establishment of the

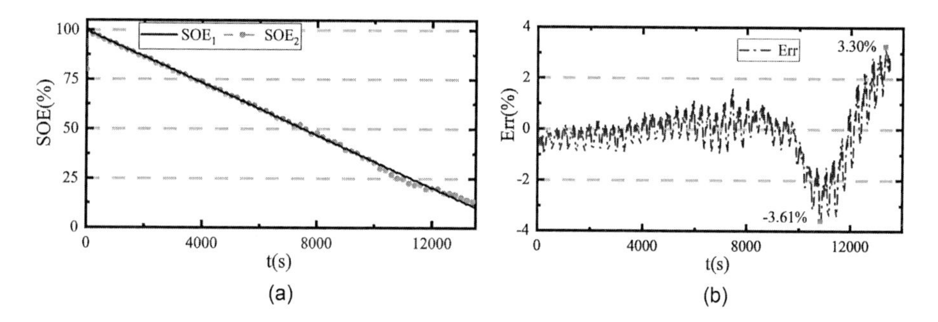

FIGURE 6.5 SOE estimation results under BBDST working conditions. (a) SOE estimation results under BBDST. (b) SOE estimation errors under BBDST.

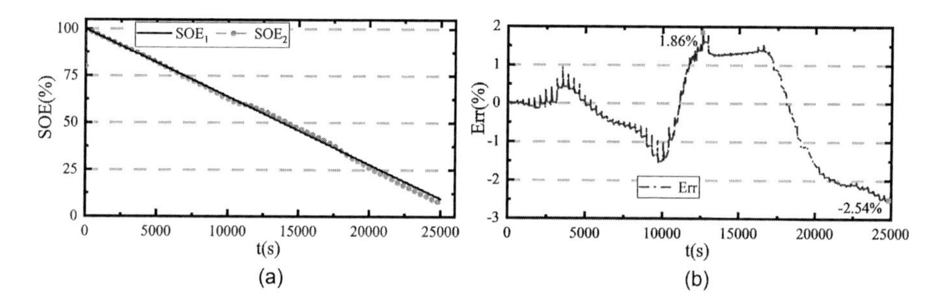

FIGURE 6.6 SOE estimation results under DST working conditions. (a) SOE estimation results under DST. (b) SOE estimation error under DST.

discrete state-space equation of the system, the determination of the initial SOE value, and the iterative SOE estimation. The control model established by the UKF algorithm needs to know the prior knowledge of the mathematical model and noise statistics, but in the actual system, because the battery model and noise statistics are unknown, the noise statistic characteristics are usually set, and the noise statistics are inaccurate. It is used in the filtering calculation process. It causes the error of the state variable estimation value to increase and even leads to the divergence of the filtering. The Sage–Husa adaptive algorithm is introduced into the UKF algorithm to predict and correct the noise in real time, thereby improving the accuracy of SOE estimation. The AUKF algorithm estimates the process noise and measurement noise through a maximal posterior estimator and obtains a suboptimal unbiased estimate of the noise.

6.3 OVERVIEW OF BATTERY CYCLE LIFE

6.3.1 Definition of Basic Concepts

The lithium-ion battery is a dynamic and time-varying electrochemical system with nonlinear behavior and complex internal mechanisms [323–325]. As the number of charge–discharge cycles increases, the performance and lifetime of lithium-ion batteries also gradually decrease. They are divided into two main degradation modes.

1. Loss of lithium inventory (LLI), which is caused by side reactions that consume lithium ions, such as the growth of solid electrolyte interphase (SEI) and lithium plating [326,327].
2. Loss of active material (LAM) in the electrode leads to the loss of storage capacity, such as the peeling of graphite and the loss of positive and negative active materials. The decrease in LAM is usually caused by a variety of factors, one of which is the deterioration of the electrode structure due to the change in the cubature of the active material during the cycle. These can cause mechanical stress, break particles, and reduce the density of lithium-ion batteries. LLI often occurs simultaneously with LAM [328].

Other factors include chemical decomposition and dissolution reactions of transition metals into the electrolyte and changes in the performance of the SEI membrane. The increase in battery resistance is caused by the formation of parasitic phases on the electrode surface and the loss of electrical contact inside the porous electrode. More specifically, the LAM is mainly caused by graphite peeling, binder decomposition, electrical contact loss caused by the corrosion of the current collector, and electrode breakage.

It is worth mentioning that these degradation mechanisms are highly related to materials. For instance, the working voltage of graphite anodes is lower than the electrochemical window of ordinary electrolytes, which can lead to the formation of SEI films. However, since the potential of lithium titanium oxide (LTO) is located within the electrochemical window of the electrolyte, no SEI film is formed in the LTO anode [329,330]. Another instance is that the cubature change of the lithium iron phosphate (LFP) cathode is smaller than that of the lithium manganese oxide cathode, so its structure is less deformed. In addition to the difference in materials, the degradation mechanism is also very different under different working conditions and different battery designs. For example, lithium plating is most likely to occur during rapid or low-temperature charging but rarely occurs during discharge [331,332]. In battery design, a cathode with a small area results in less stress and cracking of the small particles. However, a cathode with a large specific surface area results in a higher solubility of the cathode material.

The aging stress factor is an expression to describe the conditions that increase the rate of aging. In large battery systems, the BMS is usually responsible for controlling operating conditions to extend service life and ensure safe operation. The overcharge/overdischarge protection is achieved through voltage regulation of the BMS. When a battery in its pack reaches a fully charged or discharged state, the charge and discharge process should stop. The thermal management system can effectively heat/cool the battery to ensure that its temperature is within the minimum degradation range. The development of an optimal charging protocol to balance the relationship between battery capacity decay and charging time is necessary. Also, understanding the influence of aging factors is crucial for the development of reliable health diagnoses and prognoses. The data-driven method is mainly based on the quantity and quality of the aging test data, but the experimental conditions cannot cover all possible working conditions. For some application conditions, certain stress factors play a leading role. In actual working conditions, the quantitative relationship between stress factors and the aging process must be focused on the most influential factors, which should be considered when designing the experimental test plan.

To ensure the safety, reliability, and prolonged life span of a battery-powered system, the battery must be replaced when the capacity drops to 80% of the initial capacity. As an essential energy storage system in electrified transportation systems and smart grids, its failure can cause these battery-powered systems to fail. Certain serious battery failures such as thermal runaway can cause a strong release of energy, leading to fire or explosion. However, due to the uncertainty of the environment and load conditions, it is difficult to predict the degradation rate of the batteries. Due to the complexity of the battery degradation process, predicting the RUL has become a very challenging task. So, it is essential to ensure the reliable operation of the BMS, timely maintenance, and battery life applications. The internal variables directly related to the aging state of the battery are difficult to measure with sensors. Therefore, it is important to study degradation behavior and construct degradation models to estimate the RUL and state of health (SOH). The RUL is generally defined as the number of charge–discharge cycles remaining until the failure threshold. The RUL prediction is designed to use historical and current performance data to predict future conditions and provide warnings before battery failures occur. Reliable and accurate RUL prediction is of great significance to ensure the stability and safety of battery-powered systems.

The RUL prediction has attracted more attention in recent years. Some studies reviewed existing RUL prediction methods and classified them according to the models and algorithms used. According to the existing literature, RUL prediction methods include model-based, data-driven, and hybrid methods. The model-based approach aims to establish a mathematical model to describe the degradation trajectory based on battery dynamics. The model is usually a set of algebraic and differential or empirical equations. An important feature of this method is that the prediction models are designed for specific systems. The prediction models on batteries and bearings are different. However, how to balance the relationship between model complexity and prediction accuracy is challenging.

The data-driven method attempts to extract the hidden correlation from a large amount of data onto statistical theory or machine learning (ML) technology to deduce the prediction model from the measured data, rather than establishing a specific physics-based model. Also, there is no battery mathematical model for RUL prediction. Different from the model-based method, the data-driven method is more suitable for various applications. The same data-driven algorithm can be applied to both bearings and batteries by only recalibrating the parameters. The data-driven approach does not require analysis of system mechanisms, which is feasible and practical when large amounts of data are available. In recent years, hybrid methods that combine model-based and data-driven methods have also been proposed to combine the advantages of different methods.

The highly improved computing power of the central processing unit and the latest software development provide an opportunity to incorporate complex data-driven methods into battery RUL predictions. A hybrid method with online update and error correction functions has also been developed. Accurate SOH estimation is very important because it provides useful information for RUL degradation. In lithium-ion batteries, the conversion between electrical energy and chemical energy is achieved through repeated insertion and extraction of lithium ions. Side chemical reactions cause irreversible loss of electrolyte, active electrode material, and lithium stock. Therefore, the capacity decreases with the cycle period. The RUL predicts the needed information for the health indicator (HI), which can reflect the ability of the battery to provide specified performance compared with a new battery and quantify the aging state of the battery. Multiple attributes can be used as HI. The HI has multiple attributes. In existing research, capacity or battery internal resistance is usually used as HI. By predicting the future trends of these two HIs, the time interval between the current state and the predefined fault threshold can be regarded as the RUL. For example, when the battery capacity drops to 80% of the initial value or the internal resistance becomes 1.3 times the initial value, the battery reaches the retirement standard.. Since these HIs directly reflect the physical degradation of the battery, they are also called direct HIs (DHIs).

The battery health monitor has two main challenges: (1) Online measurement of battery capacity requires a complete charge–discharge process, which is difficult to achieve in many applications. (2) The predictability of DHI is very low, therefore, many relative HIs (RHIs) have been proposed. The first type of RHI extracted from the battery voltage, current, and temperature data can be easily obtained online, such as the average voltage drop and the time interval of the equal discharge voltage difference sequence. Comprehensive analysis of available data has better predictability than DHI, such as summary statistics of discharge voltage curve changes, such as minimum, mean, and variance, which is combined with the sample entropy of the measured voltage series. After extracting the RHI from the original battery measurement, a degraded modeling process is required. Degradation modeling has two main purposes: (a) assessing the correlation between RHI and DHI (usually battery capacity) and verifying the validity of the proposed RHI and (b) establishing the relationship between RHI and DHI. Therefore, after the aging modeling, the battery RUL can be effectively predicted based on the future RHI trajectory.

In recent years, significant progress has been made in the development of RUL forecasts. As mentioned earlier, RUL prediction methods are generally divided into model-based, data-driven, and hybrid methods. However, the classification standards and naming rules differ slightly. For example, in some literature, filtering methods are grouped as data-driven methods [333]. Meanwhile, some researchers classify filtering methods as model-based methods. In addition, there is no comprehensive assessment of the prognosis of battery RUL, including the latest research progress.

Lithium-ion batteries undergo sudden, instantaneous, and accidental capacity regeneration during the aging process. This phenomenon is related to the electrochemical performance of the battery and is affected by temperature and load conditions. Since the degradation trend is affected by the capacity regeneration, the capacity regeneration affects the performance predicted by RUL. Therefore, the original capacity data are usually regarded as a mixed signal, and the useful information is extracted by decomposing the original signal at different scales. For example, the empirical mode decomposition and wavelet decomposition are used to divide the original signal into different parts, including global degradation and local regeneration. Then, data-driven methods, including autoregressive integrated moving average (ARIMA) and Gaussian process regression (GPR), are used to obtain the battery RUL. Although the above data processing methods are mainly used in combination with the data-driven RUL algorithm, they are also used to detect and separate different degradation modes in the original data of the model-based RUL algorithm.

By separating the degradation mode, it can be applied to different mechanical models, and the predicted value of each model can be aggregated to obtain the RUL value, thereby reducing the difficulty of constructing the mechanical model. The combination of data-driven methods for data processing and model-based RUL prediction methods is a hybrid method. In addition, the acquired battery capacity data usually contain various types of noise and measurement errors. Using such raw data to build capacity attenuation models or training data-driven models may reduce the accuracy of predictions and even increase the errors. Therefore, the original data must be preprocessed to eliminate the noise.

Some studies have proposed an improved or discrete gray-scale model to smoothen the original data and reduce the relative randomness. Other research attempts to eliminate the noise in the original data. For instance, using wavelet denoising and variational mode decomposition to process data effectively eliminates the nonlinear and non-stationary noise. Next, the processed data can be used as input for model-based battery RUL prediction. In some studies, a relevance vector machine (RVM) is used to obtain the correlation vectors, i.e., data with less noise and measurement errors from the original historical capacity data. These representative vectors contain the battery capacity and the corresponding number of cycles and are calculated as parameters of the empirical capacity degradation model through nonlinear fitting calculations.

The degree of understanding and mathematical explanation of the battery aging process and its causes determines the feasibility of health assessment and prediction methods. Many studies have focused on identifying the basic cause and effect of performance loss. Here is a summary of the most common aging mechanisms in batteries and an outline of the main stress factors.

6.3.2 Factors Affecting Battery Life

Even when the battery is not being used, it deteriorates, which is called the calendar aging of the battery. Conversely, cyclic aging refers to the aging caused by continuous charge–discharge cycles of the battery. To better design and implement SOH estimation and RUL prediction, it is important to understand the characteristics and influencing factors [334–336]. High storage SOC and high temperature are the main driving factors for calendar aging. A high SOC means a low lithium-ion content in the active material of the positive electrode (cathode). It increases the tendency of the electrode to chemically decompose the electrolyte composition. Then, the same chemical driving force generates a higher battery voltage under a higher SOC, providing a higher driving force for lithium ions to reenter the electrode.

Regardless of the operating condition, calendar aging inevitably occurs throughout the battery life and is affected by all the factors in the calendar aging cycle. Also, the latter is affected by other factors, such as overcharge/overdischarge, current rate, and cycle depth. These factors are nonlinearly related, which makes the aging process quite complicated. The main reasons for the calendar aging of the battery are described as follows:

1. High temperature: Accelerate side reactions, including (i) the growth rate of the SEI film on the anode increases, leading to faster LLI and increasing battery resistance; (ii) metal dissolution on the cathode; and (iii) electrolyte decomposition, (ii) and (iii) leading to LAM and LLI. Extremely high temperatures may cause a thermal runaway situation, which is the most serious threat.
2. Low temperature: Slow down the transmission of lithium ions in the two electrodes and electrolyte. Where the electrolyte meets the graphite electrode, trying to charge quickly at low-temperature conditions may cause lithium-ion crowding. This may result in the graphite (partial) lithium coating that comes with the LLI. Continuous and uneven lithium plating eventually leads to the growth of lithium dendrites, which may penetrate the separator and short-circuit conditions of the batteries.
3. Overcharge/overdischarge: When the battery is overcharged, the cathode is empty and has no active lithium available. Meanwhile, the anode is filled with lithium and has no 'room' for lithium. When the lithium is excessively removed, the cathode material undergoes irreversible structural changes, followed by the dissolution of transition metal ions, such as Mn^{2+}, and the decomposition of active materials. During the overcharging process, it causes the decomposition of the electrolyte and a significant increase in the total internal resistance. Due to Joule heat and the heat generated by a series

of side reactions on the two electrodes, a large amount of heat is generated when the battery is overcharged. During the overdischarge process, the anode potential increases abnormally, which causes the copper Cu current collector anode to dissolve and form Cu^{2+} ions. After charging, the reverse reaction forms copper dendrites, which may cause internal short circuits.

4. High current: Excessive charge–discharge currents cause local overcharge and overdischarge, resulting in the same degradation reaction as general overcharge and overdischarge. High currents produce more heat waste, which may increase the battery temperature and consequently increase the speed of the aging process. Lithium-ion batteries using organic electrolytes, compared with water-based batteries, have a relatively low heat capacity, making them particularly prone to quick heat up when current flows. For graphite anodes, due to the limited ability of graphite to accept lithium ions at a high rate, rapid charging also forms a metallic Li coating, leading to LLI.
5. Mechanical stress: The battery is subjected to stress from different sources, such as manufacturing (externally applied stack pressure), expansion of electrode materials during operation, mechanical restraint of gas escape from the battery, and external loads during service. The highest stress tends to be generally generated in the electrode particles near the separator, and when the stress exceeds a certain limit, the electrode material cracks, resulting in a significant decrease in battery performance.

6.4 EXPERIMENT-BASED BATTERY LIFE ESTIMATION

Lithium-ion batteries are widely used in consumer electronics, electric vehicles, aerospace systems, and other fields. However, the inevitable problem is that due to the long-term use of the battery, its performance continues to decline in the process until it cannot meet the normal use requirement and is discarded. Therefore, the SOH of the battery must be estimated. The standard definition of battery SOH is the ratio of the capacity of a power battery discharged from a fully charged state to a cutoff voltage at a constant rate and its corresponding nominal capacity under standard conditions. This ratio is a manifestation of the current performance of the battery. In addition, the degradation of the battery performance cannot be directly measured [337]. It is often estimated through various methods to decide whether to replace the battery and prevent avoidable accidents. This chapter uses different methods to perform SOH experimental analysis on ternary lithium-ion batteries.

6.4.1 Capacity Method

The capacity method is also called the complete discharge method. The capacity measurement is mainly to conduct a complete charge–discharge experiment through the capacity test equipment. It is to obtain the battery capacity according to the charge end voltage and discharge cutoff voltage indicated by the manufacturer. Although the manufacturer gives the rated capacity with the extension of service time, the aging effect occurs, and the maximum available capacity of the battery gradually decreases. The change in capacity can characterize the health of the batteries [338,339]. The following precautions should be noted when measuring the capacity of lithium-ion batteries:

1. The discharge termination voltage is an important parameter related to the service life of lithium-ion batteries. The current nominal voltage of a single lithium-ion battery is generally designed to be 3.7 V, and the cutoff voltage is 2.75 V. Two situations that influence the manufacturer to increase the upper limit of the cutoff voltage to 3.0 V are described as follows: (1) The electrochemical properties are determined by the material of the lithium-ion battery, which is often from the perspective of safer use of batteries. (2) The cutoff voltage is designed to be as low as 2.5 V or 2.4 V, but not less than 2.4 V. Thus, when the voltage reaches the cutoff voltage, although there is remaining power, the discharge cannot continue. Otherwise, the battery is damaged.
2. The voltage at which the battery is fully charged is called the termination or terminal voltage. According to the battery manufacturer's standard, the terminal voltage is about 1.2 times the rated voltage.
3. The lithium-ion battery should not be overcharged and overdischarged, both of which cause irreversible damage.

This chapter takes a ternary lithium-ion battery as an example for capacity testing. The rated capacity is 70 Ah, the charge cutoff voltage is 4.2 V, and the discharge cutoff voltage is 2.75 V. An activation experiment is required before the capacity test to make the experimental results close to the real value.

1. The battery is charged at a constant current of 0.05 C (3.5 A) to the cell charging cutoff voltage of 4.2 V and rests for 15 minutes.
2. The battery is discharged at a constant current of 0.05 C to the cell discharge cutoff voltage of 2.75 V, which is shelved for 15 minutes, and the discharge capacity is recorded.
3. Steps (1) and (2) tests are repeated until the difference between the discharge capacities is less than ±1%.
4. The battery is charged with a 0.4 C constant current rate to the cell's terminal voltage of 4.2 V, transferred to constant voltage charging until the current drops to 0.05 C, which is then left to rest for 40 minutes.
5. The battery is discharged at a constant current of 0.4 C and rests for 40 minutes after reaching the cell discharge cutoff voltage of 2.75 V.

Steps (4) and (5) tests are repeated until the discharge capacities are less than ±1%, and the last discharge capacity is used as the actual battery capacity.

The SOH is defined from the perspective of battery capacity, as shown in Equation (6.28).

$$\mathrm{SOH} = \frac{C_M}{C_N} \tag{6.28}$$

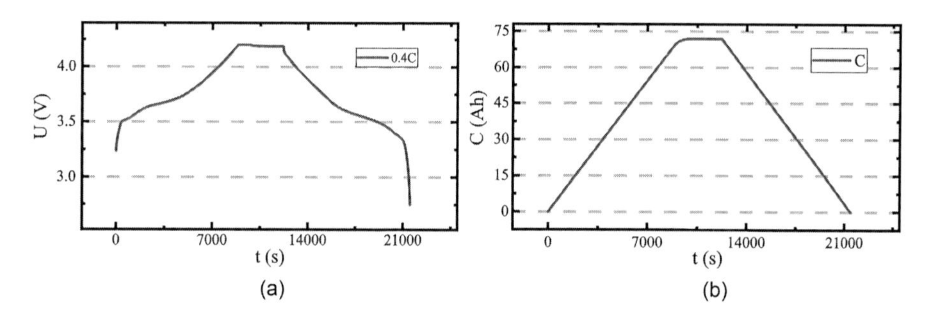

FIGURE 6.7 Constant current–constant voltage charge–discharge curve. (a) Voltage curve. (b) Capacity curve.

In Equation (6.28), C_M is the measured capacity, and C_N is the nominal capacity of the batteries. The voltage and capacity curves are shown in Figure 6.7.

In Figure 6.7, the actual capacity of the battery is 71.28 Ah.

6.4.2 Cycle Number Method

The SOH expresses the current battery capacity as a percentage. For a new battery, the SOH value is 100%. As the battery is being used, the battery will age and the SOH will gradually decrease. According to IEEE Standard 1188–1996, when the capacity of the power battery drops to 80% of the rated capacity, that is, when the SOH is 80%, the battery should be replaced [340,341].

The theoretical life of a lithium-ion battery is about 800 cycles, which is moderate among commercial rechargeable lithium-ion batteries. For LFP batteries, it is about 2000 cycles. At present, mainstream battery manufacturers promise more than 500 cycles in battery cell specifications. After the cells are described and assembled into a battery pack, due to inconsistencies, the voltage and internal resistance do not remain the same, and its cycle life reduces by 400 cycles. The above definitions are the number of cycles when the battery is fully charged and discharged.

The number of cycles of the battery is not equal to the number of charge–discharge. The number of cycles of the battery means that the battery undergoes a complete charge–discharge cycle. When the number of cycles is increased by 1, the number of charge–discharge is increased by 1. Because lithium-ion batteries have no memory effect, there is a big difference in defining the two-state parameters. For instance, a lithium-ion battery is discharged to 40% on the first day, then fully charged, and discharged to 60% the following day, and fully charged. It is a complete charge–discharge, that is, a cycle. Generally, the number of test cycles for lithium-ion batteries is directly given by the manufacturer. If the battery is fully charged each time and then discharged to end of life (EOL), the cyclic SOH calculation method can be implemented as shown in Equation (6.29).

$$\mathrm{SOH} = 1 - \frac{\mathrm{Cyc}_i}{\mathrm{Cyc}_N} \tag{6.29}$$

In Equation (6.32), Cyc_i is the current number of cycles, and Cyc_N is the nominal number of cycles of the batteries. The nominal number of cycles from the manufacturer is the number of cycles when the full capacity of the lithium-ion battery decays to 80% of the rated capacity. The capacity attenuation definitions for different batteries and different manufacturers may vary. When the SOH in Equation (6.31) drops to 0, the lithium-ion battery becomes a scrap. When the actual working condition is not a complete charge–discharge cycle, the influence of different cycle conditions, cycle states, and other factors on the service life should be considered. Under this condition, it is not advisable to use Equation (6.32) for calculation. It needs to determine the battery life based on experience and standard parameters. Taking a 45Ah lithium-ion battery as an example, the cycle aging experiment is carried out. During the cycle of discharge, the capacity test is performed every 100 times, and the aging characteristic curve is obtained, as shown in Figure 6.8.

In Figure 6.8, the battery capacity decayed to 80% of the rated capacity after 460 cycles of charge–discharge, and the battery has completely aged. The shorter the time for the battery to reach the charge–discharge cutoff voltage, the further its capacity decreases.

6.4.3 Weighted Ampere-Hour Method

From a new to an old battery, the total ampere-hours that can handle power during the entire process of charge–discharge should be a fixed value. When the accumulated ampere-hour power reaches a certain level, it is considered that the battery has reached the end of its life, and this is the ampere-hour method [342–344]. According to the weighted ampere-hour method, when the battery discharges the same amount of power under different conditions, the damage to the battery life can be dangerous. Therefore, when the cumulative ampere-hour after the power discharge is multiplied by a weighted coefficient and reaches a certain value of 80%, the battery is considered to have reached the end of its service life, as shown in Equation (6.30).

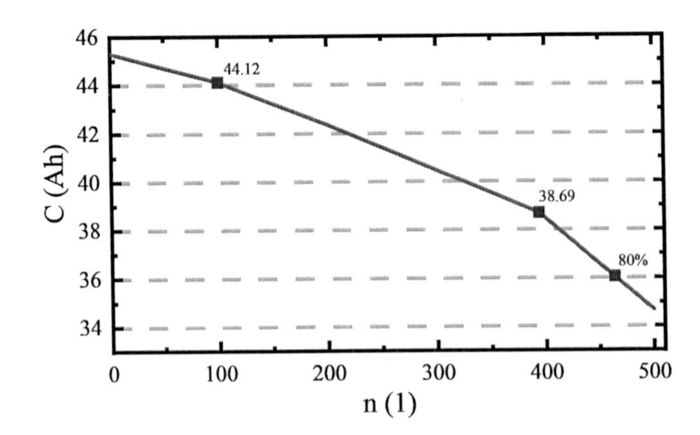

FIGURE 6.8 The aging characteristic curve.

$$\begin{cases} \mathrm{SOC}_i = \mathrm{SOC}_0 - \dfrac{\eta}{C_N}\displaystyle\int_0^t I\,d\tau \\ \mathrm{SOH} = \dfrac{\sum_{i=1}^{n} \mathrm{SOC}_i}{\mathrm{SOC}_{\mathrm{all}}} \end{cases} \tag{6.30}$$

In Equation (6.33), SOC_i is the charge–discharge capacity of the battery during the ith charge–discharge, SOC_0 is the initial value of power, C_N is the rated capacity of the battery, η is the weight, and SOH is the total power that has been charged and discharged in the previous period, and the ratio of useable power to total power in the whole life cycle of the battery. The choice of η requires a large number of experimental results to be determined jointly by experience. The proposed method has high requirements on the accuracy of the power measurement with the initial value being accurate; otherwise, the error accumulates over time.

6.4.4 Internal Resistance Method

When current flows through the battery, the internal resistance reduces the working voltage of the batteries. Due to the internal resistance of the battery, the terminal voltage of the battery is lower than the EMF and open-circuit voltage when the battery is discharged. Conversely, the terminal voltage is higher than the EMF and the open-circuit voltage when charged [345]. The internal resistance of the battery is an essential parameter of the chemical power supply. It directly affects the working voltage, working current, output energy, and power of the battery. For a practical chemical power supply, the smaller the internal resistance, the better [346,347]. The internal resistance of a lithium-ion battery is different from ordinary resistance. The resistance is generally measured in milliampere, but the commonly used Ohm's law cannot accurately measure the internal resistance. Before measuring its internal resistance, it is necessary to study the characteristics of the internal resistance in the estimation of the battery's health and the internal resistance model [348].

The internal resistance is divided into ohmic resistance (R_0) and polarization internal resistance (R_p). The polarization internal resistance is divided into electrochemical and concentration polarization internal resistance. The internal resistance is one of the key parameters of power batteries indicating the SOC, SOH, and battery capacity. It reflects the difficulty of ion and electron transmission between positive and negative electrodes in the battery during the chemical reaction processes. Different types of power batteries have different internal resistances. Even batteries of the same chemical composition have larger and stronger discharge capacity, while batteries with larger internal resistance have a relatively smaller and weaker discharge capacity.

The internal resistance of the battery is small and is often neglected under many working conditions. But power batteries are often in a state of high current and deep discharge, and the voltage drop caused by the internal resistance is large [349,350]. So, the impact of the internal resistance on the entire circuit cannot be ignored.

Regarding the composition of the internal resistance, three types of polarization of batteries are described as follows:

1. Ohmic polarization
   During the charge–discharge process, to overcome the ohmic resistance, the extra voltage must be applied to promote ion migration. The voltage is thermally converted to the environment, resulting in the ohmic polarization phenomenon. As the charging current increases sharply, ohmic polarization causes the battery temperature to rise during the charging process.
2. Concentration polarization
   When current flows through the battery, to maintain a normal reaction, the ideal situation is that the reactants on the electrode surface can be replenished in time, and the products can deplete. The diffusion rate of the product and the reactant is far less than the rate of the chemical reaction, resulting in a change in the concentration of the electrolyte solution near the electrode plate. That is to say, from the electrode surface to the middle of the solution, the electrolyte concentration distribution is non-uniform. This phenomenon is called concentration polarization.
3. Electrochemical polarization
   This polarization is caused by the fact that the speed of the electrochemical reaction to the electrode lags behind the speed of the electron movement on the electrode. For instance, before discharging the negative electrode of the battery, its surface is negatively charged, but the nearby solution is positively charged, which ensures equilibrium. During the discharging process, electrons are immediately released to the external circuit. The negative charges on the electrode surface decrease, and the oxidation reaction of metal dissolution progresses slowly [351]. The reduction of electrons on the electrode surface is supplemented in time, and the charged state of the electrode changes. The state of reduced negative surface charge promotes the electrons in the metal to leave the electrode, and the metal ions are transferred into the solution, accelerating the $Li - e^- \rightarrow Li^+$ reaction to achieve a new dynamic equilibrium. However, by comparison, before the discharge, the number of negative charges on the electrode surface decreases, and the corresponding electrode's potential becomes positive. That is, the electrochemical polarization voltage becomes higher, which seriously hinders the normal charging current. Also, when the positive electrode of the battery is discharged, the number of positive charges on the surface of the electrode decreases, and the electrode potential becomes negative.

The internal resistance of the battery increases as the battery life decreases. When a single lithium-ion battery is charged and discharged, the pulse current method is used to achieve an internal resistance of 0–200 total intervals, as shown in Figure 6.9.

In Figure 6.9, the ohmic resistance of the battery gradually increases during the entire aging process of the battery, while the polarization internal resistance remains relatively constant during the entire process. It can be observed that the change in the internal resistance of the battery is mainly caused by the ohmic resistance.

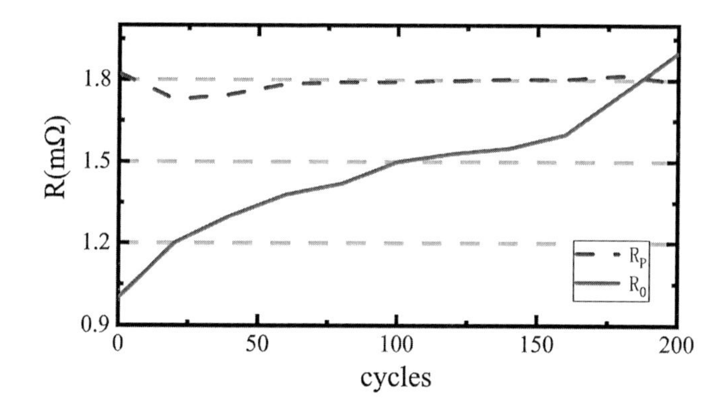

FIGURE 6.9 Variation trend of ohmic and polarization resistance with life attenuation.

Therefore, the ohmic resistance of the battery is mainly involved in the SOH estimation. The principle of the DC internal resistance test is to apply a large amount of current to charge or discharge the battery for a short period.

When the internal composition of the battery is not fully polarized, the internal resistance of the battery is calculated proportional to the voltage changes and applied current before and after the application of current. Four parameters are selected to test the DC internal resistance: current with multiplying factor used, pulse time, SOC, and test environment temperature. The changes in these parameters have a high impact on the DC internal resistance. The DC internal resistance does not include only the ohmic resistance but also the partial polarization resistance of the battery pack. The polarization of the battery is highly affected by current and time. Currently, there are three commonly used DC internal resistance test methods, which are highlighted as follows:

1. The hybrid pulse power characterization (HPPC) test method is in the 'Freedom CAR Battery Test Manual' in the United States. The test duration is 10 seconds, the applied discharge current is 5 C or higher, and the charging current is 0.75 times the discharge current. The specific current selection is based on the characteristics of the batteries, such as dynamic, operation, physical, dimensional, and aging.
2. The JEVSD7132003 (JEVS) test method from Japan is mainly for Ni/MH batteries but can also be applied to lithium-ion batteries: First, establish the current–voltage characteristic curve of the battery under 0%–100% SOC, and the used discharge rates are described as 1 C, 2 C, 5 C, and 10 C in succession. The DC internal resistance of the battery is calculated by alternating the charge or discharge under a specific SOC value. The charging or discharging time should be 10 seconds, respectively.
3. The test method proposed in the '863' major polarization new energy vehicle special project 'High-power Lithium-ion Power Battery Performance Test Specification for HEV'. The test duration is 5 seconds, and the charging test current rate is set to be 3 C and 9 C.

The two test methods of Japan electric vehicle association standards (JEVS) and HPPC have different characteristics. The JEVS adopts a 0–10 C series current, which can avoid the deviation of results caused by using a single currency. It assumes that the main component of the battery's internal resistance is approximately constant ohmic impedance, so the discharge rate is relatively low with high reliability. When the battery is charged and discharged at a high rate, the rate control step of the entire battery reaction is changed from the charge transfer process control under a small rate to the mass transfer process control. The impedance composition of the battery does not include only the ohmic impedance but also the impedance formed by polarization reaction, which varies with current and pulse time.

The HPPC method uses low and high current rates to test the output voltage and power of the battery. At the same time, the voltage response characteristics of the battery are considered under different charge–discharge current conditions. This method can obtain the battery output response and aging characteristics under the standard discharge current. But it lacks comprehensiveness, and the test results are incomplete and biased due to the different voltage responses of different batteries under a specified current [352]. For the '863' test method, different charge and discharge current conditions are used, which creates a wide variation between them. Any of these test methods can be used as a benchmark method to test and compare different battery systems. However, batteries differ from each other and the internal resistance changes along with the measured current and time. As the battery ages and its capacity decreases, the internal resistance of the battery gradually increases. The corresponding relationship between the internal resistance and the SOH is established based on the actual situation. The SOH of the battery is predicted by the accurate measurement and estimation of the internal resistance.

The internal resistance is mainly composed of ohmic resistance and polarized internal resistance. The ohmic resistance accounts for a larger proportion than polarized internal resistance in the total internal resistance. The existence of internal resistance in the internal materials of the battery leads to the formation of ohmic resistance, which sums up the internal resistance of the battery pack. It is highly affected by temperature and increases with decreasing temperature. Polarization internal resistance is produced when the chemical substances in the battery participate in a chemical reaction, and polarization reaction occurs during charge–discharge. Compared with the ohmic resistance, the polarized internal resistance is less affected by temperature. Also, the internal resistance is one of the characterization parameters of the SOH. With the aggravation of battery aging, the battery's internal resistance increases. According to the SOH equation defined by internal resistance, the battery SOH decreases along with the current increasing. The definition of SOH in terms of battery internal resistance is in Equation (6.31).

$$\mathrm{SOH}_{R_0} = \frac{R_{\mathrm{EOL}} - R}{R_{\mathrm{EOL}} - R_{\mathrm{NEW}}} \tag{6.31}$$

In Equation (6.34), R_{EOL} is the internal resistance at the end of the battery life, R_{NEW} is the resistance of the battery when it leaves the factory, and R is the internal resistance of the battery in the current state. This estimation method does not need to consider the change in battery capacity. As long as the internal resistance of the battery can be obtained, the change of SOH is estimated accordingly. The internal resistance of the battery changes with the service life of the batteries.

6.4.5 Adaptive Filtering Method

As battery degradation continues, model parameters need to be updated to maintain prediction accuracy. The update is usually done through filtering algorithms. However, there are some problems related to the initialization and sampling process in the filtering algorithm [353,354]. Initialization problems usually exist in the model parameter estimation methods, such as least squares, Kalman filter (KF), and particle filter (PF) algorithms [355,356]. Traditionally, the initial value is set based on experience or even random selection. However, inaccurate initial values slow down the convergence speed of the algorithm and may lead to divergence. The Dempster–Shafer theory is used to initialize the parameters of the PF algorithm, thereby shortening the convergence time and improving the prediction accuracy of the early battery life.

The traditional PF uses resampling technology to avoid particle degradation by eliminating small particles and replicating large particles. However, they produce a large number of large-weight particles, and the number of small-weight particles is almost zero, which causes the sample to deteriorate. The support vector regression (SVR) algorithm is used in the resampling step to maintain the diversity of particles to avoid this problem [357,358]. The basic idea is to reconstruct the posterior distribution and obtain re-weighted particles based on SVR, where the training data are the particles and their corresponding weights. Another recent study is to use the Markov Chain Monte Carlo method to solve the particle degradation problem after the standard resampling process [359].

6.4.6 Multidimensional Extended Kalman Filter

A multidimensional EKF algorithm to reduce the computational complexity of BMS is proposed. It identifies slow time-varying system parameters on a macro-scale and fast time-varying system states on a micro-scale, to provide more suitable system parameters and state estimation. The use of a multidimensional discrete-time state-space model makes the model more specific and without losing generality. It is assumed that the system has two-time scales, including macro-and micro-time scales. The system quantity on the macro-time scale changes slowly with time, but the system quantity on the micro-time scale changes quickly with time. The former is the model parameters, and the latter is the state of the system, respectively.

The DEKF method estimates the state and parameters on the same time scale. However, for a system that exhibits separation of time scales, it is natural and ideal to adjust slow time-varying parameters on the macro-time scale while maintaining the estimation of the fast time-varying state on the micro-time scale. The multi-scale framework reduces the computational workload and provides more suitable model parameter estimations [360]. Different from the dual extended Kalman estimation, the derived multidimensional extended Kalman estimation allows the time scale of state and parameter estimation to be separated. It should be noted that the macro-time scale refers to the time scale where the system state changes slowly, while the micro-time scale refers to the time scale where the system model parameters tend to change rapidly. For example, in a battery system, the SOC as a system state changes every second, which indicates that the micro-time scale is approximately one. Conversely, the cell capacity, which is a parameter of the system model, usually degrades by 1.0% or less within a month under normal usage, resulting in a macro-time scale of one day.

6.4.7 Multidimensional Particle Filter

The PF method has become a popular algorithm to solve the optimal estimation problem of nonlinear and non-Gaussian state-space models. It is because the change of the battery parameters is slower than the SOC. The method based on dual PFs can estimate battery parameters online from a macro-scale, thereby reducing the calculation cost of the BMS. However, due to the rapid change of battery parameters, the state is estimated on a micro-scale. The idea of multidimensional particle filtering to improve the accuracy of the algorithm and the ability to converge to the initial state offset is proposed. The flowchart of the multidimensional PF is shown in Figure 6.10.

Taking into account that the characteristics of system parameter change slower than state during the estimation process improves the accuracy and robustness of SOC estimation. Therefore, a multidimensional PF is proposed to reduce the calculation amount of the control system.

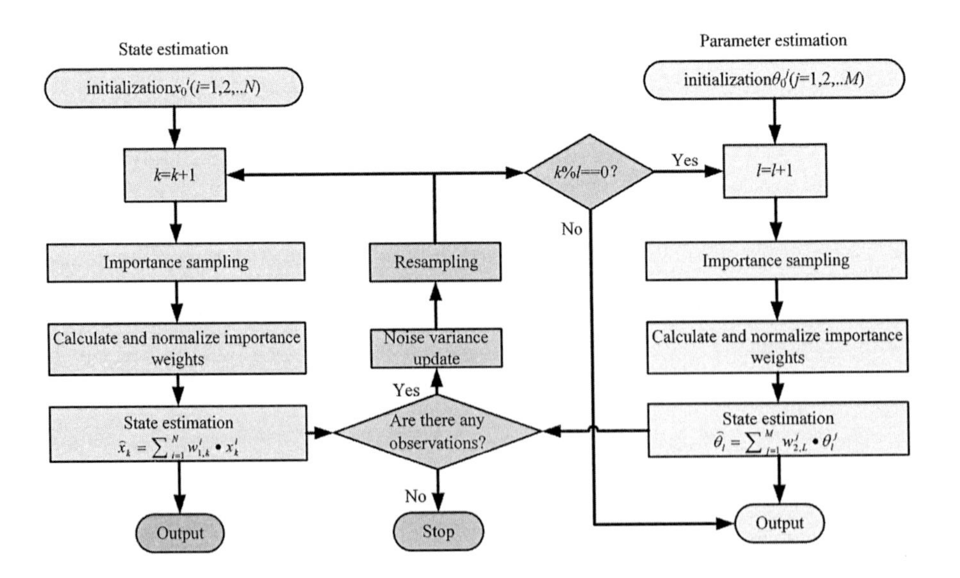

FIGURE 6.10 Flowchart of multidimensional particle filter.

6.5 DATA-DRIVEN FORECASTING

6.5.1 Machine Learning

ML methods estimate the RUL of a battery based on various measurement data, such as current, voltage, and temperature under various operating conditions. However, as the training data increase, the computational complexity also increases. In the early stage, the classification model is used to obtain a rough RUL value. When the battery is close to its EOL threshold, the regression model is used to estimate the accurate RUL value. By taking an additional classification step, the time-consuming regression model is activated only when the battery life is about to end. Therefore, the developed method highly reduces the calculation cost, thereby speeding up the calculation.

ML methods applied to battery state prediction include artificial neural network (ANN) [126,361], support vector machine (SVM), RVM [362,363], autoregressive (AR) model, and deep belief network (DBN) [364,365]. Also, these ML methods are integrated with other methods, which are used for more complex applications or to obtain more accurate results. The SVM as a method of ML has also been applied to battery state prediction. The SVM solves nonlinear problems by transforming the data onto a higher feature space into a linear problem. When SVM is applied to regression tasks, it is called SVR. The SVR is one of the popular regression methods because it describes the nonlinear correlation between input and output data, making it suitable for predictions. As a sparse Bayesian method of kernel regression, the RVM performs regression in a probabilistic manner. The sparsity feature of the RVM regression model allows for the prediction of discoveries in an effective way.

The RVM has the same functional form as the SVM and can provide probabilistic classification. Based on the health characteristics generated from the charging curve (voltage, current, and temperature curves), the remaining capacity of the lithium-ion battery is estimated using RVM. Gray rational analysis and principal component analysis have been used to extract and optimize relevant health characteristics [336–368]. AR is a point estimation by establishing a certain relationship between input and output, and it is a probabilistic method. Also, other ML methods, such as dynamic Bayesian networks (DBNs), are used for battery state prediction [369,370]. Researchers regard SOC as a hidden state in DBN and use the terminal voltage of a lithium-ion battery as an observation value, which is very convenient in practical applications. However, the above methods cannot establish a potential probability model and cannot give an expression of uncertainty in the estimated value. But for RUL estimation, it is very important not only to predict the RUL value but also to give the degree of uncertainty of the prediction. Given this, GPR derived from the Bayesian framework has received widespread attention due to its non-parametric and uncertain expression characteristics [30,371,372].

It is reasonable to learn or adopt algorithms applied in other fields to perform the data-driven prognostic and health management [373]. The RUL prediction is carried out through the deep neural network (DNN). A DNN based on a noise-reducing autoencoder is applied to classify the signal of the monitor into several degradation stages. Then, the corresponding degradation stage, a regression model based on a shallow neural network, is proposed. Finally, the regression results from different models are smoothed to obtain the RUL. The battery health prediction based on the ML method does not assume any mathematical model to describe the battery aging behavior and mainly depends on the quality of the historical test data set. Non-probabilistic ML methods only provide the estimated regression points. However, ideally, due to various sources of uncertainties, such as measurement, state estimation, model error, and uncertainty in future loads, capturing the conditional distribution of predicted uncertainty levels is a challenge.

Probabilistic methods predict data points and retain the confidence bounds. Therefore, the ML method of probability estimation is desirable because the uncertainty of estimation benefits battery users. However, the development of probabilistic ML methods is still in its infant stage. Most of the existing ML models are based on the data obtained from the training set under the same conditions, which makes users question the robustness of these models in practical applications, and their operating conditions may be very different. Therefore, it is recommended to improve probabilistic techniques by training models under complex aging conditions. In addition, the performance of these technologies is highly sensitive to structure and parameters. Appropriate structure and parameter optimization strategies should also be explored to enhance its performance for future adaptive SOH or RUL predictions.

Finding a method that can accurately predict battery life in the early stages is essential for accelerating the development, manufacturing, and optimization of emerging battery technologies. Although great efforts are made in the development of data-driven diagnostic and prognostic technologies, there are still some major challenges in the area, which are described as follows:

1. Identification of aging mechanism: Some pure data-driven methods, especially ML technology, cannot provide in-depth information about the battery aging mechanism. Therefore, it is desirable to find a method that combines the identification of aging mechanisms with online health assessment methods. Combining the ML method with the physical mechanism of degradation is indeed the right direction for future research. This technology can reveal the mechanism of battery degradation. Therefore, it is recommended at low current rates to discover the aging mechanism. This method is highly feasible because it can help find the most sensitive index of capacity loss and then use it in ML technology for SOH estimation or RUL prediction [374].
2. Self-improvement model based on online data: The degradation behavior of lithium-ion batteries is sensitive to operating conditions. Under conditions different from the training data set, the aging characteristic is still difficult to be predicted. The deviation between the experimental conditions used to develop the model and the actual operating conditions limits the practical applicability of the data-driven approach. These factors are corrected in two ways through improved experimental testing and further algorithm development. When the size of the experimental data set increases, it covers the aging information under a wide range of operating conditions, and the prediction ability of this method increases accordingly. However, this leads to a larger experimental cost. On the other hand, improving the dynamic

update capability of data-driven methods developed off-line is worthy of further research because it paves the way for self-improving models.

3. Health diagnosis and prognosis at the module and battery pack level: At present, most studies on battery health diagnosis and prognosis are conducted at the battery level. However, these are usually connected in series and/or parallel in practical applications to construct battery packs for specific energy and power requirements. Understanding the aging process of battery packs requires knowledge beyond the battery level as well as other influencing factors, such as inconsistencies in battery characteristics, imbalances between batteries, and temperature gradients. All these problems complicate accurate aging estimation and prediction models of battery packs. It is foreseeable that advancements in artificial intelligence and deep learning algorithms introduce some solutions to these problems. DNN is particularly suitable for highly complex nonlinear fitting so that it can achieve higher accuracy. Several newly developed deep learning neural networks, such as convolutional neural networks and generative adversarial networks, have been successfully applied to speech recognition and image segmentation due to strong self-learning capabilities. However, no attempt has been made for battery systems, such as module or battery pack-based health diagnosis. For RUL prediction, it is also recommended to use ANN or similar self-learning methods.

6.5.2 Support Vector Regression

In traditional regression models, the loss function is the error between $f(x)$ and its actual value y. Conversely, the SVR jump error is less than ϵ. By constructing the Lagrangian loss function, the convexity solution of SVR is obtained. If the data have a nonlinear trend, a mapping function is used to transform the low-dimensional input space into a high-dimensional feature space [375]. Therefore, the nonlinear regression problem in the input space becomes linear in the feature space.

By considering the correlation between current, voltage, operating time, energy efficiency, and operating temperature, these factors are determined by dimensionality reduction technology, which reduces the computational load while preserving all the physical characteristics of the batteries. Based on these factors, the RUL prediction model is established by SVR [376]. Based on the actual situation, the dwell time and rainwater flow counting methods are used to construct training data. The training data are then used to train the SVR model for RUL prediction. By considering battery degradation at different temperatures, the PF algorithm is used to dynamically update the parameters of the SVR model based on online data.

When a new measurement is taken, it is added to the training data to retrain the SVR model. However, the retraining of the SVR model is time-consuming. Therefore, a method to update the online SVR model with an incremental learning algorithm instead of retraining is proposed, which quickly completes the model update [377]. For instance, the equal charging time interval and discharging voltage difference are extracted as two RHIs and used to construct the training set of SVR. Then, a data-driven method combining feature vector selection and SVR is proposed.

In the feature vector selection method, a subset of the feature vector is selected to represent the original data set. This method eliminates redundant training data and narrows the search range of the SVR model.

SVM is a kernel-based non-parametric ML technology. Non-parametric models mean that the number of parameters increases with the amount of training data. It has the advantage of flexibility and can model arbitrarily complex systems while providing enough data. It performs classification by searching for a hyperplane that separates the category of interest with the highest margin. By transforming a nonlinear problem in a low-dimensional space into a linear problem in a high-dimensional feature space, the kernel function is usually used in SVM to help solve the nonlinear problem. Typically, the prediction is based on certain functions defined in the input space. Learning is the process of inferring the parameters of the function. Computational programs for SVM are particularly attractive due to their ability to process small training data sets [378]. However, when the size of the training data set increases, the number of support vectors increases. Decrement and increment strategies have been adopted to integrate relevant data samples of SVR training while ignoring irrelevant parts to improve the stability and robustness of SVR through large-scale training samples. However, the process also increases the computational cost.

6.5.3 RUL Prediction Based on Shallow Learning

1. Remaining lifetime prediction method based on neural network
   As a mathematical processing method that simulates the structure and function of the biological nervous system, the neural network has the ability of automatic learning and summarization. It mainly includes an input layer, a hidden layer, and an output layer, which often solves problems such as classification and regression. After years of research and exploration, it has shown a strong advantage in the field of remaining life prediction [379]. The neural network-based RUL prediction method uses the original measurement data or the features extracted based on the original measurement data as the input [380,381]. It continuously adjusts the structure and parameters of the network through a certain training algorithm and uses the optimized network online prediction equipment. There is no need for any prior information in the process of predicting the remaining life, and it is completely based on the prediction results obtained from the monitoring data. The current neural network-based methods mainly include multi-layer perceptron (MLP), radial basis function (RBF), and extreme learning machines (ELMs) [382].
   a. Prediction method of remaining life based on MLP neural network
      MLP is a kind of feedforward neural network (FFNN) with a hidden layer, and the neuron models of the hidden and output layers are consistent. Since MLP can approximate any form of nonlinear function, it has received extensive attention from scholars in the field of remaining useful life prediction. The MLP mostly uses the BP algorithm for training. It should be noted that the method of using the BP algorithm to

train MLP neural networks in academia is usually called the BP neural network-based method.

b. Remaining life prediction method based on RBF neural network
The RBF neural network is a neural network structure proposed in the 1980s. It is a three-layer feedforward network with a single hidden layer and approximates any continuous nonlinear function with arbitrary precision. The big difference between RBF neural network and MLP neural network in structure is that the independent variable of the excitation function is the product of the distance and deviation between the input and weight vectors, rather than the weighted sum between these vectors. Because RBF neural network has excellent characteristics such as output independent of initial weight and short training time, it has been successfully applied in signal processing, system modeling, process control, fault diagnosis, and lifetime prediction in recent years with extraordinary advantages [383].

By comparing remaining life prediction methods, on the one hand, the RBF neural network contains only one hidden layer, while the BP neural network contains multiple hidden layers. The number of hidden layers directly determines the fitting accuracy. Therefore, the determination of the number of BP neural network layers requires additional time or practical engineering experience [384]. On the other hand, the BP neural network is easy to fall into the local optimum, and the learning process converges too slowly. The RBF neural network-based remaining lifetime prediction method can overcome many problems. It can realize the dynamic determination of the network structure and the data center of the hidden layer and can adaptively predict the remaining life.

c. Prediction method of remaining life based on ELMs
ELM is a new learning algorithm for a single hidden layer feedforward neural network (SLFN). The basic idea of the ELM training process randomly selects input weights and hidden layer deviation values. It manually selects the number of hidden layer neurons according to practical engineering experience and identifies the output weight by the least squares method to rapidly determine the network structure and parameters. Because the ELMs have the characteristics of fast learning speed and strong generalization ability, they have attracted widespread attention from scholars and engineers in the fields of engineering equipment fault diagnosis, life prediction, and reliability evaluation [385].

Some advantages of the remaining life prediction method based on ELMs are described as follows: (i) It achieves a fast RUL prediction by effectively reducing model training time. (ii) The activation function uses discontinuous functions. (iii) It avoids the sensitivity and easy selection of learning parameters in the gradient descent learning algorithm. Although the method based on ELMs has many advantages, it also has certain shortcomings. Since the input weight and the hidden layer deviation value are randomly generated, the network training effect of ELMs cannot be guaranteed, resulting in a high level of uncertainty. At the same time, the number of hidden layer nodes needs to be selected based on experience and experimental methods, and it is difficult to guarantee the optimal model. In addition, since the output weight is calculated by the least squares method, the ELMs face the expansion of the problem of large outliers and the influence of noise.

2. SVM-based remaining life prediction method
As a new ML method, the SVM is developed based on the dimension theory and the principle of structural risk minimization. The SVM was first proposed by Cortes and Vapnik. It is mainly used to solve ML classification and regression problems, suitable for analyzing small samples and multidimensional data [386]. Then, the SVM began to receive widespread attention from experts and scholars in the field of reliability engineering. The main idea of the research on the remaining life prediction method based on SVM is to use the condition monitoring data obtained in the actual project to train the model to determine the model parameters (insensitivity coefficient, penalty factor, kernel function parameters, etc.). Based on the trained SVM model, the remaining useful life of the battery system is predicted.
Although the SVM effectively solves the remaining life problem of some engineering equipment, it also has many deficiencies. For example, (a) As the sample set increases, its linearity increases, which increases the overfitting and calculation time. (b) It is difficult to get a probabilistic prediction; that is, it is impossible to evaluate the uncertainty of the remaining life prediction. (c) The kernel function must satisfy the Mercer condition.
3. Future research direction
   a. Research on the remaining life prediction of equipment under multiple failure modes.
   Most of the existing research focuses on the remaining life prediction of equipment under a single failure mode, ignoring that the failure of engineering equipment is caused by the coupling effect of multiple failure modes. For example, mechanical or electronic equipment usually degenerates and falls under the influence of external factors. Under the impact of electrical stress, the equipment fails to achieve normal functions in sudden occurrences. Therefore, in the remaining life prediction of equipment under multiple failure modes, the research is worthy of further study [387].
   b. Residual life prediction of multi-component equipment considering interaction.
   At present, complex engineering equipment is usually composed of several parts through a certain connection mode. The degradation process and failure mode among these parts often influence each other, which jointly determines the remaining lifetime of engineering equipment. For example, the inertial navigation system is mainly composed of multiple gyroscopes and accelerometers. The installation mode, connection form, and interaction mechanism of these parts affect the remaining life of the inertial navigation system. However, the remaining life prediction

method based on ML fails to effectively consider the interaction of these components. Therefore, the remaining lifetime prediction research of multi-component equipment considering the interaction is more practical and needs more attention.

c. Research on intelligent feature extraction and residual life prediction
On the premise of accurately predicting the remaining life, extract the required massive monitoring data information. Both the traditional statistical data-driven and shallow ML methods need to rely on a large number of signal processing technology and expert knowledge. They manually extract characteristic information for dealing with complex engineering equipment's massive monitoring data, but they are severely limited. Deep learning can overcome such problems to a certain extent. For example, both DBN and CNN have the ability of intelligent feature extraction and residual life prediction, but the researches on intelligent feature extraction and residual life prediction are still scarce and need more effort.

4. Research on the combination of ML and traditional statistical data-driven methods
Represented neural networks and deep learning can extract the needed information contained in the monitoring data. It portrays the character information and nonlinear relation between the residual life. In the field of remaining lifetime prediction, it has a certain universality but cannot get the analytical probability distribution of residual life, and difficult to apply the arrangement of the repair strategy and develop. The traditional statistical data-driven methods represented by the Wiener and gamma process can estimate the parameters of the degradation model based on the degradation trajectory and derive the analytic probability distribution of the residual life, but the accuracy of the residual life prediction is highly affected by the degradation model selected. Therefore, it is necessary to consider how to integrate the advantages of the two methods in the subsequent research.

6.5.4 Artificial Neural Networks

The ANN method is inspired by biological neural networks such as the human brain. Generally, an ANN model consists of an input layer, one or more hidden layers, and an output layer, and its basic components are artificial neurons. The artificial neurons and processing units are arranged in the input, output, and hidden layers. The input layer obtains the preprocessed data and acts as a pipeline for the hidden layer. In the hidden layer, each neuron contains a mathematical model that is used to determine its output based on its input and can be represented by a weighted linear combination wrapped in an activation function. The higher the weight of a neuron, the more sensitive it is to the particular input.

The prediction data are obtained from the model in the output layer. In the learning process, the model parameters are adjusted by considering the number of hidden layers, the number of neurons in each layer, the interconnection weight between neurons, and the type of activation function [388]. For example, for a layer of a neural network with n neurons, the activation of each of these neurons depends on the activation of the neurons in the previous layer. The output of the previous layer of neurons is multiplied by the corresponding weight and then added. The sum is used as the input of the neurons in the current layer. A bias is added to the sum, and then, a nonlinear activation function is applied to generate an output for the neuron. After constructing the ANN, an algorithm is used to determine the weight of the interconnection based on the training data.

ANN learning is a method of producing the desired output by reacting to a given input. Under the premise of fully understanding the relationship between input and output, the ANN is used to predict the output. The learning process aims to determine the weight and bias parameters between input and output, as well as the transfer function between two adjacent layers. The ANN has also been used in combination with other battery prediction methods. Other learning algorithms of ANN, such as ELMs, are also applied to battery prediction. ELM has the function of randomly selecting hidden units and analyzing and determining the output weight of a SLFN, which provides better generalization performance at a high learning speed. The ELM has been applied to single-step and multi-step forecasting. Compared with BP neural network, ELM has a faster learning speed and higher accuracy.

Multi-layer FNN trained by the BP learning algorithm is the widely used type of neural network [389]. The BP-FNN training process includes signal forward propagation and error BP steps, which can adjust the forward propagation of the signal. Through an iterative process, the connection weights between neurons are updated to minimize the loss of function. The supervised training process continues until the error reaches an acceptable value. In an earlier study, the output layer of BP-FNN contained two neurons representing the discharge and charge capacities, while the input layer contained five neurons. The results show that the long-term performance is not encouraging. In different aging stages, the battery showed different charge–discharge voltage curves. Therefore, importance sampling is used to select representative voltage data from the voltage curve during charging, thereby improving calculation efficiency. Then, the selected voltage data are used by FNN for RUL prediction. In addition to BP-FNN, other ANN structures are also used for battery RUL prediction. The inputs to the neural network have been rearranged to include the latest online temperature, current, SOC, and historical capacity data.

However, the recurrent neural network (RNN) model can only perform short-term capacity prediction. Therefore, it cannot achieve long-term RUL predictions. In a study, to improve the prediction accuracy, an adaptive RNN is developed that uses the recursive Levenberg–Marquardt method to adaptively update the weights. In another example, a long- and short-term memory neural network is used for battery RUL prediction. The elastic mean square BP technique designed for small batch training is used to train the model and used pressure drop technology to prevent overfitting. In another study, the time interval of the equal discharge voltage difference sequence is extracted from the online monitoring data as RHI. Using the proposed RHI, the integrated learning strategy is combined with the monotonic echo state network to improve the prediction accuracy and stability. Recent studies quantify the uncertainty in RUL predictions by assuming that its value follows the Weibull distribution.

FFNN and RNN have been successfully applied to battery RUL prediction. In FFNN, the input data only propagate in one direction. When FFNN is expanded to include feedback connections, it is called RNN. The RNN retains and updates previous information for a period, making it a promising tool for capturing correlations in battery capacity degradation data. The battery degradation process usually covers hundreds of cycles, and the capacity degradation information between these cycles is highly correlated. Therefore, it is meaningful to extract and consider these correlations to make accurate RUL predictions. Because the RNN learns long-term dependencies in data, it is an efficient type of NN that can capture and update information in degraded data. Based on the analysis of the terminal voltage of the charging curve under different cycles, the FFNN is used to simulate the relationship between the battery charging curve and the RUL under constant current. The total number of cycles when the battery reaches the EOL comes from experiments using FFNN to estimate the current number of cycles of the batteries. Then, the RUL is calculated by subtracting the current number of cycles from the total number of cycles.

A unique feature of ANN is that it can learn and train from experience and examples to adapt to changing conditions. Also, it can be automatically established through training without identifying model parameters and coefficients. However, it requires a large amount of input data for training and verification, and training methods and data seriously affect accuracy. For the large-scale application of RUL prediction, the computational cost is still the bottleneck, and the structure of ANN plays an essential role in its performance. In addition, the recognition and optimization of ANN model topology are still open technical challenges. Usually, the structure is achieved through a time-consuming trial and error phase.

6.5.5 The Autoregressive Integrated Moving Average Model

The ARIMA model is different from the neural network algorithm. It only needs a small amount of data to predict the RUL value and gives the confidence bounds of the prediction results. This research is based on the improved rainflow counting method and the optimized test ARIMA model to predict the lithium-ion battery SOH to identify as few model parameters as possible and improve the prediction accuracy and efficiency, as shown in Figure 6.11.

In Figure 6.11, the SOH prediction is divided into three parts. First, import the data. Second, the obtained SOH sequence was imported into the ARIMA model and processed with differential letter paper, sequence determination, and residual testing. Finally, the original and model data are integrated and predicted.

1. Second-order stationarity test
   Second-order stationarity requires that the input sample time series undergo curve fitting, and the current trend continues for some time in the future. The first-order and second-order time points of the second-order stationary time series do not change with time. Different operations are carried out to make SOC time series meet the requirements of second-order stability. The operation equation is shown in Equation (6.32).

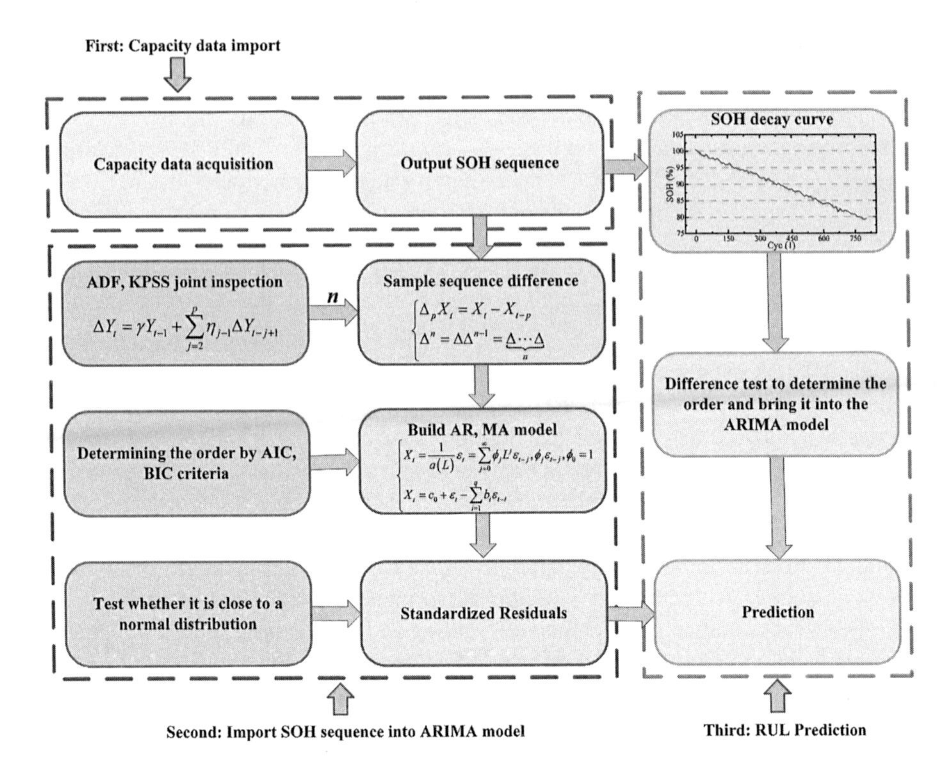

FIGURE 6.11 SOH prediction process.

$$\begin{cases} \Delta_p X_t = X_t - X_{t-p} \\ \Delta^n = \Delta\Delta^{n-1} = \underbrace{\Delta \cdots \Delta}_{n} \end{cases} \tag{6.32}$$

In Equation (6.32), Δ is the difference operator, and Δ_p is the p-step difference. The time series $\{X_t\}$ obtained by the first difference of the time series is the first-order difference. If the difference is performed multiple times, the time series $\{\Delta^n\}$ is the nth-order difference. Differential data can improve signal accuracy and remove common error interference.

Although the difference is simple and effective, if the trend is stable, the difference in the sequence causes excessive difference challenges, resulting in the loss of effective signals. If the difference order is too low, the data cannot meet the stationarity requirements and the prediction results are biased. Therefore, this research uses the combined test of augmented Dickey–Fuller (ADF) to determine the reasonable difference order. In the real situation, most time series have high-order autocorrelation, so the ADF method is used to investigate high-order autoregression in the model. The time-series data Y_t are represented by the superposition of the weight of the historical data and the random disturbance. Its expression is shown in Equation (6.32).

$$\begin{cases} Y_t = \sum_{j=1}^{p} \alpha_j Y_{t-j} + \varepsilon_t \\ \Delta Y_t = \gamma Y_{t-1} + \sum_{j=2}^{p} \eta_{j-1} \Delta Y_{t-j+1} \\ \gamma = \sum_{i=1}^{p} \alpha_i - 1 \\ \eta_i = -\sum_{i=1}^{p-1} \alpha_{i+1} \end{cases} \tag{6.33}$$

In Equation (6.32), α_j is the AR coefficient, and ε_t is the random disturbance term. Subtract Y_{t-1} from both ends of the equal sign of the first sub-expression. Then, it is written as the second sub-expression after the differential transformation. The first sub-equation is a linear differential equation. When $\gamma = 0$, the corresponding characteristic equation has at least one unit root. At this time, the stationarity of the sequence $\{Y_t\}$ is in a critical state; that is, it is a non-stationary sequence. Therefore, it is necessary to continue to differentiate the sequence until $\gamma < 0$ of the new sequence, which is stable. Therefore, based on (6.33), the original hypothesis and alternative hypothesis of ADF are H_0: $\gamma = 0$ and H_1: $\gamma < 0$, respectively. The original hypothesis is that its sequence is non-stationary, and the alternative hypothesis is that its sequence is stationary.

Although ADF has good adaptability in advanced inspections, it is a single-ended inspection, which inevitably leads to excessively high inspection order, resulting in an excessive difference in the series, which in turn leads to the loss of valid data. Therefore, this research introduces the Kwiatkowski-Phillips-Schmidt-Shin test (KPSS) test and uses KPSS and Augmented Dickey-Fuller test (ADF) to test the sequence together [390]. Only when two inspections pass at the same time can the difference order be determined. The KPSS test is used to eliminate the intercept and trend term from the sequence to be tested to construct the statistic learning model. The expression of the inspection process is shown in Equation (6.34).

$$\begin{cases} y_t = x_t\delta + \varepsilon_t, y_0 = 0, \varepsilon_t \sim i.i.d.\left(0, \sigma_\varepsilon^2\right) \\ \hat{u}_t = y_t - x_t\hat{\delta} \\ \text{LM} = \sum_t \dfrac{S(t)^2}{\left(T^2 f_0\right), S(t) = \sum_{i=1}^{t} \hat{u}_i} \end{cases} \tag{6.34}$$

In Equation (6.34), x_t is a vector sequence of exogenous variables, including the intercept term and the trend term of the tested sequence y_t. The second sub-equation is used to estimate the residual sequence by introducing the least squares method to regress and judge whether the original sequence has a unit root by checking whether the residual has a unit root. The third sub-equation is the LM statistic, and f_0 is the residual error under the condition of zero frequency. The null hypothesis is H_0: $\gamma = 0$, and the alternative hypothesis is H_1: $\gamma < 0$ of the KPSS test. The assumptions of KPSS and ADF are opposite, the original hypothesis is that the sequence is stationary, and the alternative hypothesis is that the sequence is non-stationary. When the LM statistic is less than 3 critical values, the null hypothesis is rejected; that is, its sequence has a unit root.

2. ARIMA model establishment

The AR model takes itself as the process of regression variables, which uses the linear combination of random variables at a certain time in the previous period to describe the linear regression process of random variables. Compared with other linear regressions, autoregression does not use x to predict y but uses x to predict x itself. The set time series $\{X_t\}$ is shown in Equation (6.35).

$$X_t = \varepsilon_t + a_0 + \sum_{i=1}^{p} a_i X_{t-i} \tag{6.35}$$

In Equation (6.38), $\{\varepsilon_t\}$ is the white noise sequence; $a_0, a_1, \ldots, a_p$ is $p+1$ real number. This model is called the p-order AR model, which is recorded as the AR(p) model, and $\{X_t\}$ suitable for this model is AR(p) sequence. The AR coefficient polynomial defined in AR(p) is the AR coefficient polynomial of the AR(p) model, as shown in Equation (6.36).

$$\begin{cases} a(u) = 1 - a_0 - \sum_{i=1}^{p} a_i u^i \\ \partial(L) = 1 - \sum_{i=1}^{p} a_i L^i, \partial(L) X_t = \varepsilon_t \\ X_t = \dfrac{1}{a(L)} \varepsilon_t = \sum_{j=0}^{\infty} \phi_j L^j \varepsilon_{t-j}, \phi_j \varepsilon_{t-j}, \phi_0 = 1 \end{cases} \tag{6.36}$$

The first sub-equation in Equation (6.36) is the AR coefficient polynomial. L is the lag operator; at this time, the operator expression of the AR(p) model is expressed as ε_t. The rule is obtained by bringing in multiple different p values, and finally, the solution of AR(p) is shown in the third sub-equation. ϕ_j is the sum of the coefficients of all items in the $\{X_t\}$ sequence. The function of the moving average (MA) model is to add constraints to the

AR model, and a finite number of parameters b are proposed to limit the parameter and avoid the divergence of the AR model. The model of MA is shown in Equation (6.37).

$$\begin{cases} X_t = c_0 + 1 - \varepsilon_t \sum_{i=1}^{p} b_1 L^i, q > 0 \\ X_t = a_0 + \sum_{i=1}^{p} a_i X_{t-i} + \varepsilon_t - \sum_{j=1}^{q} b_j \varepsilon_{t-i} \end{cases} \tag{6.37}$$

In Equation (6.37), c_0 is a constant, and $\{\varepsilon_t\}$ is the white noise sequence. The absolute value of b_1 must be less than 1 to make the $\{X_t\}$ sequence stationary; otherwise, the $\{b_i\}$ sequence diverges. Therefore, the combination of the AR model and the MA model becomes the ARMA model, as shown in the second sub-Equation (6.37). Combining the second-order smoothing process of the data with the ARMA model forms the ARIMA model. There are three main parameters in the ARIMA model: the difference order d, the order p of the AR model, and the order q of the MA model. From Equations (6.32) to (6.37), the value of d can be determined. The values of p and q are determined using the Akaike information criterion (AIC), as shown in Equation (6.38).

$$\begin{cases} \text{AIC} = 2k - 2\ln(L) \\ \text{AIC}(p_0, q_0) = \min_{1 \le p,q \le M(N)} \text{AIC}(p,q) \end{cases} \tag{6.38}$$

In Equation (6.38), k is the number of parameters, and the number depends on p and q in Equation (6.38). When the accuracy is guaranteed, the smaller the value of k, the better. L is the maximum likelihood value of the model. Because the maximum likelihood requires high data cubature and is difficult to solve, the residual variance obtained by least squares estimation is often used as an approximate substitute in practice. The AIC conducts a comprehensive analysis of k and $\ln(L)$. When the order of the model increases, $\ln(L)$ usually decreases. When the number of observation data is given, the value of k increases as the order of the model increases. When the model order is gradually increased with the fitted data, the value of AIC shows a decreasing trend. At this time, the residual variance of the model decreases rapidly, and $\ln(L)$ plays a decisive role. When the order rises to a certain point, the AIC value reaches a minimum. Subsequently, no matter how the model order increases, the residual variance is almost unchanged. At this time, the influence of $\ln(L)$ is weakened, and k plays a decisive role. Given the highest order $M(N)$ in advance, m_0 and n_0 are taken as the best order of the model.

3. Residual test

A residual test is also needed to ensure that the order of p and q in ARMA is appropriate. The residual is the residual signal after subtracting the signal fitted by the model from the original signal. If the residuals are randomly distributed and are white noise sequences, it means that all the effective signals have been extracted into the ARMA model. Therefore, the significance test of the model is the white noise test of the residual sequence. Its original hypothesis, alternative hypothesis, and Ljung–Box statistics are shown in Equation (6.39).

$$\begin{cases} H_0 : \varepsilon_1 = \varepsilon_2 = \cdots = \varepsilon_m = 0, \ \forall m \geq 1 \\ H_1 : \text{at least exists } \varepsilon_k \neq 0, \ \forall m \geq 1, k \leq m \\ \text{LB} = n(n+2) \sum_{k=1}^{m} \left(\frac{\hat{\varepsilon}_k^2}{n-k} \right) \sim \chi^2(m), \ \forall m > 0 \end{cases} \tag{6.39}$$

In Equation (6.39), n is the number of samples, and $\hat{\varepsilon}_k^2$ is the correlation coefficient of the k-order lag of the sample. This statistic obeys the χ^2 distribution with m degrees of freedom. Given the significance level α, the rejection domain is $\text{LB} > X_{1-a,m}^2$. If the null hypothesis is rejected, there is still relevant information in the residual sequence, and the fitting model is not significant. If the null hypothesis is not rejected, the fitted model is considered to be significantly effective.

6.6 CONCLUSION

This chapter introduces the related research content of battery SOE estimation. SOE is a parameter required by the lithium-ion BMS, which can reflect the residual energy of the batteries. When the battery is charged to the cutoff voltage, the SOE is 1, and when the battery is discharged to the cutoff voltage, the SOE is 0. Traditional SOE estimation methods include static initial estimation methods and dynamic period estimation methods. The influence of environment temperature, discharge rate, and battery health status on the total discharge is analyzed comprehensively, and the calculation equation of SOE is modified. In the method of SOE estimation based on UKF, the principle of UKF filtering is used to realize SOE estimation, including establishing the discrete state-space equation of the system, determining the initial SOE value, and iteratively estimating the SOE. In the SOE estimation based on the AUKF algorithm, which is used to estimate the process noise and measurement noise, the suboptimal unbiased estimation of noise is obtained. These two methods can estimate the state energy of lithium-ion batteries well. The main factors affecting the battery life and the test methods are analyzed to predict the battery life. Due to the complexity of the battery degradation process, predicting the RUL of the battery has become a very challenging task.

It is essential to study degradation behavior and establish a degradation model to estimate SOH and RUL, which refers to the number of charge–discharge cycles remaining before reaching the failure threshold. The factors that affect battery life are extreme temperature variations (high and low), overcharge/overdischarge, high current, and mechanical stress. The capacity, cycle number, weighted ampere-hour, and internal resistance methods are used to estimate the SOH of ternary lithium-ion batteries. The weighted ampere-hour method is highly dependent on the accuracy of power measurement and the initial value. When measuring the SOH of lithium-ion batteries, both the multidimensional EKF and multidimensional PF can effectively reduce the computational complexity of the BMS. This also mainly introduces the estimation method of battery cycle life. The lithium-ion battery is a dynamic and time-varying electrochemical system with nonlinear behavior and complicated internal mechanisms. As the number of charge and discharge cycles increases, the performance and life of lithium-ion batteries gradually decline. The adaptive filtering method uses a multidimensional EKF and multidimensional PF to expand. Data-driven forecasting uses ML, SVR, SVM, etc., for the estimation.

References

1. Luiso, S. and P. Fedkiw, Lithium-ion battery separators: Recent developments and state of art. *Current Opinion in Electrochemistry*, 2020. 20: p. 99–107.
2. Park, K.-H. and K.-H. Lee, Space qualification of small satellite Li-ion battery system for the secured reliability. *Journal of The Korean Society Aeronautical and Space Sciences*, 2014. 42(4): p. 351–359.
3. Yuan, Q., et al., Multitime scale analysis of surface temperature distribution of lithium-ion batteries in quantity-quality change under local high-temperature heat source. *Journal of Energy Engineering*, 2020. 146(6): p. 1–12.
4. Poullikkas, A., A comparative overview of large-scale battery systems for electricity storage. *Renewable & Sustainable Energy Reviews*, 2013. 27: p. 778–788.
5. Walter, M., M.V. Kovalenko, and K.V. Kravchyk, Challenges and benefits of post-lithium-ion batteries. *New Journal of Chemistry*, 2020. 44(5): p. 1677–1683.
6. Xiao, J., et al., Understanding and applying coulombic efficiency in lithium metal batteries. *Nature Energy*, 2020. 5(8): p. 561–568.
7. Zhang, J., et al., Microstructure engineering of solid-state composite cathode via solvent-assisted processing. *Joule*, 2021. 5(7): p. 1845–1859.
8. Adam, A., et al., Fast-charging of automotive lithium-ion cells: In-situ lithium-plating detection and comparison of different cell designs. *Journal of the Electrochemical Society*, 2020. 167(13): p. 1–16.
9. Abdullah, S., et al., An energy-efficient message scheduling algorithm with joint routing mechanism at network layer in internet of things environment. *Wireless Personal Communications*, 2020. 111(3): p. 1821–1835.
10. Hashemi, S.R., et al., Machine learning-based model for lithium-ion batteries in BMS of electric/hybrid electric aircraft. *International Journal of Energy Research*, 2021. 45(4): p. 5747–5765.
11. Kumar, B., N. Khare, and P.K. Chaturvedi, FPGA-based design of advanced BMS implementing SoC/SoH estimators. *Microelectronics Reliability*, 2018. 84: p. 66–74.
12. Ee, Y.J., et al., Lithium-ion battery State of Charge (SoC) estimation with non-electrical parameter using uniform Fiber Bragg Grating (FBG). *Journal of Energy Storage*, 2021. 40: p. 1–9.
13. Xu, X., et al., A hybrid observer for SOC estimation of lithium-ion battery based on a coupled electrochemical-thermal model. *International Journal of Green Energy*, 2019. 16(15): p. 1527–1538.
14. Babaeiyazdi, I., A. Rezaei-Zare, and S. Shokrzadeh, State of charge prediction of EV Li-ion batteries using EIS: A machine learning approach. *Energy*, 2021. 223: p. 1–9.
15. Chen, J., et al., Neural network-based state of charge observer design for lithium-ion batteries. *IEEE Transactions on Control Systems Technology*, 2018. 26(1): p. 313–320.
16. Guo, F., et al., State of charge estimation in electric vehicles at various ambient temperatures. *International Journal of Energy Research*, 2020. 44(9): p. 7357–7370.
17. Wang, Q.K., et al., State of charge-dependent polynomial equivalent circuit modeling for electrochemical impedance spectroscopy of lithium-ion batteries. *IEEE Transactions on Power Electronics*, 2018. 33(10): p. 8449–8460.
18. Hasan, R. and J. Scott, Comments on "state of charge-dependent polynomial equivalent circuit modeling for electrochemical impedance spectroscopy of lithium-ion batteries". *IEEE Transactions on Power Electronics*, 2020. 35(4): p. 4448–4448.

19. Hekmat, S. and G.R. Molaeimanesh, Hybrid thermal management of a Li-ion battery module with phase change material and cooling water pipes: An experimental investigation. *Applied Thermal Engineering*, 2020. 166: p. 1–139.
20. Zhang, J., et al., Variable-order equivalent circuit modeling and state of charge estimation of lithium-ion battery based on electrochemical impedance spectroscopy. *Energies*, 2021. 14(3): p. 1–20.
21. Zhang, L., et al., High-safety separators for lithium-ion batteries and sodium-ion batteries: Advances and perspective. *Energy Storage Materials*, 2021. 41: p. 522–545.
22. Gu, F., et al., An investigation of the current status of recycling spent lithium-ion batteries from consumer electronics in China. *Journal of Cleaner Production*, 2017. 161: p. 765–780.
23. Fei, H., W. Feng, and T. Xu, Zinc naphthalenedicarboxylate coordination complex: A promising anode material for lithium and sodium-ion batteries with good cycling stability. *Journal of Colloid and Interface Science*, 2017. 488: p. 277–281.
24. Wang, X.W., et al., Critical design features of thermal-based radioisotope generators: A review of the power solution for polar regions and space. *Renewable & Sustainable Energy Reviews*, 2020. 119: p. 1–19.
25. Dillon, S.J. and K. Sun, Microstructural design considerations for Li-ion battery systems. *Current Opinion in Solid State & Materials Science*, 2012. 16(4): p. 153–162.
26. Wang, Z., et al., Evaluating the thermal failure risk of large-format lithium-ion batteries using a cone calorimeter. *Journal of Fire Sciences*, 2019. 37(1): p. 81–95.
27. Hu, X.S., et al., Advanced fault diagnosis for lithium-ion battery systems: A review of fault mechanisms, fault features, and diagnosis procedures. *IEEE Industrial Electronics Magazine*, 2020. 14(3): p. 65–91.
28. Lu, J.Y., et al., Mathematical modeling and frequency-domain characteristics of a periodic pulse-discharged lithium-ion battery system. *IEEE Transactions on Industry Applications*, 2021. 57(2): p. 1801–1809.
29. Mondal, B., et al., Vortex generators for active thermal management in lithium-ion battery systems. *International Journal of Heat and Mass Transfer*, 2018. 124: p. 800–815.
30. Li, X.Y., et al., State of health estimation for Li-ion battery using incremental capacity analysis and Gaussian process regression. *Energy*, 2020. 190: p. 1–11.
31. Ye, M., et al., A double-scale and adaptive particle filter-based online parameter and state of charge estimation method for lithium-ion batteries. *Energy*, 2018. 144: p. 789–799.
32. Yang, Y., et al., Carbon oxides emissions from lithium-ion batteries under thermal runaway from measurements and predictive model. *Journal of Energy Storage*, 2021. 33: p. 1–10.
33. Zhang, X., et al., Recent progress in rate and cycling performance modifications of vanadium oxides cathode for lithium-ion batteries. *Journal of Energy Chemistry*, 2021. 59: p. 343–363.
34. Zhao, C.P., J.H. Sun, and Q.S. Wang, Thermal runaway hazards investigation on 18650 lithium-ion battery using extended volume accelerating rate calorimeter. *Journal of Energy Storage*, 2020. 28: p. 1–9.
35. Fei, H.F., et al., Poly(propylene carbonate)-based polymer electrolyte with an organic cathode for stable all-solid-state sodium batteries. *Acta Physico-Chimica Sinica*, 2020. 36(5): p. 1–6.
36. Li, Z., et al., Stabilizing atomic Pt with trapped interstitial F in alloyed PtCo nanosheets for high-performance zinc-air batteries. *Energy & Environmental Science*, 2020. 13(3): p. 884–895.
37. Sharma, S., M.V. Aware, and A. Bhowate, Symmetrical six-phase induction motor-based integrated driveline of electric vehicle with predictive control. *IEEE Transactions on Transportation Electrification*, 2020. 6(2): p. 635–646.

38. Chawla, N., N. Bharti, and S. Singh, Recent advances in non-flammable electrolytes for safer lithium-ion batteries. *Batteries-Basel*, 2019. 5(1): p. 1–25.
39. Vicente, R., et al., Impacts of laser cooling for low earth orbit observation satellites: An analysis in terms of size, weight and power. *Cryogenics*, 2020. 105: p. 1–10.
40. Zhang, K.X., et al., Exergy analysis of electric vehicle heat pump air conditioning system with battery thermal management system. *Journal of Thermal Science*, 2020. 29(2): p. 408–422.
41. Shi, X., et al., The influence of cutoff charge-voltage on performance of lithium manganite/lithium titanate lithium-ion battery for power applications. *Materials Technology*, 2016. 31(11): p. 642–645.
42. Shi, J.P., Prediction study on the degeneration of lithium-ion battery based on fuzzy inference system. *Modern Physics Letters B*, 2017. 31(19–21): p. 1–16.
43. Landa-Medrano, I., et al., Sodium-oxygen battery: Steps toward reality. *Journal of Physical Chemistry Letters*, 2016. 7(7): p. 1161–1166.
44. Johnson, I.D., B.J. Ingram, and J. Cabana, The quest for functional oxide cathodes for magnesium batteries: A critical perspective. *ACS Energy Letters*, 2021. 6(5): p. 1892–1900.
45. Yan, X.D., et al., Aminoalkyldisiloxane as effective electrolyte additive for improving high temperature cycle life of nickel-rich LiNi0.6Co0.2Mn0.2O2/graphite batteries. *Journal of Power Sources*, 2020. 461: p. 1–10.
46. Zheng, Q.F., et al., A cyclic phosphate-based battery electrolyte for high voltage and safe operation. *Nature Energy*, 2020. 5(4): p. 291–298.
47. Zhang, Y.Z., et al., State of charge-dependent aging mechanisms in graphite/Li(NiCoAl)O-2 cells: Capacity loss modeling and remaining useful life prediction. *Applied Energy*, 2019. 255: p. 1–8.
48. Zhu, J.G., et al., Investigation of capacity fade for 18650-type lithium-ion batteries cycled in different state of charge (SoC) ranges. *Journal of Power Sources*, 2021. 489: p. 1–12.
49. Zhu, X.H., et al., Electrochemical impedance study of commercial LiNi0.80Co0.15Al0.05O2 electrodes as a function of state of charge and aging. *Electrochimica Acta*, 2018. 287: p. 10–20.
50. Schipper, F. and D. Aurbach, A brief review: Past, present and future of lithium ion batteries. *Russian Journal of Electrochemistry*, 2016. 52(12): p. 1095–1121.
51. Akyurek, A.S. and T.S. Rosing, Optimal distributed nonlinear battery control. *IEEE Journal of Emerging and Selected Topics in Power Electronics*, 2017. 5(3): p. 1045–1054.
52. Rinkel, B.L.D., et al., Electrolyte oxidation pathways in lithium-ion batteries. *Journal of the American Chemical Society*, 2020. 142(35): p. 15058–15074.
53. Sprocati, R., et al., Modeling electrokinetic transport and biogeochemical reactions in porous media: A multidimensional Nernst-Planck-Poisson approach with PHREEQC coupling. *Advances in Water Resources*, 2019. 127: p. 134–147.
54. Su, Y.B., et al., A more stable lithium anode by mechanical constriction for solid state batteries. *Energy & Environmental Science*, 2020. 13(3): p. 908–916.
55. Yoo, K. and J. Kim, Thermal behavior of full-scale battery pack based on comprehensive heat-generation model. *Journal of Power Sources*, 2019. 433: p. 1–10.
56. May, G.J., A. Davidson, and B. Monahov, Lead batteries for utility energy storage: A review. *Journal of Energy Storage*, 2018. 15: p. 145–157.
57. Gong, Y., et al., Evaluation of lithium-ion batteries through the simultaneous consideration of environmental, economic and electrochemical performance indicators. *Journal of Cleaner Production*, 2018. 170: p. 915–923.
58. Guo, M.M., et al., Recovery of cathode materials from spent lithium-ion batteries and their application in preparing multi-metal oxides for the removal of oxygenated VOCs: Effect of synthetic methods. *Environmental Research*, 2021. 193: p. 1–11.

59. Rudyi, A.S., et al., A solid-state lithium-ion battery: Structure, technology, and characteristics. *Technical Physics Letters*, 2020. 46(3): p. 215–219.
60. Li, J., et al., Design and test of a new droop control algorithm for a SMES/battery hybrid energy storage system. *Energy*, 2017. 118: p. 1110–1122.
61. Lawder, M.T., et al., Battery Energy Storage System (BESS) and Battery Management System (BMS) for grid-scale applications. *Proceedings of the IEEE*, 2014. 102(6): p. 1014–1030.
62. Lasrado, D., S. Ahankari, and K. Kar, Nanocellulose-based polymer composites for energy applications-A review. *Journal of Applied Polymer Science*, 2020. 137(27): p. 1–14.
63. Shanmukaraj, D., et al., Towards efficient energy storage materials: Lithium intercalation/organic electrodes to polymer electrolytes-a road map. *Journal of the Electrochemical Society*, 2020. 167(7): p. 1–11.
64. Diaz, C., et al., Particle-filtering-based prognostics for the state of maximum power available in lithium-ion batteries at electromobility applications. *IEEE Transactions on Vehicular Technology*, 2020. 69(7): p. 7187–7200.
65. Mai, R.K., et al., Variable parameter and variable frequency-based IPT charging system with configurable charge current and charge voltage. *IET Power Electronics*, 2020. 13(4): p. 751–757.
66. He, H.W., J.F. Cao, and X. Cui, Energy optimization of electric vehicle's acceleration process based on reinforcement learning. *Journal of Cleaner Production*, 2020. 248: p. 1–12.
67. Hein, S., T. Danner, and A. Latz, An electrochemical model of lithium plating and stripping in lithium ion batteries. *ACS Applied Energy Materials*, 2020. 3(9): p. 8519–8531.
68. Ahmed, M.S., S.A. Raihan, and B. Balasingam, A scaling approach for improved state of charge representation in rechargeable batteries. *Applied Energy*, 2020. 267: p. 1–12.
69. Chen, R., et al., A flexible and safe aqueous zinc-air battery with a wide operating temperature range from −20 to 70 degrees C. *ACS Sustainable Chemistry & Engineering*, 2020. 8(31): p. 11501–11511.
70. Ponte, S., et al., An embedded platform for positioning and obstacle detection for small unmanned aerial vehicles. *Electronics*, 2020. 9(7): p. 1–18.
71. Anjana, P.K., et al., Lithium-ion-based electrochemical energy storage in a layered vanadium formate coordination polymer. *Chempluschem*, 2020. 85(6): p. 1137–1144.
72. Bag, S., et al., Electrochemical studies on symmetric solid-state Na-ion full cell using Na3V2(PO4)(3) electrodes and polymer composite electrolyte. *Journal of Power Sources*, 2020. 454: p. 1–10.
73. Pramudita, J.C., et al., Graphene and selected derivatives as negative electrodes in sodium- and lithium-ion batteries. *Chemelectrochem*, 2015. 2(4): p. 600–610.
74. Abdelshafy, A.M., et al., Optimized energy management strategy for grid connected double storage (pumped storage-battery) system powered by renewable energy resources. *Energy*, 2020. 192: p. 1–16.
75. de la Torre, S., et al., Optimal battery sizing considering degradation for renewable energy integration. *IET Renewable Power Generation*, 2019. 13(4): p. 572–577.
76. Du, J.Y., et al., Battery degradation minimization oriented energy management strategy for plug-in hybrid electric bus with multi-energy storage system. *Energy*, 2018. 165: p. 153–163.
77. Liu, S.Q., et al., Study about thermal runaway behavior of high specific energy density Li-ion batteries in a low state of charge. *Journal of Energy Chemistry*, 2021. 52: p. 20–27.
78. Liu, Y., et al., Storage aging mechanism of LiNi0.8Co0.15Al0.05O2/Graphite Li-ion batteries at high state of charge. *Journal of Inorganic Materials*, 2021. 36(2): p. 175–180.

79. Bi, H.J., et al., Low-temperature thermal pretreatment process for recycling inner core of spent lithium iron phosphate batteries. *Waste Management & Research*, 2020. 39(1): p. 146–155.
80. Nguyen, T.T., et al., An accurate state of charge estimation method for lithium iron phosphate battery using a combination of an unscented Kalman filter and a particle filter. *Energies*, 2020. 13(17): p. 1–15.
81. Sun, L., et al., Comparative study on thermal runaway characteristics of lithium iron phosphate battery modules under different overcharge conditions. *Fire Technology*, 2020. 56(4): p. 1555–1574.
82. Cheng, C.S., et al., Extraction of intrinsic parameters of lead-acid batteries using energy recycling technique. *IEEE Transactions on Power Electronics*, 2019. 34(5): p. 4765–4779.
83. Krivik, P., et al., Determination of state of charge of lead-acid battery by EIS. *Journal of Energy Storage*, 2019. 21: p. 581–585.
84. Zou, C.F., et al., A review of fractional-order techniques applied to lithium-ion batteries, lead-acid batteries, and supercapacitors. *Journal of Power Sources*, 2018. 390: p. 286–296.
85. Feng, D.J., et al., Mixed lithium salts electrolyte improves the high-temperature performance of nickel-rich based lithium-ion batteries. *Journal of the Electrochemical Society*, 2020. 167(11): p. 1–8.
86. Kato, A., et al., High-temperature performance of all-solid-state lithium-metal batteries having Li/Li3PS4 interfaces modified with Au thin films. *Journal of the Electrochemical Society*, 2018. 165(9): p. 1950–1954.
87. Zhang, H.Y., et al., Thermal analysis of a 6s4p lithium-ion battery pack cooled by cold plates based on a multi-domain modeling framework. *Applied Thermal Engineering*, 2020. 173: p. 1–14.
88. Sun, B., et al., Acoustic response characteristics of lithium cobaltate/graphite battery during cycling. *Journal of the Electrochemical Society*, 2022. 169(3): p. 1–12.
89. Liao, X.-Q., et al., Dimethyl trimethylsilyl phosphite as a novel electrolyte additive for high voltage layered lithium cobaltate-based lithium ion batteries. *New Journal of Chemistry*, 2021. 45(6): p. 3160–3168.
90. Okamoto, E., et al., Analysis of heat generation of lithium ion rechargeable batteries used in implantable battery systems for driving undulation pump ventricular assist device. *Artificial Organs*, 2007. 31(7): p. 538–541.
91. Inui, Y., S. Sakamoto, and T. Tanaka, Investigation of degradation and voltage response of lithium-ion battery based on impedance and electromotive force measurement. *Electrical Engineering in Japan (English translation of Denki Gakkai Ronbunshi)*, 2017. 201(3): p. 14–24.
92. Shan, H., et al., Investigation of self-discharge properties and a new concept of open-circuit voltage drop rate in lithium-ion batteries. *Journal of Solid State Electrochemistry*, 2022. 26(1): p. 163–170.
93. Ohzuku, T. and R.J. Brodd, An overview of positive-electrode materials for advanced lithium-ion batteries. *Journal of Power Sources*, 2007. 174(2): p. 449–456.
94. Nicoll, K.J., et al., Thigh burns from exploding e-cigarette lithium ion batteries: First case series. *Burns*, 2016. 42(4): p. 42–46.
95. Pan, B., et al., Aging mechanism diagnosis of lithium ion battery by open circuit voltage analysis. *Electrochimica Acta*, 2020. 362: p. 1–8.
96. Stroe, A.I., V. Knap, and D.I. Stroe, Comparison of lithium-ion battery performance at beginning-of-life and end-of-life. *Microelectronics Reliability*, 2018. 88–90: p. 1251–1255.
97. Nayak, P.K., et al., From lithium-ion to sodium-ion batteries: Advantages, challenges, and surprises. *Angewandte Chemie-International Edition*, 2018. 57(1): p. 102–120.

98. Shi, H.T., et al., Improved splice-electrochemical circuit polarization modeling and optimized dynamic functional multi-innovation least square parameter identification for lithium-ion batteries. *International Journal of Energy Research*, 2021. 45(10): p. 15323–15337.
99. Nishio, K., et al., High rate capability of all-solid-state lithium batteries using quasi-solid-state electrolytes containing ionic liquids. *Journal of the Electrochemical Society*, 2020. 167(4): p. 1–6.
100. Shrivastava, P., et al., Overview of model-based online state-of-charge estimation using Kalman filter family for lithium-ion batteries. *Renewable & Sustainable Energy Reviews*, 2019. 113: p. 1–23.
101. Shen, Y., et al., Variable structure battery-based fuel cell hybrid power system and its incremental fuzzy logic energy management strategy. *International Journal of Hydrogen Energy*, 2020. 45(21): p. 12130–12142.
102. Celik, M., et al., SOC estimation for Li-ion batteries using extended Kalman filter with PID controlled process noise according to the voltage error. *2019 11th International Conference on Electrical and Electronics Engineering* (*ELECO 2019*), Izmir, Turkey, 2019: p. 810–814.
103. Cen, Z.H. and P. Kubiak, Lithium-ion battery SOC/SOH adaptive estimation via simplified single particle model. *International Journal of Energy Research*, 2020. 44(15): p. 12444–12459.
104. Liu, Z. and X.J. Dang, A new method for state of charge and capacity estimation of lithium-ion battery based on dual strong tracking adaptive h infinity filter. *Mathematical Problems in Engineering*, 2018. 2018: p. 1–19.
105. Ng, M.F., et al., Predicting the state of charge and health of batteries using data-driven machine learning. *Nature Machine Intelligence*, 2020. 2(3): p. 161–170.
106. Qaisar, S.M., A proficient Li-ion battery state of charge estimation based on event-driven processing. *Journal of Electrical Engineering & Technology*, 2020. 15(4): p. 1871–1877.
107. Luo, Y., et al., Dual anode materials for lithium- and sodium-ion batteries. *Journal of Materials Chemistry A*, 2018. 6(10): p. 4236–4259.
108. Xiao, D.X., et al., Reduced-coupling coestimation of SOC and SOH for lithium-ion batteries based on convex optimization. *IEEE Transactions on Power Electronics*, 2020. 35(11): p. 12332–12346.
109. Xu, J., et al., Coupling effect of state-of-health and state-of-charge on the mechanical integrity of lithium-ion batteries. *Experimental Mechanics*, 2018. 58(4): p. 633–643.
110. Yu, J.S., et al., Online state-of-health prediction of lithium-ion batteries with limited labeled data. *International Journal of Energy Research*, 2020. 44 (14): p. 11345–11353.
111. Yun, Z.H., et al., State-of-health prediction for lithium-ion batteries based on a novel hybrid approach. *Energies*, 2020. 13(18): p. 1–22.
112. Ren, P., et al., Novel co-estimation strategy based on forgetting factor dual particle filter algorithm for the state of charge and state of health of the lithium-ion battery. *International Journal of Energy Research*, 2022. 46(2): p. 1094–1107.
113. Li, X., et al., Intelligent two-step estimation approach for vehicle mass and road grade. *IEEE Access*, 2020. 8: p. 218853–218862.
114. Bhowmik, P., S. Chandak, and P.K. Rout, State of charge and state of power management among the energy storage systems by the fuzzy tuned dynamic exponent and the dynamic PI controller. *Journal of Energy Storage*, 2018. 19: p. 348–363.
115. Lai, X., et al., Co-estimation of state of charge and state of power for lithium-ion batteries based on fractional variable-order model. *Journal of Cleaner Production*, 2020. 255: p. 1–14.
116. Li, X., et al., A S/N-doped high-capacity mesoporous carbon anode for Na-ion batteries. *Journal of Materials Chemistry A*, 2019. 7(19): p. 11976–11984.

117. Liu, X.T., et al., An improved state of charge and state of power estimation method based on genetic particle filter for lithium-ion batteries. *Energies*, 2020. 13(2): p. 1–13.
118. Liu, H.L., et al., Research on energy-saving characteristics of battery-powered electric-hydrostatic hydraulic hybrid rail vehicles. *Energy*, 2020. 205: p. 1–16.
119. Shrivastava, P., et al., Combined state of charge and state of energy estimation of lithium-ion battery using dual forgetting factor-based adaptive extended Kalman filter for electric vehicle applications. *IEEE Transactions on Vehicular Technology*, 2021. 70(2): p. 1200–1215.
120. Zhang, Z.F., et al., Estimation of state-of-energy for lithium batteries based on dual adaptive particle filters considering variable current and noise effects. *International Journal of Energy Research*, 2021. 45 (11): p.15921–15935.
121. Ma, L., C. Hu, and F. Cheng, State of charge and state of energy estimation for lithium-ion batteries based on a long short-term memory neural network. *Journal of Energy Storage*, 2021. 37: p. 1–10.
122. Chen, Y., et al., Remaining useful life prediction and state of health diagnosis of lithium-ion battery based on second-order central difference particle filter. *IEEE Access*, 2020. 8: p. 37305–37313.
123. Duan, B., et al., Remaining useful life prediction of lithium-ion battery based on extended Kalman particle filter. *International Journal of Energy Research*, 2020. 44(3): p. 1724–1734.
124. He, B., L. Liu, and D. Zhang, Digital twin-driven remaining useful life prediction for gear performance degradation: A review. *Journal of Computing and Information Science in Engineering*, 2021. 21(3): p. 1–14.
125. Jia, J.F., et al., SOH and RUL prediction of lithium-ion batteries based on Gaussian process regression with indirect health indicators. *Energies*, 2020. 13(2): p. 1–15.
126. Qin, W., et al., Remaining useful life prediction for lithium-ion batteries using particle filter and artificial neural network. *Industrial Management & Data Systems*, 2020. 120(2): p. 312–328.
127. Dong, X.L., et al., Low-temperature charge/discharge of rechargeable battery realized by intercalation pseudocapacitive behavior. *Advanced Science*, 2020. 7(14): p. 1–13.
128. Qu, H.N., et al., Application of ac impedance as diagnostic tool – Low temperature electrolyte for a Li-ion battery. *Electrochimica Acta*, 2019. 322: p. 1–8.
129. Wang, Y.Q., et al., Low temperature performance enhancement of high-safety lithium-sulfur battery enabled by synergetic adsorption and catalysis. *Electrochimica Acta*, 2020. 353: p. 1–11.
130. Antonucci, V., et al., Li-ion battery modeling and state of charge estimation method including the hysteresis effect. *Electronics*, 2019. 8(11): p. 1–20.
131. Azzollini, I.A., et al., Lead-acid battery modeling over full state of charge and discharge range. *IEEE Transactions on Power Systems*, 2018. 33(6): p. 6422–6429.
132. Hu, M.H., et al., Lithium-ion battery modeling and parameter identification based on fractional theory. *Energy*, 2018. 165: p. 153–163.
133. Stubler, T., A. Lahyani, and A.A. Zayoud, Lithium-ion battery modeling using CC-CV and impedance spectroscopy characterizations. *SN Applied Sciences*, 2020. 2(5): p. 1–8.
134. Miniguano, H., et al., General parameter identification procedure and comparative study of Li-ion battery models. *IEEE Transactions on Vehicular Technology*, 2020. 69(1): p. 235–245.
135. Sun, C., et al., Improved parameter identification and state-of-charge estimation for lithium-ion battery with fixed memory recursive least squares and sigma-point Kalman filter. *Electrochimica Acta*, 2021. 387: p. 1–14.
136. Khalid, A. and A.I. Sarwat, Unified univariate-neural network models for lithium-ion battery state-of-charge forecasting using minimized Kalman information criterion algorithm. *IEEE Access*, 2021. 9: p. 39154–39170.

137. Lipu, M.S.H., et al., State of charge estimation for lithium-ion battery using recurrent NARX neural network model based lighting search algorithm. *IEEE Access*, 2018. 6: p. 28150–28161.
138. Yan, K., et al., Simplified mechanism modeling and discharge characteristic analysis of high C-rate LiFePO4 battery. *Diangong Jishu Xuebao/Transactions of China Electrotechnical Society*, 2022. 37(3): p. 599–609.
139. Tian, Y., et al., Parallel-connected battery module modeling based on physical characteristics in multiple domains and heterogeneous characteristic analysis. *Energy*, 2022. 239: p. 1–16.
140. Baccouche, I., et al., Li-ion battery modeling and characterization: An experimental overview on NMC battery. *International Journal of Energy Research*, 2022. 46(4): p. 3843–3859.
141. Chayambuka, K., et al., An experimental and modeling study of sodium-ion battery electrolytes. *Journal of Power Sources*, 2021. 516: p. 1–11.
142. Bonkile, M.P. and V. Ramadesigan, Effects of sizing on battery life and generation cost in PV wind battery hybrid systems. *Journal of Cleaner Production*, 2022. 340: p. 1–14.
143. Liang, X., et al., A new high-capacity and safe energy storage system: Lithium-ion sulfur batteries. *Nanoscale*, 2019. 11(41): p. 19140–19157.
144. Zheng, Y., et al., State-of-charge inconsistency estimation of lithium-ion battery pack using mean-difference model and extended Kalman filter. *Journal of Power Sources*, 2018. 383: p. 50–58.
145. Li, J., et al., Preparation of LiCoO2 cathode materials from spent lithium-ion batteries. *Ionics*, 2009. 15(1): p. 111–113.
146. Ding, X., et al., An improved Thevenin model of lithium-ion battery with high accuracy for electric vehicles. *Applied Energy*, 2019. 254: p. 1–8.
147. Lyu, Z.Q. and R.J. Gao, Li-ion battery state of health estimation through Gaussian process regression with Thevenin model. *International Journal of Energy Research*, 2020. 44(13): p. 10262–10281.
148. Pai, K.J., A reformatory model incorporating PNGV battery and three-terminal-switch models to design and implement feedback compensations of LiFePO4 battery chargers. *Electronics*, 2019. 8(2): p. 1–28.
149. Lei, Z., Y. Zhang, and X. Lei, Temperature uniformity of a heated lithium-ion battery cell in cold climate. *Applied Thermal Engineering*, 2018. 129: p. 148–154.
150. Vasak, M. and G. Kujundzic, A battery management system for efficient adherence to energy exchange commands under longevity constraints. *IEEE Transactions on Industry Applications*, 2018. 54(4): p. 3019–3033.
151. Wang, S., et al., Improved covariance matching electrical equivalent modeling for accurate internal state characterization of packing lithium-ion batteries. *International Journal of Energy Research*, 2022. 46(3): p. 3602–3620.
152. Yang, Y., et al., Battery energy storage system size determination in renewable energy systems: A review. *Renewable & Sustainable Energy Reviews*, 2018. 91: p. 109–125.
153. Wang, L.M., et al., Study on electrochemical and thermal characteristics of lithium-ion battery using the electrochemical-thermal coupled model. *International Journal of Energy Research*, 2019. 43(6): p. 2086–2107.
154. Ling, L.Y. and Y. Wei, State-of-charge and state-of-health estimation for lithium-ion batteries based on dual fractional-order extended Kalman filter and online parameter identification. *IEEE Access*, 2021. 9: p. 47588–47602.
155. Lee, S.-J., et al., Lithium-ion battery module temperature monitoring by using planer home-made micro thermocouples. *International Journal of Electrochemical Science*, 2013. 8(3): p. 4131–4141.
156. Meng, J.H., et al., A novel multiple correction approach for fast open circuit voltage prediction of lithium-ion battery. *IEEE Transactions on Energy Conversion*, 2019. 34(2): p. 1115–1123.

157. Kulova, T.L., et al., A brief review of post-lithium-ion batteries. *International Journal of Electrochemical Science*, 2020. 15(8): p. 7242–7259.
158. Huang, B., et al., Recycling of lithium-ion batteries: Recent advances and perspectives. *Journal of Power Sources*, 2018. 399: p. 274–286.
159. Guha, A. and A. Patra, Online estimation of the electrochemical impedance spectrum and remaining useful life of lithium-ion batteries. *IEEE Transactions on Instrumentation and Measurement*, 2018. 67(8): p. 1836–1849.
160. Wang, Y.J., et al., Model based insulation fault diagnosis for lithium-ion battery pack in electric vehicles. *Measurement*, 2019. 131: p. 443–451.
161. Chen, Y., et al., Online state of charge estimation for battery in electric vehicles based on forgetting factor recursive least squares. *Shanghai Jiaotong Daxue Xuebao/Journal of Shanghai Jiaotong University*, 2020. 54(12): p. 1340–1346.
162. Liu, X.L., et al., Online identification of power battery parameters for electric vehicles using a decoupling multiple forgetting factors recursive least squares method. *CSEE Journal of Power and Energy Systems*, 2020. 6(3): p. 735–742.
163. Elsergany, A.M., et al., An adaptive autotuned polynomial-based extended Kalman filter for sensorless surface temperature estimation of Li-ion battery cells. *IEEE Access*, 2022. 10: p. 14038–14048.
164. Partovibakhsh, M. and G. Liu, An adaptive unscented Kalman filtering approach for online estimation of model parameters and state-of-charge of lithium-ion batteries for autonomous mobile robots. *IEEE Transactions on Control Systems Technology*, 2015. 23(1): p. 357–363.
165. Li, M., et al., A battery SOC estimation method based on AFFRLS-EKF. *Sensors*, 2021. 21(17): p. 1–12.
166. Liu, F., et al., SOC estimation based on data driven exteaded Kalman filter algorithm for power battery of electric vehicle and plug-in electric vehicle. *Journal of Central South University*, 2019. 26(6): p. 1402–1415.
167. Sun, F., et al., Adaptive unscented Kalman filtering for state of charge estimation of a lithium-ion battery for electric vehicles. *Energy*, 2011. 36(5): p. 3531–3540.
168. Zou, C., et al., Electrochemical estimation and control for lithium-ion battery health-aware fast charging. *IEEE Transactions on Industrial Electronics*, 2018. 65(8): p. 6635–6645.
169. Zheng, D., et al., Real-time estimation of battery state of charge with metabolic grey model and LabVIEW platform. *IEEE Access*, 2018. 6: p. 13170–13180.
170. Harun, M.J.H. and S.A.H. Ali, Numerical simulation of concentration profile in cathode of lithium ion cell. *Materials Research Innovations*, 2011. 15: p. 187–192.
171. Zhang, Z., et al., Active cell balancing of lithium-ion battery pack based on average state of charge. *International Journal of Energy Research*, 2020. 44(4): p. 2535–2548.
172. Zhang, X., et al., A novel method for lithium-ion battery state of energy and state of power estimation based on multi-time-scale filter. *Applied Energy*, 2018. 216: p. 442–451.
173. Wu, X., X. Han, and J. Du, Investigating capacity degradation of LiFePO4 battery for electric vehicles under different overcharge conditions. *International Journal of Vehicle Design*, 2020. 84(1–4): p. 219–237.
174. Wang, L., et al., Research on multiple states joint estimation algorithm for electric vehicles under charge mode. *IEEE Access*, 2018. 6: p. 40143–40153.
175. Shi, Y., et al., The optimization of state of charge and state of health estimation for lithium ions battery using combined deep learning and Kalman filter methods. *International Journal of Energy Research*, 2021. 45(7): p. 11206–11230.
176. Shen, P., et al., The co-estimation of state of charge, state of health, and state of function for lithium-ion batteries in electric vehicles. *IEEE Transactions on Vehicular Technology*, 2018. 67(1): p. 92–103.

177. Otoyama, M., et al., Investigation of state-of-charge distributions for LiCoO2 composite positive electrodes in all-solid-state lithium batteries by Raman imaging. *Chemistry Letters*, 2016. 45(7): p. 810–812.
178. Ng, K.S., et al., Enhanced coulomb counting method for estimating state-of-charge and state-of-health of lithium-ion batteries. *Applied Energy*, 2009. 86(9): p. 1506–1511.
179. Adaikkappan, M. and N. Sathiyamoorthy, Modeling, state of charge estimation, and charging of lithium-ion battery in electric vehicle: A review. *International Journal of Energy Research*, 2022. 46(3): p. 2141-2165.
180. Bayda-Smykaj, M., et al., White light from dual intramolecular charge-transfer emission in a silylene-bridged styrylcarbazole and pyrene dyad. *Journal of Physical Chemistry C*, 2021. 125(23): p. 12488–12495.
181. Mastali, M., et al., Battery state of the charge estimation using Kalman filtering. *Journal of Power Sources*, 2013. 239: p. 294–307.
182. Beelen, H., H.J. Bergveld, and M.C.F. Donkers, Joint estimation of battery parameters and state of charge using an extended Kalman filter: A single-parameter tuning approach. *IEEE Transactions on Control Systems Technology*, 2021. 29(3): p. 1087–1101.
183. Lei, K.-B. and Z.-Q. Chen, Estimation of state of charge of battery based on improved multi-innovation extended Kalman filter SOC. *Zhejiang Daxue Xuebao (Gongxue Ban)/Journal of Zhejiang University (Engineering Science)*, 2021. 55(10): p. 1978–1985 and 2001.
184. Nejad, S. and D.T. Gladwin, Online battery state of power prediction using PRBS and extended Kalman filter. *IEEE Transactions on Industrial Electronics*, 2020. 67(5): p. 3747–3755.
185. Zhang, L., et al., Intelligent computing for extended Kalman filtering SOC algorithm of lithium-ion battery. *Wireless Personal Communications*, 2018. 102(2): p. 2063–2076.
186. Reichbach, N. and A. Kuperman, Recursive-least-squares-based real-time estimation of supercapacitor parameters. *IEEE Transactions on Energy Conversion*, 2016. 31(2): p. 810–812.
187. Xu, L., J. Wang, and Q. Chen, Kalman filtering state of charge estimation for battery management system based on a stochastic fuzzy neural network battery model. *Energy Conversion and Management*, 2012. 53(1): p. 33–39.
188. Qian, K., et al., Modified dual extended Kalman filters for SOC estimation and online parameter identification of lithium-ion battery via modified gray wolf optimizer. *Proceedings of the Institution of Mechanical Engineers Part D-Journal of Automobile Engineering*, 2021. 236(8): p. 1761–1774.
189. Xu, W., et al., Novel reduced-order modeling method combined with three-particle nonlinear transform unscented Kalman filtering for the battery state-of-charge estimation. *Journal of Power Electronics*, 2020. 20(6): p. 1541–1549.
190. Xie, F., et al., A novel battery state of charge estimation based on the joint unscented Kalman filter and support vector machine algorithms. *International Journal of Electrochemical Science*, 2020. 15(8): p. 7935–7953.
191. Wang, T.P., et al., Model-based unscented Kalman filter observer design for lithium-ion battery state of charge estimation. *International Journal of Energy Research*, 2018. 42(4): p. 1603–1614.
192. Zhu, F. and J. Fu, A novel state-of-health estimation for lithium-ion battery via unscented Kalman filter and improved unscented particle filter. *IEEE Sensors Journal*, 2021. 21(22): p. 25449–25456.
193. Seung, J., S. Yoo, and K. Chong, Experiments on state and unmeasured-parameter estimation of two degree-of-freedom system for precise control based on JAUKF. *International Journal of Precision Engineering and Manufacturing*, 2019. 20(7): p. 1159–1168.

194. Shi, H., et al., Adaptive iterative working state prediction based on the double unscented transformation and dynamic functioning for unmanned aerial vehicle lithium-ion batteries. *Measurement & Control*, 2020. 53(9–10): p. 1760–1773.
195. Ahwiadi, M. and W. Wang, An adaptive particle filter technique for system state estimation and prognosis. *IEEE Transactions on Instrumentation and Measurement*, 2020. 69(9): p. 6756–6765.
196. Corno, M. and G. Pozzato, Active adaptive battery aging management for electric vehicles. *IEEE Transactions on Vehicular Technology*, 2020. 69(1): p. 258–269.
197. He, Z., et al., Adaptive state of charge estimation for Li-ion batteries based on an unscented Kalman filter with an enhanced battery model. *Energies*, 2013. 6(8): p. 4134–4151.
198. Rauh, A., S.S. Butt, and H. Aschemann, Nonlinear state observers and extended Kalman filters for battery systems. *International Journal of Applied Mathematics and Computer Science*, 2013. 23(3): p. 539–556.
199. Jiang, C., et al., A novel adaptive extended Kalman filtering and electrochemical-circuit combined modeling method for the online ternary battery state-of-charge estimation. *International Journal of Electrochemical Science*, 2020. 15(10): p. 9720–9733.
200. Goud, J.S., R. Kalpana, and B. Singh, An online method of estimating state of health of a Li-ion battery. *IEEE Transactions on Energy Conversion*, 2021. 36(1): p. 111–119.
201. Zhang, X., et al., A novel method for identification of lithium-ion battery equivalent circuit model parameters considering electrochemical properties. *Journal of Power Sources*, 2017. 345: p. 21–29.
202. Kwon, S.-J., et al., Research of adaptive extended Kalman filter-based SOC estimator for frequency regulation ESS. *Applied Sciences-Basel*, 2019. 9(20): p. 1–21.
203. Zhang, Q., et al., A novel fractional variable-order equivalent circuit model and parameter identification of electric vehicle Li-ion batteries. *ISA Transactions*, 2020. 97: p. 448–457.
204. Yuan, S., et al., A transfer function type of simplified electrochemical model with modified boundary conditions and Pade approximation for Li-ion battery: Part 2. Modeling and parameter estimation. *Journal of Power Sources*, 2017. 352: p. 258–271.
205. Wei, Z., et al., Signal-disturbance interfacing elimination for unbiased model parameter identification of lithium-ion battery. *IEEE Transactions on Industrial Informatics*, 2021. 17(9): p. 5887–5897.
206. Chen, Z., Y. Fu, and C.C. Mi, State of charge estimation of lithium-ion batteries in electric drive vehicles using extended Kalman filtering. *IEEE Transactions on Vehicular Technology*, 2013. 62(3): p. 1020–1030.
207. Wang, B., et al., Fractional-order modeling and parameter identification for lithium-ion batteries. *Journal of Power Sources*, 2015. 293: p. 151–161.
208. Song, Z., et al., Parameter identification of lithium-ion battery pack for different applications based on Cramer-Rao bound analysis and experimental study. *Applied Energy*, 2018. 231: p. 1307–1318.
209. Ren, B., et al., Parameter identification of a lithium-ion battery based on the improved recursive least square algorithm. *IET Power Electronics*, 2020. 13(12): p. 2531–2537.
210. Shi, H., et al., On-line adaptive asynchronous parameter identification of lumped electrical characteristic model for vehicle lithium-ion battery considering multi-time scale effects. *Journal of Power Sources*, 2022. 517: p. 1–14.
211. Hentunen, A., T. Lehmuspelto, and J. Suomela, Time-domain parameter extraction method for Thevenin-equivalent circuit battery models. *IEEE Transactions on Energy Conversion*, 2014. 29(3): p. 558–566.
212. Li, J., et al., A novel state estimation approach based on adaptive unscented Kalman filter for electric vehicles. *IEEE Access*, 2020. 8: p. 185629–185637.

213. Hou, J., Y. Yang, and T. Gao, A normal-gamma-based adaptive dual unscented Kalman filter for battery parameters and state-of-charge estimation with heavy-tailed measurement noise. *International Journal of Energy Research*, 2020. 44(5): p. 3510–3525.
214. Faisal, M., et al., Fuzzy-based charging-discharging controller for lithium-ion battery in microgrid applications. *IEEE Transactions on Industry Applications*, 2021. 57(4): p. 4187–4195.
215. Deng, Z., et al., Online available capacity prediction and state of charge estimation based on advanced data-driven algorithms for lithium iron phosphate battery. *Energy*, 2016. 112: p. 469–480.
216. Choi, S.W., et al., Interleaved isolated single-phase PFC converter module for three-phase EV charger. *IEEE Transactions on Vehicular Technology*, 2020. 69(5): p. 4957–4967.
217. Jia, J.F., et al., Multi-scale prediction of RUL and SOH for lithium-ion batteries based on WNN-UPF combined model. *Chinese Journal of Electronics*, 2021. 30(1): p. 26–35.
218. Li, Y., et al., A novel fusion model for battery online State of Charge (SOC) estimation. *International Journal of Electrochemical Science*, 2021. 16(1): p. 1–10.
219. Li, Y., et al., Safety modeling and protection for lithium-ion batteries based on artificial neural networks method under mechanical abuse. *Science China-Technological Sciences*, 2021. 64(11): p. 2373–2388.
220. Xiong, R., et al., A novel fractional order model for state of charge estimation in lithium ion batteries. *IEEE Transactions on Vehicular Technology*, 2019. 68(5): p. 4130–4139.
221. Qiao, J., et al., A novel bias compensation recursive least square-multiple weighted dual extended Kalman filtering method for accurate state-of-charge and state-of-health co-estimation of lithium-ion batteries. *International Journal of Circuit Theory and Applications*, 2021. 49(11): p. 3879–3893.
222. Pourjafar, S., et al., A bidirectional multiport DC-DC converter applied for energy storage system with hybrid energy sources. *International Journal of Circuit Theory and Applications*, 2021. 49(8): p. 2453–2478.
223. Wang, H., Y. Zheng, and Y. Yu, Joint estimation of SOC of lithium battery based on dual Kalman filter. *Processes*, 2021. 9(8): p. 1–10.
224. Ren, Z., et al., A comparative study of the influence of different open circuit voltage tests on model-based state of charge estimation for lithium-ion batteries. *International Journal of Energy Research*, 2021. 45(9): p. 13692–13711.
225. Wassiliadis, N., et al., Revisiting the dual extended Kalman filter for battery state-of-charge and state-of-health estimation: A use-case life cycle analysis. *Journal of Energy Storage*, 2018. 19: p. 73–87.
226. Xu, W.H., et al., A novel adaptive dual extended Kalman filtering algorithm for the Li-ion battery state of charge and state of health co-estimation. *International Journal of Energy Research*, 2021. 45(12): p. 14592–14602.
227. Xu, X.M., et al., State estimation of lithium batteries for energy storage based on dual extended Kalman filter. *Mathematical Problems in Engineering*, 2020. 2020: p. 1–14.
228. Cui, X.Y., et al., State-of-charge estimation of power lithium-ion batteries based on an embedded micro control unit using a square root cubature Kalman filter at various ambient temperatures. *International Journal of Energy Research*, 2019. 43(8): p. 3561–3577.
229. Fu, S., et al., State of charge estimation of lithium-ion phosphate battery based on weighted multi-innovation cubature Kalman filter. *Journal of Energy Storage*, 2022. 50: p. 1–11.
230. Linghu, J.Q., et al., Estimation for state-of-charge of lithium-ion battery based on an adaptive high-degree cubature Kalman filter. *Energy*, 2019. 189(5): p. 1–12.
231. Tian, Y., et al., A combined method for state-of-charge estimation for lithium-ion batteries using a long short-term memory network and an adaptive cubature Kalman filter. *Applied Energy*, 2020. 265: p. 1–14.

232. Shen, J., et al., State of charge estimation framework for lithium-ion batteries based on square root cubature Kalman filter under wide operation temperature range. *International Journal of Energy Research*, 2021. 45(4): p. 5586–5601.
233. Zhao, L., Z. Liu, and G. Ji, Lithium-ion battery state of charge estimation with model parameters adaptation using H extended Kalman filter. *Control Engineering Practice*, 2018. 81: p. 114–128.
234. Sun, T.-W. and T.-H. Tsai, A battery management system using interleaved pulse charging with charge and temperature balancing based on NARX network. *IEEE Transactions on Circuits and Systems I: Regular Papers*, 2022. 69(4): p. 1811–1819.
235. Wang, Q., et al., State of charge estimation for lithium-ion battery based on NARX recurrent neural network and moving window method. *IEEE Access*, 2021. 9: p. 83364–83375.
236. Wei, M., et al., State of charge estimation of lithium-ion batteries using LSTM and NARX neural networks. *IEEE Access*, 2020. 8: p. 189236–189245.
237. Zhang, Y., et al., Novel feedback-Bayesian BP neural network combined with extended Kalman filtering for the battery state-of-charge estimation. *International Journal of Electrochemical Science*, 2021. 16(6): p. 1–12.
238. Xiong, W., Y. Mo, and C. Yan, Lithium-ion battery parameters and state of charge joint estimation using bias compensation least squares and the alternate algorithm. *Mathematical Problems in Engineering*, 2020. 2020: p. 1–16.
239. Wu, M., et al., State of charge estimation of power lithium-ion battery based on a variable forgetting factor adaptive Kalman filter. *Journal of Energy Storage*, 2021. 41: p. 1–8.
240. Wang, Y., et al., A state of charge estimation method of lithium-ion battery based on fused open circuit voltage curve. *Applied Sciences-Basel*, 2020. 10(4): p. 1–15.
241. Cao, Y. and J.A. Abu Qahouq, Evaluation of bi-directional single-inductor multi-input battery system with state-of-charge balancing control. *IET Power Electronics*, 2018. 11(13): p. 2140–2150.
242. Al-Shalabi, M., et al., Energy efficient multi-hop path in wireless sensor networks using an enhanced genetic algorithm. *Information Sciences*, 2019. 500: p. 259–273.
243. Du, X., et al., An information appraisal procedure: Endows reliable online parameter identification to lithium-ion battery model. *IEEE Transactions on Industrial Electronics*, 2022. 69(6): p. 5889–5899.
244. Cui, Y., et al., State of health diagnosis model for lithium ion batteries based on real-time impedance and open circuit voltage parameters identification method. *Energy*, 2018. 144: p. 647–656.
245. Zhang, T., et al., A systematic framework for state of charge, state of health and state of power co-estimation of lithium-ion battery in electric vehicles. *Sustainability*, 2021. 13(9): p. 1–19.
246. Yang, F.Q., Modeling analysis for the growth of a Li sphere and Li whisker in a solid-state lithium metal battery. *Physical Chemistry Chemical Physics*, 2020. 22(24): p. 13737–13745.
247. Zhou, S.D., et al., Adaptive model parameter identification for lithium-ion batteries based on improved coupling hybrid adaptive particle swarm optimization- simulated annealing method. *Journal of Power Sources*, 2021. 482: p. 1–13.
248. Jiang, Y., et al., State of health estimation for lithium-ion battery using empirical degradation and error compensation models. *IEEE Access*, 2020. 8: p. 123858–123868.
249. Jiaqiang, E., et al., Effects analysis on active equalization control of lithium-ion batteries based on intelligent estimation of the state-of-charge. *Energy*, 2022. 238: p. 1–14.
250. Khayamy, M., A. Nasiri, and O. Okoye, Development of an equivalent circuit for batteries based on a distributed impedance network. *IEEE Transactions on Vehicular Technology*, 2020. 69(6): p. 6119–6128.

251. Chang, J.Q., M.S. Chi, and T. Shen, Model based state-of-energy estimation for LiFePO4 batteries using unscented particle filter. *Journal of Power Electronics*, 2020. 20(2): p. 624–633.
252. Chen, A., et al., A temperature and current rate adaptive model for high-power lithium-titanate batteries used in electric vehicles. *IEEE Transactions on Industrial Electronics*, 2020. 67(11): p. 9492–9502.
253. Chen, B., et al., Robust state-of-charge estimation for lithium-ion batteries based on an improved gas-liquid dynamics model. *Energy*, 2022. 238: p. 1–12.
254. Wang, S., et al., A novel safety assurance method based on the compound equivalent modeling and iterate reduce particle-adaptive Kalman filtering for the unmanned aerial vehicle lithium ion batteries. Energy Science & Engineering, 2020. 8(5): p. 1484–1500.
255. Li, J., et al., Joint estimation of state of charge and state of health for lithium-ion battery based on dual adaptive extended Kalman filter. *International Journal of Energy Research*, 2021. 45(9): p. 13307–13322.
256. Berecibar, M., et al., Critical review of state of health estimation methods of Li-ion batteries for real applications. *Renewable & Sustainable Energy Reviews*, 2016. 56: p. 572–587.
257. Chen, L., et al., State of charge estimation of lithium-ion batteries based on fuzzy fractional-order unscented Kalman filter. *Fractal and Fractional*, 2021. 5(3): p. 1–17.
258. Chen, L., et al., A novel combined estimation method of online full-parameter identification and adaptive unscented particle filter for Li-ion batteries SOC based on fractional-order modeling. *International Journal of Energy Research*, 2021. 45(10): p. 15481–15494.
259. Tang, J., et al., A health monitoring method based on multiple indicators to eliminate influences of estimation dispersion for lithium-ion batteries. *IEEE Access*, 2019. 7: p. 122302–122314.
260. Roman, D., et al., Machine learning pipeline for battery state-of-health estimation. *Nature Machine Intelligence*, 2021. 3(5): p. 447–456.
261. Suhail, M., et al., Development of progressive fuzzy logic and ANFIS control for energy management of plug-in hybrid electric vehicle. *IEEE Access*, 2021. 9: p. 62219–62231.
262. Oji, T., et al., Data-driven methods for battery SOH estimation: Survey and a critical analysis. *IEEE Access*, 2021. 9: p. 126903–126916.
263. Ebbesen, S., P. Elbert, and L. Guzzella, Battery state-of-health perceptive energy management for hybrid electric vehicles. *IEEE Transactions on Vehicular Technology*, 2012. 61(7): p. 2893–2900.
264. Hu, X.S., et al., Battery health prediction using fusion-based feature selection and machine learning. *IEEE Transactions on Transportation Electrification*, 2021. 7(2): p. 382–398.
265. Ngoc-Tham, T., A.B. Khan, and W. Choi, State of charge and state of health estimation of AGM VRLA batteries by employing a dual extended Kalman filter and an ARX model for online parameter estimation. *Energies*, 2017. 10(1): p. 1–18.
266. Moura, S.J., J.L. Stein, and H.K. Fathy, Battery-health conscious power management in plug-in hybrid electric vehicles via electrochemical modeling and stochastic control. *IEEE Transactions on Control Systems Technology*, 2013. 21(3): p. 679–694.
267. Tan, Y. and G. Zhao, Transfer learning with long short-term memory network for state-of-health prediction of lithium-ion batteries. *IEEE Transactions on Industrial Electronics*, 2020. 67(10): p. 8723–8731.
268. Feng, X.N., et al., Online state-of-health estimation for Li-ion battery using partial charging segment based on support vector machine. *IEEE Transactions on Vehicular Technology*, 2019. 68(9): p. 8583–8592.
269. Li, J.B., et al., A novel state of charge approach of lithium ion battery using least squares support vector machine. *IEEE Access*, 2020. 8: p. 195398–195410.

270. Meng, J.H., et al., Lithium-ion battery state of health estimation with short-term current pulse test and support vector machine. *Microelectronics Reliability*, 2018. 88–90: p. 1216–1220.
271. Chung, D.-W., J.-H. Ko, and K.-Y. Yoon, State-of-charge estimation of lithium-ion batteries using LSTM deep learning method. *Journal of Electrical Engineering & Technology*, 2022. 17(3): p. 1931–1945.
272. Rahman, F., et al., An assistive model for visually impaired people using YOLO and MTCNN. *Proceedings of 2019 the 3rd International Conference on Cryptography, Security and Privacy (ICCSP 2019) with Workshop 2019 the 4th International Conference on Multimedia and Image Processing (ICMIP 2019)*, Kuala Lumpur, Malaysia, 2019: p. 225–230.
273. Zraibi, B., et al., Remaining useful life assessment for lithium-ion batteries using CNN-LSTM-DNN hybrid method. *IEEE Transactions on Vehicular Technology*, 2021. 70(5): p. 4252–4261.
274. Yun, Z. and W. Qin, Remaining useful life estimation of lithium-ion batteries based on optimal time series health indicator. *IEEE Access*, 2020. 8: p. 55447–55461.
275. Liu, X., et al., A new dynamic SOH estimation of lead-acid battery for substation application. *International Journal of Energy Research*, 2017. 41(4): p. 579–592.
276. Klass, V., M. Behm, and G. Lindbergh, A support vector machine-based state-of-health estimation method for lithium-ion batteries under electric vehicle operation. *Journal of Power Sources*, 2014. 270: p. 262–272.
277. He, Z., et al., Online state-of-health estimation of lithium-ion batteries using Dynamic Bayesian Networks. *Journal of Power Sources*, 2014. 267: p. 576–583.
278. Hametner, C., S. Jakubek, and W. Prochazka, Data-driven design of a cascaded observer for battery state of health estimation. *IEEE Transactions on Industry Applications*, 2018. 54(6): p. 6258–6266.
279. Lyu, Z., R. Gao, and L. Chen, Li-ion battery state of health estimation and remaining useful life prediction through a model-data-fusion method. *IEEE Transactions on Power Electronics*, 2021. 36(6): p. 6228–6240.
280. Kariem, H., E. Touti, and T. Fetouh, The efficiency of PSO-based MPPT technique of an electric vehicle within the city. *Measurement & Control*, 2020. 53(3–4): p. 461–473.
281. Li, R., et al., State of charge prediction algorithm of lithium-ion battery based on PSO-SVR cross validation. *IEEE Access*, 2020. 8: p. 10234–10242.
282. Telmoudi, A.J., et al., Modeling and state of health estimation of nickel-metal hydride battery using an EPSO-based fuzzy c-regression model. *Soft Computing*, 2020. 24(10): p. 7265–7279.
283. Li, D., et al., Battery thermal runaway fault prognosis in electric vehicles based on abnormal heat generation and deep learning algorithms. *IEEE Transactions on Power Electronics*, 2022. 37(7): p. 8513–8525.
284. Hu, C., et al., State of charge estimation for lithium-ion batteries based on TCN-LSTM neural networks. *Journal of the Electrochemical Society*, 2022. 169(3): p. 1–11.
285. He, J.T., et al., State-of-health estimation of lithium-ion batteries using incremental capacity analysis based on voltage-capacity model. *IEEE Transactions on Transportation Electrification*, 2020. 6(2): p. 417–426.
286. Lin, C.P., et al., SOH estimation and SOC recalibration of lithium-ion battery with incremental capacity analysis & cubic smoothing spline. *Journal of the Electrochemical Society*, 2020. 167(9): p. 1–16.
287. Farjah, A. and T. Ghanbari, Early ageing detection of battery cells in battery management system. *Electronics Letters*, 2020. 56(12): p. 616–618.
288. Lim, J.-Y., et al., Remaining useful life prediction for lithium-ion batteries using EMD-CNN-LSTM hybrid method. *The Transactions of the Korean Institute of Power Electronics*, 2022. 27(1): p. 48–55.

289. Li, K., W. Ni, and F. Dressler, LSTM-characterized deep reinforcement learning for continuous flight control and resource allocation in UAV-assisted sensor network. *IEEE Internet of Things Journal*, 2022. 9(6): p. 4179–4189.
290. Kulikov, G.Y. and M.V. Kulikova, Hyperbolic-singular-value-decomposition-based square-root accurate continuous-discrete extended-unscented Kalman filters for estimating continuous-time stochastic models with discrete measurements. *International Journal of Robust and Nonlinear Control*, 2020. 30(5): p. 2033–2058.
291. Lei, X., et al., A novel temperature-hysteresis model for power battery of electric vehicles with an adaptive joint estimator on state of charge and power. *Energies*, 2019. 12(19): p. 1–24.
292. Yu, H., et al., Porous carbon derived from metal-organic framework@graphene quantum dots as electrode materials for supercapacitors and lithium-ion batteries. *RSC Advances*, 2019. 9(17): p. 9577–9583.
293. Li, S., et al., State-of-charge estimation of lithium-ion batteries in the battery degradation process based on recurrent neural network. *Energies*, 2021. 14(2): p. 1–21.
294. Li, Y., et al., Optimization of charging strategy for lithium-ion battery packs based on complete battery pack model. *Journal of Energy Storage*, 2021. 37: p. 1–16.
295. Pan, H., et al., Establishing a dynamic model of lithium-ion battery charging internal resistance based on multiple factors. *Diangong Jishu Xuebao/Transactions of China Electrotechnical Society*, 2021. 36(10): p. 2199–2206.
296. Chen, L., et al., Estimation the internal resistance of lithium-ion-battery using a multi-factor dynamic internal resistance model with an error compensation strategy. *Energy Reports*, 2021. 7: p. 3050–3059.
297. Zheng, F., et al., Temperature dependent power capability estimation of lithium-ion batteries for hybrid electric vehicles. *Energy*, 2016. 113: p. 64–75.
298. Zhang, W., et al., Joint state-of-charge and state-of-available-power estimation based on the online parameter identification of lithium-ion battery model. Ieee Transactions on Industrial Electronics, 2022. 69(4): p. 3677–3688.
299. Zhang, W., W. Shi, and Z. Ma, Adaptive unscented Kalman filter based state of energy and power capability estimation approach for lithium-ion battery. *Journal of Power Sources*, 2015. 289: p. 50–62.
300. Abdelaty, H., et al., Machine learning prediction models for battery-electric bus energy consumption in transit. *Transportation Research Part D-Transport and Environment*, 2021. 96: p. 1–27.
301. Zhang, H., et al., Cost-effective lebesgue sampling long short-term memory networks for lithium-ion batteries diagnosis and prognosis. *IEEE Transactions on Industrial Electronics*, 2022. 69(2): p. 1958–1967.
302. White, G. and S. Clarke, Short-term QoS forecasting at the edge for reliable service applications. *IEEE Transactions on Services Computing*, 2022. 15(2): p. 1089–1102.
303. Xiong, R., et al., A robust state-of-charge estimator for multiple types of lithium-ion batteries using adaptive extended Kalman filter. *Journal of Power Sources*, 2013. 243: p. 805–816.
304. Tovar Rosas, M.A., M. Robles Perez, and E.R. Martinez Perez, Itineraries for charging and discharging a BESS using energy predictions based on a CNN-LSTM neural network model in BCS, Mexico. *Renewable Energy*, 2022. 188: p. 1141–1165.
305. Oyewole, I., A. Chehade, and Y. Kim, A controllable deep transfer learning network with multiple domain adaptation for battery state-of-charge estimation. *Applied Energy*, 2022. 312: p. 1–15.
306. Esfandyari, M.J., et al., A hybrid model predictive and fuzzy logic based control method for state of power estimation of series-connected Lithium-ion batteries in HEVs. *Journal of Energy Storage*, 2019. 24: p. 1–14.

307. Moradi-Sepahvand, M., T. Amraee, and S.S. Gougheri, Deep learning based hurricane resilient coplanning of transmission lines, battery energy storages, and wind farms. *IEEE Transactions on Industrial Informatics*, 2022. 18(3): p. 2120–2131.
308. Li, B., et al., A linear recursive state of power estimation method based on fusion model of voltage and state of charge limitations. *Journal of Energy Storage*, 2021. 40: p. 1–12.
309. Hussein, A.A., Adaptive artificial neural network-based models for instantaneous power estimation enhancement in electric vehicles' Li-ion batteries. *IEEE Transactions on Industry Applications*, 2019. 55(1): p. 840–849.
310. Farmann, A. and D.U. Sauer, A comprehensive review of on-board state-of-available-power prediction techniques for lithium-ion batteries in electric vehicles. *Journal of Power Sources*, 2016. 329: p. 123–137.
311. Fan, Y., et al., The power state estimation method for high energy ternary lithium-ion batteries based on the online collaborative equivalent modeling and adaptive correction - unscented Kalman filter. *International Journal of Electrochemical Science*, 2021. 16(1): p. 1–28.
312. Trevey, J.E., et al., Electrochemical investigation of all-solid-state lithium batteries with a high capacity sulfur-based electrode. *Journal of the Electrochemical Society*, 2012. 159(7): p. 1019–1022.
313. Sobon, J. and B. Stephen, Model-free non-invasive health assessment for battery energy storage assets. *IEEE Access*, 2021. 9: p. 54579–54590.
314. Mukherjee, N. and D. De, A new state-of-charge control derivation method for hybrid battery type integration. *IEEE Transactions on Energy Conversion*, 2017. 32(3): p. 866–875.
315. An, F.L., et al., State of Energy Estimation for Lithium-Ion Battery Pack via Prediction in Electric Vehicle Applications. IEEE Transactions on Vehicular Technology, 2022. 71(1): p. 184–195.
316. Chawla, N., Recent advances in air-battery chemistries. *Materials Today Chemistry*, 2019. 12: p. 324–331.
317. Zhao, B., et al., Estimation of the SOC of energy-storage lithium batteries based on the voltage increment. *IEEE Access*, 2020. 8: p. 198706–198713.
318. Zhao, T., et al., A study on half-cell equivalent circuit model of lithium-ion battery based on reference electrode. *International Journal of Energy Research*, 2021. 45(3): p. 4155–4169.
319. Fiorenti, S., et al., Modeling and experimental validation of a hybridized energy storage system for automotive applications. *Journal of Power Sources*, 2013. 241: p. 112–120.
320. Zhang, S.Z., X. Guo, and X.W. Zhang, An improved adaptive unscented Kalman filtering for state of charge online estimation of lithium-ion battery. *Journal of Energy Storage*, 2020. 32: p. 1–13.
321. Lawder, M.T., V. Viswanathan, and V.R. Subramanian, Balancing autonomy and utilization of solar power and battery storage for demand based microgrids. *Journal of Power Sources*, 2015. 279: p. 645–655.
322. Xu, Y.D., et al., State of charge estimation for lithium-ion batteries based on adaptive dual Kalman filter. *Applied Mathematical Modelling*, 2020. 77: p. 1255–1272.
323. Zhang, X., Y. Han, and W.-p. Zhang, A review of factors affecting the lifespan of lithium-ion battery and its health estimation methods. *Transactions on Electrical and Electronic Materials*, 2021. 22(5): p. 567–574.
324. Zhang, X., et al., Fuzzy adaptive filtering-based energy management for hybrid energy storage system. *Computer Systems Science and Engineering*, 2021. 36(1): p. 117–130.
325. Zhang, Z., B. Xue, and J. Fan, Noise adaptive moving horizon estimation for state-of-charge estimation of Li-ion battery. *IEEE Access*, 2021. 9: p. 5250–5259.
326. Seo, G., et al., Rapid determination of lithium-ion battery degradation: High C-rate LAM and calculated limiting LLI. *Journal of Energy Chemistry*, 2022. 67: p. 663–671.

327. Casino, S., et al., Al2O3 protective coating on silicon thin film electrodes and its effect on the aging mechanisms of lithium metal and lithium ion cells. *Journal of Energy Storage*, 2021. 44: p. 1–10.
328. Chouchane, M. and A.A. Franco, Deconvoluting the impacts of the active material skeleton and the inactive phase morphology on the performance of lithium ion battery electrodes. *Energy Storage Materials*, 2022. 47: p. 649–655.
329. Kick, M., et al., Mobile small polarons qualitatively explain conductivity in lithium titanium oxide battery electrodes. *Journal of Physical Chemistry Letters*, 2020. 11(7): p. 2535–2540.
330. Ahamed, P., et al., Role of gelatin and chitosan cross-linked aqueous template in controlling the size of lithium titanium oxide. *SN Applied Sciences*, 2022. 4(4): p. 1–13.
331. Wang, H., Y. Sun, and Y. Jin, Simulation study on overcharge thermal runaway propagation of lithium-iron-phosphate energy storage battery clusters. *Jixie Gongcheng Xuebao/Journal of Mechanical Engineering*, 2021. 57(14): p. 32–39.
332. Lee, M.-H., et al., Electrochemically induced crystallite alignment of lithium manganese oxide to improve lithium insertion kinetics for dye-sensitized photorechargeable batteries. *ACS Energy Letters*, 2021. 6(4): p. 1198–1204.
333. Hasib, S.A., et al., A comprehensive review of available battery datasets, RUL prediction approaches, and advanced battery management. *IEEE Access*, 2021. 9: p. 86166–86193.
334. Pastor-Flores, P., et al., Analysis of Li-ion battery degradation using self-organizing maps. *45th Annual Conference of the IEEE Industrial Electronics Society (IECON 2019)*, Lisbon, Portugal, 2019: p. 4525–4530.
335. Pugalenthi, K. and N. Raghavan, A holistic comparison of the different resampling algorithms for particle filter based prognosis using lithium ion batteries as a case study. *Microelectronics Reliability*, 2018. 91: p. 160–169.
336. Qiao, J., X. Liu, and Z. Chen, Prediction of the remaining useful life of lithium-ion batteries based on empirical mode decomposition and deep neural networks. *IEEE Access*, 2020. 8: p. 42760–42767.
337. Zhang, N., et al., A novel method for estimating state-of-charge in power batteries for electric vehicles. *International Journal of Precision Engineering and Manufacturing*, 2019. 20(5): p. 845–852.
338. Ospina Agudelo, B., W. Zamboni, and E. Monmasson, Application domain extension of incremental capacity-based battery SoH indicators. *Energy*, 2021. 234: p. 1–14.
339. Park, J., et al., Integrated approach based on dual extended Kalman filter and multivariate autoregressive model for predicting battery capacity using health indicator and SOC/SOH. *Energies*, 2020. 13(9): p. 1–20.
340. Zhao, X., et al., Preparation of gel polymer electrolyte with high lithium ion transference number using GO as filler and application in lithium battery. *Ionics*, 2020. 26(9): p. 4299–4309.
341. Zhang, T., et al., Facile in situ chemical cross-linking gel polymer electrolyte, which confines the shuttle effect with high ionic conductivity and Li-ion transference number for quasi-solid-state lithium-sulfur battery. *ACS Applied Materials and Interfaces*, 2021. 13(37): p. 44497–44508.
342. Sun, Y., et al., Estimation method of state-of-charge for lithium-ion battery used in hybrid electric vehicles based on variable structure extended Kalman filter. *Chinese Journal of Mechanical Engineering (English Edition)*, 2016. 29(4): p. 717–726.
343. Liu, M., et al., Reliability evaluation of large scale battery energy storage systems. *IEEE Transactions on Smart Grid*, 2017. 8(6): p. 2733–2743.
344. Wang, C., et al., Coordinated predictive control for wind farm with BESS considering power dispatching and equipment ageing. *IET Generation, Transmission and Distribution*, 2018. 12(10): p. 2406–2414.

345. Razavi-Far, R., et al., Extreme learning machine based prognostics of battery life. *International Journal on Artificial Intelligence Tools*, 2018. 27(8): p. 1–11.
346. Hua, Y., et al., A comprehensive review on inconsistency and equalization technology of lithium-ion battery for electric vehicles. *International Journal of Energy Research*, 2020. 44(14): p. 11059–11087.
347. He, Z., et al., State-of-charge estimation of lithium ion batteries based on adaptive iterative extended Kalman filter. *Journal of Energy Storage*, 2021. 39: p. 1–7.
348. Guo, Y., Y. Dai, and S. Fu, Estimation of battery SOC and SOH for urban rail vehicle composite power energy storage system. *Journal of Railway Science and Engineering*, 2020. 17(11): p. 2920–2928.
349. Wang, W., et al., Failure warning at the end of service-life of lead-acid batteries for backup applications. *Applied Sciences-Basel*, 2020. 10(17): p. 1–22.
350. Wang, F.-K. and T. Mamo, A hybrid model based on support vector regression and differential evolution for remaining useful lifetime prediction of lithium-ion batteries. *Journal of Power Sources*, 2018. 401: p. 49–54.
351. Zhang, Z., et al., Lithium-ion batteries remaining useful life prediction method considering recovery phenomenon. *International Journal of Electrochemical Science*, 2019. 14(8): p. 7149–7165.
352. Gao, D., et al., A method for predicting the remaining useful life of lithium-ion batteries based on particle filter using Kendall rank correlation coefficient. *Energies*, 2020. 13(16): p. 1–13.
353. Chan, S.-C., et al., A new variable forgetting factor-based bias-compensation algorithm for recursive identification of time-varying multi-input single-output systems with measurement noise. *IEEE Transactions on Instrumentation and Measurement*, 2020. 69(7): p. 4555–4568.
354. Al-Gabalawy, M., et al., State of charge estimation of a Li-ion battery based on extended Kalman filtering and sensor bias. *International Journal of Energy Research*, 2021. 45(5): p. 6708–6726.
355. Gholizadeh, M. and A. Yazdizadeh, Systematic mixed adaptive observer and EKF approach to estimate SOC and SOH of lithiumion battery. *IET Electrical Systems in Transportation*, 2020. 10(2): p. 135–143.
356. Sakile, R. and U.K. Sinha, Estimation of state of charge and state of health of lithium-ion batteries based on a new adaptive nonlinear observer. *Advanced Theory and Simulations*, 2021. 4(11): p. 1–9.
357. Zhang, Y., et al., Motor driver-based topology of integrated on-board charging system and data-driven inductance identification method. *IEEE Journal on Emerging and Selected Topics in Circuits and Systems*, 2022. 12(1): p. 310–319.
358. Zhang, N., B.D. Leibowicz, and G.A. Hanasusanto, Optimal residential battery storage operations using robust data-driven dynamic programming. *IEEE Transactions on Smart Grid*, 2020. 11(2): p. 1771–1780.
359. Liu, Z.F., A. Ivanco, and S. Onori, Aging characterization and modeling of nickel-manganese-cobalt lithium-ion batteries for 48V mild hybrid electric vehicle applications. *Journal of Energy Storage*, 2019. 21: p. 519–527.
360. Chen, L., L. Xu, and Y. Zhou, Novel approach for lithium-ion battery on-line remaining useful life prediction based on permutation entropy. *Energies*, 2018. 11(4): p. 1–15.
361. Rao, K.K., Y. Yao, and L.C. Grabow, Accelerated modeling of lithium diffusion in solid state electrolytes using artificial neural networks. *Advanced Theory and Simulations*, 2020. 3(9): p. 1–9.
362. Shin, S.U., et al., A time-interleaved resonant voltage mode wireless power receiver with delay-based tracking loops for implantable medical devices. *IEEE Journal of Solid-State Circuits*, 2020. 55(5): p. 1374–1385.

363. Song, Y.C., et al., Satellite lithium-ion battery remaining useful life estimation with an iterative updated RVM fused with the KF algorithm. *Chinese Journal of Aeronautics*, 2018. 31(1): p. 31–40.
364. Li, S.Q., J.W. Li, and H.X. Wang, Big data driven lithium-ion battery modeling method: A cyber-physical system approach. *2019 IEEE International Conference on Industrial Cyber Physical Systems (ICPS 2019)*, Taipei, Taiwan, 2019: p. 161–166.
365. Liu, D.T., et al., Hybrid state of charge estimation for lithium-ion battery under dynamic operating conditions. *International Journal of Electrical Power & Energy Systems*, 2019. 110: p. 48–61.
366. Vyas, M., et al., State-of-charge prediction of lithium ion battery through multivariate adaptive recursive spline and principal component analysis. *Energy Storage*, 2021. 3(2): p. 1–15.
367. Banguero, E., et al., Diagnosis of a battery energy storage system based on principal component analysis. *Renewable Energy*, 2020. 146: p. 2438–2449.
368. Wang, L., et al., Development of a typical urban driving cycle for battery electric vehicles based on kernel principal component analysis and random forest. *IEEE Access*, 2021. 9: p. 15053–15065.
369. Dong, G., W. Han, and Y. Wang, Dynamic Bayesian network-based lithium-ion battery health prognosis for electric vehicles. *IEEE Transactions on Industrial Electronics*, 2021. 68(11): p. 10949–10958.
370. Khan, K., M. Jafari, and L. Gauchia, Comparison of Li-ion battery equivalent circuit modelling using impedance analyzer and Bayesian networks. *IET Electrical Systems in Transportation*, 2018. 8(3): p. 197–204.
371. Feng, J., et al., Cross trajectory Gaussian process regression model for battery health prediction. *Journal of Modern Power Systems and Clean Energy*, 2021. 9(5): p. 1217–1226.
372. Jiang, B. et al., A cell-to-pack state estimation extension method based on a multilayer difference model for series-connected battery packs. IEEE Transactions on Transportation Electrification, 2022. 8(2): p. 2037–2049.
373. Dong, G., et al., Data-driven battery health prognosis using adaptive Brownian Motion Model. IEEE Transactions on Industrial Informatics, 2020.16(7): p. 4736–4746.
374. Tian, J., et al., Capacity attenuation mechanism modeling and health assessment of lithium-ion batteries. *Energy*, 2021. 221: p. 1–14.
375. Lyu, C., Y. Jia, and Z. Xu, DRO-MPC-based data-driven approach to real-time economic dispatch for islanded microgrids. *IET Generation Transmission & Distribution*, 2020. 14(24): p. 5704–5711.
376. Kim, S., et al., A novel prognostics approach using shifting kernel particle filter of Li-ion batteries under state changes. *IEEE Transactions on Industrial Electronics*, 2021. 68(4): p. 3485–3493.
377. Liu, X., et al., A generalizable, data-driven online approach to forecast capacity degradation trajectory of lithium batteries. *Journal of Energy Chemistry*, 2022. 68: p. 548–555.
378. Liang, H. and J. Ma, Data-driven resource planning for virtual power plant integrating demand response customer selection and storage. *IEEE Transactions on Industrial Informatics*, 2022. 18(3): p. 1833–1844.
379. Li, W., et al., Data-driven systematic parameter identification of an electrochemical model for lithium-ion batteries with artificial intelligence. *Energy Storage Materials*, 2022. 44: p. 557–570.
380. Duan, J., et al., State of charge estimation of lithium battery based on improved correntropy extended Kalman filter. *Energies*, 2020. 13(16): p. 1–18.
381. Chen, X., et al., A novel fireworks factor and improved elite strategy based on back propagation neural networks for state-of-charge estimation of lithium-ion batteries. *International Journal of Electrochemical Science*, 2021. 16(9): p. 1–14.

382. Wang, Z., N. Liu, and Y. Guo, Adaptive sliding window LSTM NN based RUL prediction for lithium-ion batteries integrating LTSA feature reconstruction. *Neurocomputing*, 2021. 466: p. 178–189.
383. Wang, J., S. Zhang, and X. Hu, A fault diagnosis method for lithium-ion battery packs using improved RBF neural network. *Frontiers in Energy Research*, 2021. 9: p. 1–9.
384. Gou, B., Y. Xu, and X. Feng, State-of-health estimation and remaining-useful-life prediction for lithium-ion battery using a hybrid data-driven method. *IEEE Transactions on Vehicular Technology*, 2020. 69(10): p. 10854–10867.
385. Zhu, J., et al., RUL prediction of lithium-ion battery based on improved DGWO-ELM method in a random discharge rates environment. *IEEE Access*, 2019. 7: p. 125176–125187.
386. Dong, G. and Z. Chen, Data-driven energy management in a home microgrid based on Bayesian optimal algorithm. *IEEE Transactions on Industrial Informatics*, 2019. 15(2): p. 869–877.
387. Zhang, H., et al., Stochastic process-based degradation modeling and RUL prediction: From Brownian motion to fractional Brownian motion. *Science China Information Sciences*, 2021. 64(7): p. 1–14.
388. Zhang, Z., et al., The effects of Fe2O3 based DOC and SCR catalyst on the combustion and emission characteristics of a diesel engine fueled with biodiesel. *Fuel*, 2021. 290: p. 1–11.
389. Wu, J., C. Zhang, and Z. Chen, An online method for lithium-ion battery remaining useful life estimation using importance sampling and neural networks. *Applied Energy*, 2016. 173: p. 134–140.
390. Zou, L., et al., Online prediction of remaining useful life for Li-ion batteries based on discharge voltage data. *Energies*, 2022. 15(6): p. 2237–2237.

Index

Printed in Canada